£32.50

Occupational Ergonomics

Manufacturing Systems Engineering Series

Series editor: Hamid R. Parsaei, Department of Industrial Engineering, University of Louisville, USA

The globalization of business and industry and the worldwide competitive economy are forcing business leaders in the manufacturing and service sectors to utilize fully the best equipment and techniques available. The objective is to have efficient control of the organizational structure in order to produce high quality products at lower prices within a shorter period of time.

Since the introduction of computers in the 1950s, Manufacturing Systems Engineering has experienced tremendous growth. The development of the discipline has helped industry to become more productive and to make more efficient use of resources. Manufacturing information systems, total quality management, facility layout, material handling, value engineering and cost analysis, safety, computer-integrated manufacturing, production planning and shop floor control are just some of the areas in which manufacturing systems engineers have been traditionally involved in order to help improve understanding and awareness in the manufacturing and service sectors. The recent emphasis and concern about the environment and product recyclability and re-usability have brought new perspectives and more challenges to this ever-growing engineering discipline.

The aim of the *Manufacturing Systems Engineering Series* is to provide an outlet for state-of-the-art topics in manufacturing systems engineering. This series is also intended to provide a scientific and practical basis for researchers, practitioners and students involved in manufacturing systems areas. Issues which are addressed in this series include, but are not limited to, the following:

- Production systems design and control
- Life cycle analysis
- Simulation in manufacturing
- Manufacturing cost estimating
- Industrial safety
- Fuzzy logic and neural networks in manufacturing
- CAD/CAM/CIM

We would welcome proposals to write material for this series from colleagues and industry leaders around the world. We hope that researchers both in academia and government, as well as private organizations and individual practitioners, will find this series informative and worthwhile.

Occupational Ergonomics

Principles and applications

F. Tayyari, PE, CPE, CSP

Associate Professor of Industrial and Manufacturing Engineering and
Technology, Bradley University, USA

and

J.L. Smith, PE, CPE

Professor and Chairman of the Department of Industrial Engineering,
Texas Tech University, USA

 CHAPMAN & HALL

London · Weinheim · New York · Tokyo · Melbourne · Madras

Published by Chapman & Hall, 2–6 Boundary Row, London SE1 8HN, UK

Chapman & Hall, 2–6 Boundary Row, London SE1 8HN, UK

Chapman & Hall GmbH, Pappelallee 3, 69469 Weinheim, Germany

Chapman & Hall USA, Fourth Floor, 115 Fifth Avenue, New York, NY 10003, USA

Chapman & Hall Japan, ITP-Japan, Kyowa Building, 3F, 2-2-1 Hirakawacho, Chiyoda-ku, Tokyo 102, Japan

Chapman & Hall Australia, 102 Dodds Street, South Melbourne, Victoria 3205, Australia

Chapman & Hall India, R. Seshadri, 32 Second Main Road, CIT East, Madras 600 035, India

First edition 1997

© 1997   F. Tayyari and J.L. Smith

Typeset in 10/12 pt Palatino by Cambrian Typesetters, Frimley, Surrey

Printed in Great Britain by T.J. Press Ltd., Padstow, Cornwall

ISBN 0 412 58650 9   1001 025143

∞ Printed on permanent acid-free text paper, manufactured in accordance with ANSI/NISO Z39.48-1992 and ANSI/NISO Z39.48-1984 (Permanence of Paper)

Contents

An introduction to ergonomics

1

1.1 INTRODUCTION

Ergonomics can be defined as "that branch of science that is concerned with the achievement of optimal relationships between workers and their work environment." It deals with the assessment of the human's capabilities and limitations (biomechanics and anthropometry), work and environmental stresses (work physiology and industrial psychology), static and dynamic forces on the human body structure (biomechanics), vigilance (industrial psychology), fatigue (work physiology and industrial psychology), design simulation and training, and design of workstations and tools (anthropometry and engineering). Therefore, ergonomics draws heavily from many areas of science and engineering.

The term **ergonomics** has its roots in Ramazzini's study of the ill-effects of poor posture and poorly designed tools on the health of workers in the early 1700s (Byers *et al.*, 1978). The term ergonomics has been derived from two Greek words – *ergon* (meaning work) and *nomikos* (meaning law). Ergonomics is, then, the study of work laws. Its goal is to fit (adapt) work to individuals, as opposed to fitting workers to the work, through developing knowledge that results in efficient adaptation of work methods to the individual's physiological and psychological characteristics. Therefore, the mission of an ergonomist (ergonomics practitioner) is to identify and alleviate those work stresses that adversely affect the health, safety and efficiency of workers.

There exists, however, some confusion concerning the definition of the term ergonomics and the objectives toward which it is directed. This is mainly due to the fact that two general terms are used for naming essentially the same discipline or field of study. The term ergonomics is mostly used in Europe (especially in the UK), while the term **human factors** is widely used in the USA and Canada. It is, however, argued that the field of human factors in the USA covers a broader subject area since it involves not only the design of work tasks and equipment but also the design of consumer products. The term ergonomics is becoming more universally accepted as demonstrated by the recent (1993) name change of the Human Factors Society to the Human Factors and Ergonomics Society. Whichever term is used, the major application of the field is toward the use of bioengineering, biotechnology, and biomechanics to improve the workplace environment for workers.

1.2 CONTRIBUTING DISCIPLINES

Ergonomics is a multidisciplinary science which draws heavily from many disciplines or fields of study. Figure 1.1 illustrates the major disciplines contributing to the field of ergonomics.

- **Mathematics**: Mathematical science facilitates the study and use of numerical factors and symbolic notations dealing with quantities, magnitudes, and formulas describing relationships and attributes that are concerned with humans and their work environment. This area includes algebra, trigonometry, calculus, statistics, biometrics (the measurement of body structure), and symbolic logic.
- **Biological sciences**: From biological sciences (i.e., anatomy, physiology, and anthropometry) the subjects of body structure (e.g., muscular, skeletal and nervous systems), reproduction, and functions of cells and organisms in human biology are addressed.
- **Psychology (behavioral) sciences**: With the help of psychology or behavioral sciences human behavior both as individuals and as groups is studied. Related subjects are individual similarities and differences in learning, perception, reaction to stimuli, attitudes and habits, and group behavior.

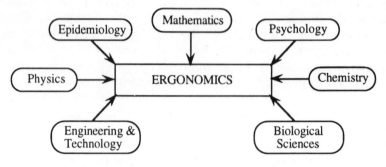

Fig. 1.1 The major disciplines contributing to ergonomics.

- **Physics**: Physics provides for the study of the phenomena associated with matter in general and the laws governing these phenomena, especially those dealing with the human body in the work environment. Subjects included within the broad scope of physics are mechanics (statics, dynamics and fluid), heat, light, sound/noise, vibration, electricity, magnetism, gases, and radiation.
- **Chemistry**: Chemical reactions and equilibrium, especially those dealing with metabolic reactions are addressed.
- **Engineering and technology**: Engineering and technology concepts and techniques are utilized in the design of workplaces and equipment for use by humans in the workplace. Engineering topics concerned with ergonomics are applied mechanics (statics, dynamics, fluid mechanics, and thermodynamics), electrical circuits and devices/machines, engineering designs, and computer science and technology. Some refer to those applications as bioengineering.
- **Epidemiology**: The concentration of epidemiology is the study of the incidences, distributions, and control of diseases in a population. In ergonomics, it involves the study of cause-and-effect and prevention of occupational injuries and illnesses.

1.3 OBJECTIVES OF ERGONOMICS

As illustrated in Figure 1.2, ergonomics deals with the interaction between human and machine in the work environment. The main objective of ergonomics is to achieve an optimal relationship between people and their work environment. The two conflicting factors in this optimization process are workers' productivity and their health and physical well-being. That is, while workers should perform their job in

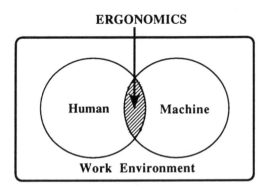

Fig. 1.2 Ergonomics deals with the interaction of the human operator and machine in the work environment.

the most efficient manner possible, they must also be protected against undue physical, biological, and psychological strain that may occur as a result of performing the required tasks.

1.4 OCCUPATIONAL ERGONOMICS

Although ergonomics was originally aimed at occupational settings to study the relationship between workers and their work environment, it has been expanded to cover other areas such as consumer product design and the auto industry. Therefore, the terms occupational, industrial and, often, applied ergonomics have been used to refer to an area of specialty in ergonomics that specifically deals with occupational situations. Occupational ergonomists strive to review work systems and modify them to minimize occupational stresses. Ergonomics principles can be used in the following industrial applications:

- design, modification, replacement and maintenance of equipment for enhanced productivity, and the work-life and product quality;
- design and modification of work spaces and workplace layout for ease and speed of operation, service and maintenance;
- design and modification of work methods, including automation and task allocation between human operators and machines;
- controlling physical factors (e.g., heat, cold, noise, vibration, and light) in the workplace for the best productivity and safety of employees.

Occupational ergonomics is a solution-oriented branch of ergonomics rather than just an evaluation of work-related problems. Its goal is to optimize worker well-being and productivity by treatment of the work stressors. Ergonomics interventions are preventive, before injuries occur, thereby avoiding future medical treatments.

1.5 WORKPLACE STRESS FACTORS

Every workplace presents its own, distinctive stress factors. The following are the major recognized stress factors in the workplace:

- complexity and number of tools used in the workplace;
- unnatural environmental conditions (e.g., thermal, noise, vibration, illumination, toxic materials, etc.);
- physical and mental workload.

The stress from these factors placed on the worker is significant. Ergonomic principles can be applied to the design of the workplace so that the strain placed on the worker is minimized.

1.6 OCCUPATIONAL FACTORS AFFECTING THE WORKER

Among the factors affecting the worker, the following are most important to be considered in optimizing the relationship between workers and their job:

- environmental conditions (e.g., temperature, illumination, noise) of the workplace;
- the physical and mental requirements of the job;
- the worker's exposure to hazardous materials;
- the interaction between the worker and the work equipment.

Heat stress due to exposure to a relatively high heat load places a physiological strain on the worker, resulting in potential health impairment. **Cold stress** can also physiologically affect the human body, resulting in impairment of the individual's health as well as job performance. Exposure to severe cold stress can result in cold injuries and illness (e.g., frostbite, hypothermia when the body temperature falls below 35°C). **Poor illumination** can reduce the worker's job performance and can be a potential strain to vision. Inadequate illumination can also cause accidents, resulting in personal injuries. **Noise** and **vibration**, at a high level, may be physiologically and/or psychologically harmful to the human body. Noise itself can make communication difficult and can be permanently harmful to the worker's hearing ability. It may mentally disturb the worker, resulting in job performance impairment and/or accidents. Prolonged exposure to vibration may cause health impairment.

Automation has eliminated many of the heavy physical activities that used to be performed manually by human beings. Despite a tremendous amount of technological advancement and automation there are, however, still many physically demanding jobs to be performed by individuals. The **physical demands** can place biomechanical and physiological stress on workers while they are performing their job. This stress can affect workers' job performance as well as their health and safety. Modernization of the industrial environment has increased the **mental demands** placed on the worker. In the past, physical demands of the job were the primary cause of worker fatigue, but today mental demands may contribute as much as or more than the physical demands to overall worker fatigue.

The wide variety of chemicals produced and used in today's industry exposes many workers to potentially hazardous materials during their work. The workers exposed to these hazardous materials can develop both acute and chronic injuries or illnesses.

It is, therefore, obvious that the worker in the industrial environment is exposed to many potentially hazardous factors. The ergonomists and industrial hygienists who are trained in ergonomics/human factors principles play significant roles in controlling and minimizing the adverse effects of the work environment to insure healthier, safer and more efficient workplaces.

1.7 WORKPLACE HAZARDS

Hazards are present everywhere, including workplaces. Single exposures to hazards are not necessarily dangerous. However, repeated exposure to a concentration of a hazard for a sufficient length of time may make it dangerous. Workplace hazards may be categorized as physical, chemical, and biological.

1.7.1 PHYSICAL HAZARDS

Physical hazards include extremes of temperature, noise, vibration, and radiation. Falling, slipping, and tripping may also be categorized as physical hazards. Exposure to high temperatures can lead to heat stress and heat stroke. Long exposure to noise of sufficient intensity could result in permanent hearing damage. Exposure to excessive vibration over time could lead to nerve and blood vessel damage. Exposure to high levels of ionizing radiation, produced by such instruments as X-ray and radioisotope machines used in health care centers, can result in cancer and leukemia. Exposure to excessive levels of non-ionizing radiation, produced by sources such as microwave ovens and lasers, could cause burns or eye injuries.

1.7.2 CHEMICAL HAZARDS

Hazardous chemicals may be in the form of liquids, solids, gases, dusts, fumes, or mists. Hazardous materials can cause external or local problems such as irritation, skin burns, or ulceration. They may enter the body when inhaled, swallowed, or absorbed through skin and cause serious long-term health problems. Typical examples of hazardous substances are methyl alcohol (i.e., methanol), lead, acids, and alkalis. Methyl alcohol can enter the body by inhalation or absorption. Lead can enter the body by inhalation and ingestion, which is very hazardous.

1.7.3 BIOLOGICAL HAZARDS

In the general industrial setting, biological hazards are not yet of important concern. However, workers in health care centers, scientific and research laboratories who handle biological and medical specimens and animals must be protected against biological hazards.

The first defense against workplace hazards is to keep workers informed. All workers have the "right to know" about the hazards present in their workplace and to be assured that those hazards are not endangering their health. Manufacturers and suppliers are required to convey hazard information by means of labeling controlled products or containers of controlled products and prescribing information on material safety data sheets (MSDSs). Employers are responsible:

- to inform their employees of potential workplace hazards and to make MSDSs available to their employees;
- to develop appropriate, safe work procedures and to train their employees in those safe work procedures;
- to educate and train their workers on the safe use and handling of hazardous materials;
- to provide their employees with protection against workplace hazards.

1.8 RESULTS OF ERGONOMICS APPLICATIONS

The following are some expected outcomes of applying the principles of ergonomics to the workplace:

- understanding the effects of a particular type of work on workers' bodies and their job performance;
- predicting the potential long-term (or cumulative) effects of work on workers' bodies;
- assessment of the fitness of the workplace and/or tools to workers in performing a job;
- improvement of productivity and well-being of workers by "fitting the task to the person," or "fitting the person to the task." The result of such efforts is to achieve the best matching between worker capabilities and job requirements;
- establishment of a knowledge base support for designers, engineers, and medical personnel for improving the productivity and well-being of individuals.

1.9 BASIC TERMINOLOGY

There are certain terms used in describing the relationships between body parts and joint movements. These terms have been developed by anatomists in order to communicate about the human body. This section presents the basic terminology that is useful for understanding the human body's physiological responses to work and biomechanical evaluations.

1.9.1 ANATOMY AND PHYSIOLOGY

- **Anatomy** is defined as that branch of science which deals with the body's structures, including muscles, bones, tendons, ligaments, and other structures.
- **Physiology** is the science of the body's functions, including metabolism, muscle mechanics, oxygen and nutrient distribution, temperature regulation, nerve transmission, cognition, and other functional activities.

1.9.2 THE ANATOMICAL POSITION

The anatomical position is standing erect, the eyes looking forward to the horizon, the arms by the sides, the palms of the hands and the toes directed forward (Fig. 1.3).

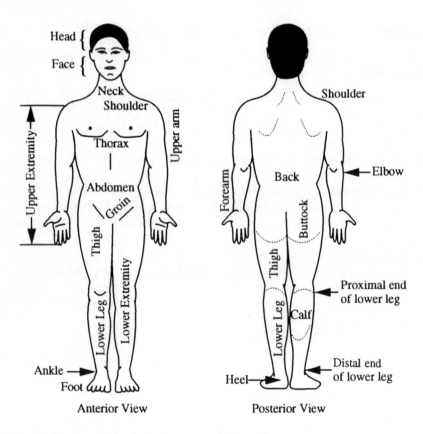

Fig. 1.3 The human body in the anatomical position.

1.9.3 REFERENCE PLANES

As shown in Figure 1.4, there are three general reference planes: the sagittal plane (there is also a special sagittal plane, called mid sagittal), coronal plane, and transverse plane.

- The **mid sagittal (median) plane** is a vertical plane dividing the body into right and left halves. A **sagittal (parasagittal) plane** refers to any vertical plane parallel to and including the median plane which divides the body into right and left parts.
- A **coronal (frontal) plane** is any vertical plane perpendicular to the median plane which divides the body into anterior (front) and posterior (back) portions.
- A **transverse** (or **horizontal) plane** is any horizontal plane at right angles to the sagittal and frontal (coronal) planes, dividing the body into superior (upper) and inferior (lower) parts.

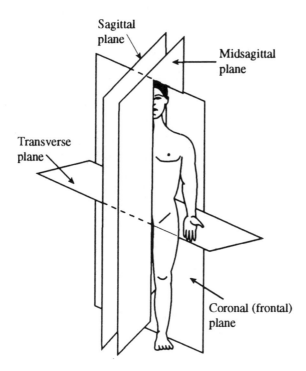

Fig. 1.4 Reference planes.

1.9.4 TERMS OF RELATIONSHIP

The terms of spatial anatomical relationships are briefly described in Table 1.1 and illustrated in Figure 1.5.

1.10 APPLICATIONS AND DISCUSSION

Ergonomics is a very broad discipline, encompassing training in several fields. Successful ergonomics programs are going to understand the broad base of the discipline and to look for ergonomic solutions that utilize the full range of the discipline. If the engineer ignores the non-engineering factors (or the psychologist ignores the non-psychological factors), the evaluation will be short-sighted and will most likely not lead to a match where human capabilities and limitations are consistent with the demands of the job.

The occupational ergonomist should look for potential contributors to the ergonomics team and take advantage of the resources available. For example, the facility might have medical, safety, human resources, engineering, training, or other departments with personnel who could contribute to the ergonomics effort. The work of a "lone ergonomist" will have much less impact than that of a multidisciplinary ergonomics team.

Table 1.1 Terms of relationship

Term	Meaning
Anterior	Refers to front, nearer the front surface of the body
Posterior	Refers to back, nearer the back surface of the body
Superior	Above, upper or higher part of body, or nearer the crown of the head
Inferior	Below, the lower part of the body, or nearer the soles of the feet
Medial	Nearer the median plane of the body (or body part) which divides the body (or body part) into right and left halves
Lateral	Farther from the median plane
Proximal	The end of a body member nearer the body
Distal	The end of a body segment farther from the body
Palmar or volar	Anterior surface of the hand or forearm
Dorsal	Pertaining to back, nearer the back (of the foot, hand and forearm, e.g., dorsal surface of the hand, opposite of palmar)
Plantar	Refers to the sole of the foot

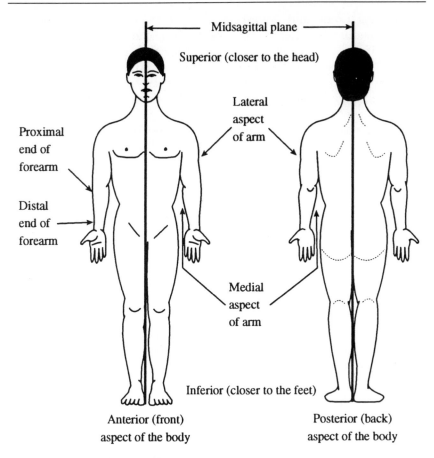

Midsagittal plane

Superior (closer to the head)

Lateral aspect of arm

Proximal end of forearm

Distal end of forearm

Medial aspect of arm

Inferior (closer to the feet)

Anterior (front) aspect of the body

Posterior (back) aspect of the body

Fig. 1.5 Terms of relationship.

REVIEW QUESTIONS

1. Development of the ergonomics field of study is traced back to Ramazzini's study of the ill-effects of two factors on the health of workers (in the early 1700s). What were those two factors?
2. What is the mission of ergonomists?
3. What aspects of work tools are considered as ergonomic-related stress factors?
4. The main objective of ergonomics is to achieve an optimal relationship between workers and their work environment. What are the two conflicting factors in this optimization process?
5. What is the first defense of the employer against workplace hazards?

6. Manufacturers and suppliers are required to convey hazard information by means of labeling on controlled products or containers of controlled products and prescribed information on MSDSs. What does the acronym MSDS stand for?
7. What are the responsibilities of employers with regard to safe-guarding their employees against hazardous materials in the workplace?
8. Ergonomics is a multidisciplinary science which draws heavily from many disciplines or fields of study. What are the six major disciplines from which ergonomics draws?
9. What are the major recognized stress factors in the workplace whose effects on workers can be minimized by the application of ergonomics principles?
10. What are the most important occupational factors affecting workers that must be considered in optimizing the relationship between workers and their job?
11. What are the three categories of workplace hazards? (Give at least one example of each.)
12. Define the anatomical position.
13. Explain what is meant by anterior and posterior.
14. Explain what is meant by superior and inferior.
15. Explain what is meant by medial and lateral.
16. Define the mid sagittal plane.
17. Explain what a sagittal plane is.
18. Explain what a coronal (frontal) plane is.
19. Explain what a transverse plane is.

REFERENCES

Byers, B.B., Hirtz, R.J. and McClintock, J.C. (1978) *Industrial Hygiene Engineering and Control: Ergonomics – Student Manual.* (Contract CDC–210–75–0076.) Division of Training and Manpower Development, National Institute for Occupational Safety and Health (NIOSH), US Department of Health and Human Services (formerly Health, Education, and Welfare–HEW), Cincinnati, Ohio.

Skeletal system

2

2.1 INTRODUCTION

The skeletal and muscular systems of the human body form the movement mechanism and perform other functions that are of tremendous importance for maintaining life. The skeletal system provides the mechanical levers whose movements are accomplished by contraction of the muscles. Movement of the body is of great interest in ergonomics. To gain an understanding of how this function is accomplished, this chapter presents an overview of the human skeletal system; the muscular system will be described in Chapter 3. Because of its importance in understanding the subject of back injuries, a special emphasis is placed on describing the structure of the spinal column and potential back injuries due to manual materials handling.

2.2 THE SKELETAL SYSTEM

The skeletal system consists of all the bones, cartilages, and the joints formed by their attachments to each other by connective tissues (Fig. 2.1). This system is made of three predominant types of tissues (Anthony and Kolthoff, 1975): bone tissues; cartilage tissues; and hemopoietic tissues, which are responsible for forming blood cells.

2.2.1 BONES

There are 206 bones in the human body; they are usually classified according to their shapes, as follows (Anthony and Kolthoff, 1975):

- **Long bones**: humerus (in the upper arm), radius and ulna (in the forearm), femur (in the thigh), tibia and fibula (in the lower leg), and phalanges (finger bones, metacarpals in the palm of the hands, toe bones and metatarsals in the feet);
- **Short bones**: carpals (wrist bones) and tarsals (ankle bones);

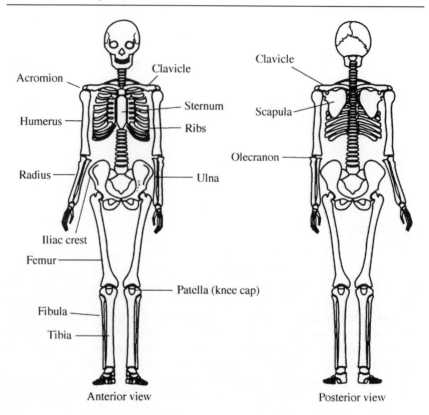

Anterior view   Posterior view

Fig. 2.1 The human skeleton.

- **Flat bones**: scapulae, ribs, and the skull;
- **Irregular bones**: vertebrae, sacrum, coccyx, mandible (the lower jaw), and the hyoid bone.

The hyoid bone is a single bone in the neck that does not articulate with any other bones. Its U shape can be felt between the mandible and upper part of the larynx, to which the tongue muscle is attached.

2.2.1.1 Gross skeletal structure

The human skeleton consists of the following two parts:

- **Axial skeleton**, consisting of the skull (the head, face, and ear bones), hyoid bone, vertebral bones in the spine, the ribs and sternum (breast bone), and the pelvis (Fig. 2.2a).
- **Appendicular skeleton**, consisting of the bones attached to the axial skeleton. It includes the upper and lower extremities (Fig. 2.2b). The

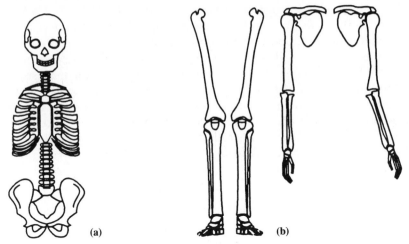

Fig. 2.2 The human skeleton consists of two groups: (a) axial skeleton and (b) appendicular skeleton.

upper extremities consist of the clavicle and scapula in the shoulder, the humerus in the upper arm, the radius and ulna in the forearm, and the wrist and hand bones. The lower extremities consist of the femur in the thigh, the patella (the knee cap), the fibula and tibia in the lower leg, and the foot bones.

2.2.1.2 Functions of the bones

The skeletal system performs the following critical functions:

- supporting the body's framework (structure), preventing the entire body from collapsing into a heap of soft tissues;
- providing shells to protect vital organs; for example, the ribcage protects the heart and lungs, and the skull protects the brain;
- allowing movement of the body. The skeletal system provides the mechanical levers for certain groups of muscles whose contraction moves the body;
- housing bone marrow which produces red blood cells;
- storing calcium and phosphorus. The bones are the greatest calcium and phosphorus reserves of the body, constantly being drawn upon and added to.

2.2.2 CARTILAGE

Cartilage is made of a transparent material formed by round cells (cartilage cells), that are embedded and bound together in a firm, gel-like, intercellular substance (Anthony and Kolthoff, 1975). Cartilage is firm, elastic, flexible and capable of rapid growth, and plays an important role in supporting the body structure so long as the load is moderate (Åstrand and Rodahl, 1986). Cartilage is found between the joints of the spinal column (as intervertebral disks), at the joint surface of the limbs, and at the end of the ribs. Cartilage is avascular; that is, there is not any canal system or blood vessel penetrating the cartilage structure (Anthony and Kolthoff, 1975). Hence, nutrients and oxygen can reach the isolated cartilage cells only by diffusion. The cartilage cells absorb these vital materials from capillaries in the fibrous covering of cartilage or, in the case of articulating cartilage, from synovial fluid. There are three types of cartilage tissues:

- **Hyaline cartilage tissue**, that covers articular surfaces (where bones join) to cushion the joints against the impact effects of jolts. The word hyaline is a Greek term, meaning glassy. It resembles milk glass in appearance.
- **Fibrous cartilage tissue** has the greatest rigidity and tensile strength of the three types of cartilage tissue. The ribcage is mainly made up of this type of cartilage tissue. The intervertebral disks are also fibrocartilages.
- **Elastic cartilage tissue** has both elasticity and firmness. The cartilage that protrudes from the body is of this type (e.g., ears and nose).

2.2.3 JOINT ARTICULATIONS

A joint is the junction between two or more bones. The types of joint movements are determined by bone shapes and joint structures. For example, where free movements are desirable, the bone shape and joint structure facilitate unhampered and smooth movements. Where only slight movements are required, the bones are so shaped and joints so constructed as to permit only slight movements. Based on their functions, joint articulations are divided into the following two types (Anthony and Kolthoff, 1975):

- **Diarthroses** (or diarthrotic joints): in a diarthrotic joint a small space exists between the articulating surfaces of the two joined bones. Since no other tissues grow in this cavity, the surfaces move freely against one another. Hence, they are functionally defined as **freely movable** joints. Specific examples of diarthrotic joints include **ball-and-socket** joints (Fig. 2.3a) and **hinge** joints (Fig. 2.3b). Other classifications of diarthrotic joints can be found in anatomy textbooks.
- **Synarthroses** (or synarthrotic joints): a synarthrotic joint does not have a joint cavity. Fibrous cartilage or bone tissues grow between the articulating surfaces of the two joined bones and make them unable to

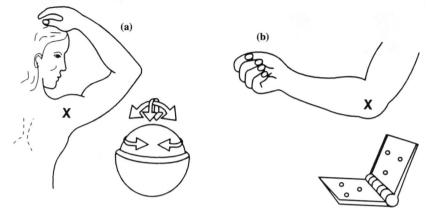

Fig. 2.3 (a) Ball-and-socket joint (e.g., in the shoulder); (b) hinge type of joint (e.g., in the elbow).

move freely against one another. Therefore, they are functionally defined as **immovable** (or slightly movable) joints that do not allow free movement. Examples of synarthrotic joints are the skull joints.

2.3 JOINT MOVEMENTS

The freely movable joints (diarthroses) can perform one or more of the following joint movements: flexion, extension, abduction, adduction, rotation, circumduction, pronation, and supination. These movements are described as follows:

- **Extension and flexion**: Flexion is a movement which decreases the angle between two bones. It is also referred to as bending or making an angle. Extension is a stretching or straightening movement which increases the angle between two bones. Figure 2.4 illustrates extension and flexion movements of joints.
- **Abduction and adduction**: Abduction means moving away laterally from the central axis of the body (e.g., the median plane). Adduction means moving toward the central axis of the body (e.g., the median plane). Figure 2.5 illustrates abduction and adduction movements of joints.
- **Circumduction** is an action which involves flexion, abduction, extension, and adduction, in sequence. The moving segment describes the surface of a cone.
- **Rotation** is a movement of a bone around its long axis, such as the rotation of the humerus in the upper arm. Figure 2.6 illustrates the rotation of the elbow in which the radius rotates around the ulna.

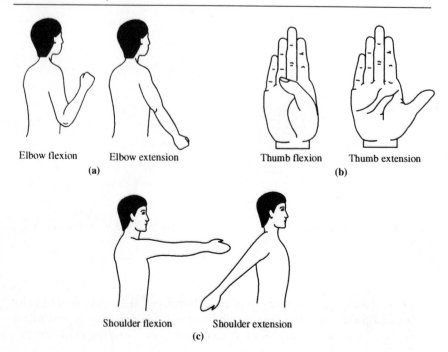

<table>
</table>

Elbow flexion   Elbow extension   Thumb flexion   Thumb extension

(a)   (b)

Shoulder flexion   Shoulder extension

(c)

Fig. 2.4 Flexion and extension of (a) the elbow; (b) the thumb; (c) the shoulder.

- **Pronation and supination**: Pronation is a medial rotation (or inward rotation) of a body member. For example, medial rotation of the forearm (Fig. 2.7a) brings the palm of the hand downward (facing the ground). Supination is a lateral (or outward) rotation of a body member. For example, lateral rotation of the forearm (Fig. 2.7b) brings the palm of the hand upward (facing up).

2.4 THE BACK STRUCTURE

The back is a complex structure that is made of:

- muscles;
- bones (vertebrae and their processes);
- intervertebral disks;
- ligaments;
- tendons;
- blood supply;
- the spinal cord and branched nerves.

Figure 2.8 illustrates the structure of the spinal column, which is a stack of 33 (in some 34) vertebrae (seven cervical, 12 thoracic, five

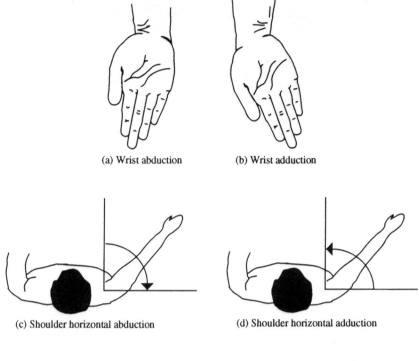

(a) Wrist abduction   (b) Wrist adduction

(c) Shoulder horizontal abduction   (d) Shoulder horizontal adduction

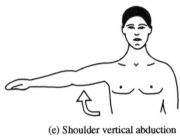

(e) Shoulder vertical abduction

Fig. 2.5 Abduction and adduction of the wrist and shoulder.

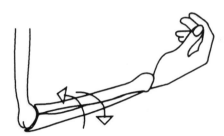

Fig. 2.6 The elbow rotation.

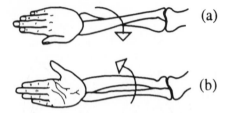

Fig. 2.7 (a) Supination and (b) pronation of the forearm.

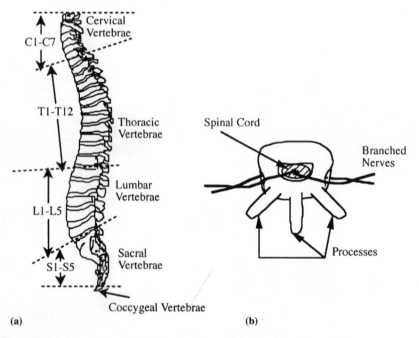

Fig. 2.8 (a) Structure of the spinal column; (b) top view of a vertebra.

lumbar, five sacral and four or five coccygeal vertebrae). This stack of vertebrae forms three natural curvatures in a **double-S** shape. When the spine is balanced or in a natural position, none of the spine curvatures are exaggerated or flattened. Figure 2.8 also shows the top view of a vertebra that has three bony appendages, called processes. The spinal cord passes through the spine and branches out nerves at each vertebra. They carry instructions and signals back and forth between the brain and the entire body.

The seven **cervical vertebrae** (C1–C7) constitute a structural framework for the neck. The next 12 bones are called **thoracic vertebrae** (T1–T12) that lie behind the ribcage (the thoracic cavity). The next five bones, below the thoracic vertebrae, are the **lumbar vertebrae** (L1–L5) that support the lower back. The next five vertebrae are the **sacral vertebrae** (S1–S5) which are fused together and named the **sacrum**. Below the sacrum lies the **coccyx**, which is made up of four or five fused coccygeal vertebrae.

The vertebrae are separated by the intervertebral disks. An intervertebral disk is a tough, fibrous ring, resembling a sac, that is filled with a viscous fluid (a gel-like substance). These disks act as shock absorbers between the vertebrae and provide flexibility for the vertebrae. They are, however, the most fragile parts of the spine, especially in the lumbar (lower-back) region. A schematic representation of an intervertebral disk, and its connection with vertebrae and the spinal cord, is given in Figure 2.9.

The interior of an intervertebral disk does not have a blood supply, and it must be fed by diffusion through the fibrous outer ring. Pressure on the disk creates a diffusion gradient from the interior to the exterior, so that the tissue fluid is squeezed out. When the pressure is reduced, this gradient is reversed, and the tissue fluid diffuses back, taking nutrients with it. Thus, to keep the disks well-nourished and in good condition, they need to be subjected to frequent changes of pressure, similar to a pumping mechanism.

The structure of the lower back is significant in discussing back injuries to workers. The area of focus is typically the intervertebral disk

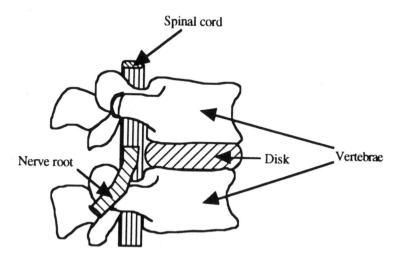

Fig. 2.9 Schematic illustration of an intervertebral disk, lying between two vertebrae, and the spinal cord and a nerve root branching out from the cord.

between the fifth lumbar (L5) and the first sacral (S1) vertebrae. A major concern of biomechanics is to study compressive and shear forces acting on this disk.

2.5 THE HAND STRUCTURE

There are a total of 28 bones in the hand and wrist, including 12 phalanges in the fingers, two phalanges in the thumb, five metacarpal bones in the palm of the hand, eight carpus (carpal bones) in the wrist, and one small bone close to the head of the metacarpal in the thumb; this is called the sesamoid bone. The hand–wrist bones and their joints are depicted in Figure 2.10. The carpal bones form a small, tunnel-like structure in the wrist which is covered on the palmar side by the transverse carpal ligaments. This tunnel is called the carpal tunnel.

2.6 APPLICATIONS AND DISCUSSION

Chapter 2 is intended to acquaint the reader with the skeletal system and some of the terminology that an ergonomist might encounter. The very broad overview of the skeletal system is not meant to be a

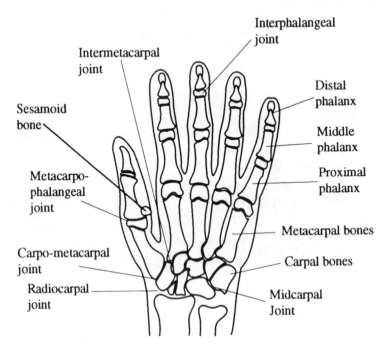

Fig. 2.10 Hand and wrist bones.

substitute for a course in basic anatomy. The ergonomist is urged to refer to an anatomy text (or a functional anatomy text such as those used by physical therapists, kinesiologists, or occupational physicians). The objective of this chapter is not to lead ergonomists to believe that they are experts in anatomy, but rather to provide a basis for them to discuss anatomical issues with ergonomics team members trained in anatomy. There are many excellent references available, which do not typically focus on ergonomics, but they can be very helpful in understanding human anatomy.

REVIEW QUESTIONS

1. What are the components of the skeletal system? What are the three predominant types of tissues of this system?
2. How many bones are there in the human skeleton?
3. The bones in the human body are usually classified into four categories according to their shapes. Name these four categories and give an example of each.
4. Name the bone in the neck that does not articulate with any other bones.
5. What critical functions are performed by the skeletal system?
6. The gross structure of the human skeleton consists of two parts. Name the two parts.
7. Describe why cartilages are said to be avascular.
8. Explain how nutrients and oxygen can reach the cartilage cells.
9. List the three types of cartilage tissues.
10. Briefly describe the diarthroses and synarthroses types of joint articulations and give an example of each.
11. Define flexion, extension, adduction, abduction, circumduction, rotation, pronation, and supination.
12. How many cervical, thoracic, lumbar, sacral, and coccygeal vertebrae are there in the spinal column?
13. How many bones are there in the fingers, thumb, palm of the hand, wrist, and in the whole hand?

REFERENCES

Anthony, C.P. and Kolthoff, N.J. (1975) *Textbook of Anatomy and Physiology*, 9th edn. C.V. Mosby, St Louis.
Åstrand, P.O. and Rodahl, K. (1986) *Textbook of Work Physiology: Physiological Bases of Exercise*, 3rd edn. McGraw-Hill, New York.

Muscular system and work

3

3.1 INTRODUCTION

The muscular system provides power for performing mechanical work. The muscles transform chemical energy stored in the body into physical activities. As in any engine, this energy transformation is performed with a mechanical efficiency. For example, about 20–25% of the utilized energy is transformed into mechanical work and the rest is dissipated as heat into the surrounding environment. The mechanical work is of various forms, such as moving parts of the body, carrying loads, and manipulating objects.

This chapter familiarizes the reader with basic phenomena concerning the muscular system. It also deals with muscle functions, muscular work, and the relevant basic laws of physics.

3.1.1 MUSCLES

The bones of our body are held together at their joints by means of ligaments. The bones can be moved by muscular activities (contraction and extension). Each muscle group (Fig. 3.1) is made up of several to thousands or millions of muscle fibers. The fibers of long muscles are usually bound together in bundles.

3.1.2 MOTOR NERVES AND MOTOR UNIT

Motor nerves supply signals from the nervous system to the muscle system. A single motor nerve fiber which innervates (supplies with nerve to) a group of (one to several thousand) muscle fibers within a muscle is called a **motor unit**. All the muscle fibers served (innervated) by the same motor nerve contract and relax at the same time, working as a unit.

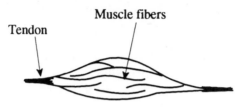

Tendon

Muscle fibers

Fig. 3.1 A muscle group.

3.1.3 MUSCLE CLASSIFICATIONS

The muscle system of the human body consists of three basic types of muscle fibers: smooth muscles (e.g., stomach muscles), cardiac muscles (i.e., the muscles of the heart) and striated or skeletal muscles (e.g., biceps brachii). **Smooth muscles**, which are automatic, contract and extend without requiring an external nerve supply and are relatively slow in contracting. Smooth-muscle fibers typically form the hollow cavities of the body (stomach, blood vessels, etc.) The size of blood vessels changes through contraction of smooth muscles. **Cardiac muscles** are also automatic and contract more rapidly than smooth muscles. Cardiac muscles also have the important property of being self-exciting. **Skeletal (striated) muscles** contract and extend voluntarily under some external stimuli and, hence, are often called voluntary muscles. Skeletal muscles are more rapid in contracting than cardiac and smooth muscles. Skeletal muscles are important in locomotion and movement, and consequently are of greatest interest to the ergonomist. This chapter will concentrate on skeletal muscle activities.

3.2 CHEMICAL REACTIONS DURING MUSCULAR CONTRACTIONS

The chemical reactions during muscular contractions are taking place either anaerobically (i.e., in the absence of oxygen) or aerobically (i.e., in the presence of oxygen). The anaerobic breakdown of adenosine triphosphate (ATP) in the muscle generates the energy for muscle contraction to take place. There are several sources of ATP available for the muscle:

- **ATP stored in the cell**. A small amount of ATP is stored in the cell and is readily available to support muscle contraction. Unfortunately, at maximal levels of muscular contraction, the stored ATP will only support approximately 1 second of muscular contraction. The general form of the chemical reaction is:

$$ATP \rightarrow ADP + P + energy$$

- **ATP-CP system.** A small amount of creatine phosphate (CP) is also stored in the cells. The creatine phosphate combines with the ADP (adenosine diphosphate generated in the breakdown of ATP) to regenerate an ATP that can be used for muscle contraction. During maximal levels of muscle contractions, the CP stores will support approximately 3–4 s of muscular activity. Chemically, this reaction is shown as:

$$CP \rightarrow P + C + energy$$

$$ADP + P + energy \rightarrow ATP$$

- **Anaerobic glycolysis.** Glucose stored in the cell or diffused into the cell from the circulatory system can be broken down anaerobically (without oxygen) to generate ATP for muscle contractions.

$$Glucose + P + ADP \rightarrow ATP + lactic \ acid$$

The body has a tolerance limit for lactic acid accumulation. For strenuous activities (maximal contractions), lactic acid tolerances will be reached in 35–45 s. Therefore, anaerobically, the body has less than 1 min to generate maximal-level muscle contractions. It is obvious that we cannot rely on anaerobic mechanisms to generate ATP for muscle contractions.

- **Aerobic glycolysis.** For most activities, glycogen (or fats or proteins) are broken down chemically with oxygen to generate ATP. This process can continue for a few seconds to several hours, depending on the level of demand of the activity being performed. For basal or low levels of activity, we rely on aerobic glycolysis for our muscle contractions to perform basic bodily functions for the duration of our lives. The basic chemical reaction can be described as:

$$Glucose + free \ fatty \ acids + P + ADP + O_2 \rightarrow CO_2 + H_2O + ATP$$

During most submaximal activities, aerobic glycolysis is utilized to generate the ATP needed for muscular contractions. The fuels normally used for aerobic glycolysis are fats and carbohydrates. Carbohydrates are more easily broken down and therefore are utilized as a primary source of fuel when fast responses are required. However, the body has a limited capability to store carbohydrates. Since carbohydrates are stored with water molecules, they are not an efficient fuel from the standpoint of storage efficiency. Fats, which have a much higher stored energy potential, are the primary source of fuel for low- to moderate-intensity activities. Table 3.1 shows the relative stored energy potential and storage capacities of fats and carbohydrates.

An individual's ability to perform at maximal or near maximal levels depends not only on the fuel sources but also on the physiologic characteristics of the person, including muscle fiber type, capillary

Table 3.1 Relative storage potentials and capacities for carbohydrates and fats

Fuel source	Stored energy potential	Stored energy capacity
Fats	9.1 kcal g^{-1}	50 000–75 000 kcal
Carbohydrates	4.3 kcal g^{-1}	1200–1500 kcal

Table 3.2 Differences between slow- and fast-twitch muscle fibers

	Muscle fiber	
Property	Slow-twitch (red fiber)	Fast-twitch (white fiber)
Speed of contraction	Slow	Fast
Strength of contraction	Low	High
Aerobic dependency	High	Low
Size	Small	Large
Capillary density	High	Low
Endurance	High	Low
Myoglobin content	High	Low
Mitochondrial enzyme activity	High	Low

density, myoglobin content, and mitochondrial enzyme activity. Muscle fibers are classified as type I (slow-twitch, oxidative, or red fibers) and type II (fast-twitch, glycolytic, or white fibers). More recently, type II fibers have been further divided into type IIa (a fast-twitch fiber with some of the characteristics of a slow-twitch fiber) and type IIb (a traditional fast-twitch fiber). The slow-twitch fibers are employed during endurance-type activities, and the fast-twitch fibers are utilized in performing short-duration, but high-intensity, activities. Table 3.2 summarizes some of the differences between fast- and slow-twitch fibers.

3.3 MUSCLE FIBER STIMULATION

Through phenomena such as selective permeability and active ion transport, membranes in the body usually have a resting potential that is not zero (that is, one side of the membrane is more positive than the other). Nerve and muscle membranes exhibit these properties. The muscle membrane maintains its resting potential until a physical, chemical, or electrical stimulus disturbs the equilibrium of the membrane and causes a depolarization of the membrane to occur. Once the depolarization has occurred, the action potential spreads along the membrane until the entire membrane has been stimulated. The depolarization and repolarization of the membranes generate the electrical activity in the muscle that can be measured with electromyography (EMG). An example of a motor unit stimulation and its depolarization and repolarization is shown in Figure 3.2.

In Figure 3.2, the resting potential of the membrane is −85 mV. When a stimulus is presented at time t, depolarization of the membrane begins. When the membrane potential in a healthy cell reaches its threshold value (in this case, −55 mV), the depolarization will continue to completion. At complete depolarization, the membrane potential has reversed to +45 mV. Then, through selective permeability and active ion transport, the membrane rapidly repolarizes to its resting potential (in this case −85 mV).

An important physiologic law, the **all-or-none law**, states that once the threshold has been reached, an action potential will continue to completion (that is, the membrane will depolarize and then repolarize). The all-or-none law also applies to muscle fibers. For muscles, the law states that once adequately stimulated, a muscle fiber will contract completely.

In order to generate the proper amount of strength to perform the task at hand, the proper muscle fibers must be recruited. Two basic strategies are utilized to achieve increased muscle strength:

- The process of **temporal summation** refers to increasing the rate of stimulation of the muscle fiber. Increased strength can be achieved by restimulating the muscle fiber before it has a chance to recover. The muscle strength will increase until tetanization (the point at which an increased stimulation rate becomes indistinguishable) is achieved.

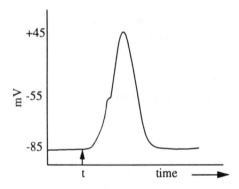

Fig. 3.2 Membrane potential and changes during an action potential.

- The process of **spatial summation** refers to recruiting more motor units to achieve increased levels of muscle strength. According to the **size principle**, smaller motor units are added first and then increasingly larger motor units are recruited until the desired strength is achieved. This recruitment strategy allows for smooth output and movement. In typical human activity, both temporal and spatial summation are used in conjunction to achieve the desired muscular output.

3.4 FUNCTIONAL CHARACTERISTICS OF MUSCLE TISSUES

Muscle tissues have the following five distinctive characteristics:

- **conductivity**: ability to transmit impulses;
- **irritability**: ability to respond to a stimulus;
- **extensibility**: ability to be stretched;
- **elasticity**: ability to return to their original length when stretching force is removed;
- **contractibility**: ability to contract or shorten.

3.5 MUSCLE CONTRACTION

A muscle can contract (shrink) to half its normal length (Murrell, 1965; Grandjean, 1988). Thus, the amount of movement by a muscle depends on the normal length of its individual fibers. On the other hand, the force that can be exerted by a muscle fiber depends on its cross-sectional area, not its length, so that the strength of the muscle depends on the number of its fibers. Therefore, the strength of a muscle is the sum of the strengths of its muscle fibers.

The maximum strength of a human muscle is between 30 and 40 N $\cdot$ cm^{-2} of the cross-sectional area of the muscle (Grandjean, 1988). Thus, a muscle of 1 cm^2 cross-sectional area can hold a weight of 3–4 kg (i.e. 30–40 N). A muscle has its greatest force at its relaxed (resting) length. As the muscle shortens or lengthens, its power declines.

Not all the muscle fibers are necessarily stimulated at the same time (as stated earlier, smaller motor units are recruited first and then progressively larger motor units until the desired strength is attained). Therefore, not all the muscle fibers are necessarily contracted at the same time. Fatigue results in failure of a muscle to contract in response to a stimulus.

In general, women can exert on average about 30% less force than men. This is due to the fact that women have narrower (less cross-sectional area) muscles (Grandjean, 1988). However, the distributions of muscular strength overlap, so there are some women with greater strength values than some men. Human strength should not be generalized by sex, but rather should reflect the strength capabilities of a particular population and strength requirements of a particular job or set of jobs.

3.5.1 TYPES OF MUSCLE CONTRACTIONS

There are several different types of muscle contractions, with the more common types of contractions listed below:

- **Isometric** (static) contraction occurs during a prolonged state of contraction, in which the muscle remains at the same length and performs "no physical work," but tension within the muscle increases.
- **Isotonic** (dynamic) contraction occurs when the muscle shortens and performs work, but the tension within the muscle remains the same.
- **Isokinetic** (dynamic) contraction occurs when the muscle (or external load) moves at a constant velocity.
- **Isoinertial** (dynamic) contraction involves the movement of a constant load (free weights).
- **Fibrillation** is an abnormal contraction of the cardiac muscle without producing an effective movement. Fibrillation is typically the result of asynchronous firing of cardiac muscle fibers.
- **Convulsion** is an abnormal or uncoordinated smooth contraction of a group of muscles.

Of the six types of muscle contractions, only the first four are of concern in ergonomics and biomechanics.

During dynamic efforts the state of the muscles alternates between contraction and extension, and therefore between tension and relaxation. During this kind of muscular effort, the muscles act as a pump for the blood-return system. Muscle activity facilitates venous return by moving the blood in the veins from a series of **one-way valves** toward the heart. For this reason, dynamic efforts are less fatiguing than static efforts.

During static contractions, the internal pressure produced by the muscle tissue restricts the size of blood vessels, which prevents blood flow through the muscle and, consequently, the removal of waste products and the supply of oxygen to the muscle is reduced. This explains why we cannot sustain a static muscular effort for very long, but we are able to carry out a dynamic effort for a relatively long period of time. Therefore, isometric contractions should be designed out of work systems, as much as possible. Table 3.3 gives some common static (isometric) components of occupational work.

Under approximately similar conditions static muscular work (effort), as compared to dynamic work, leads to a higher energy consumption and raised heart rate, and longer rest periods are needed.

Table 3.3 Some examples of static components of work

Static component of work	Situational examples
Pushing/pulling heavy objects	Pushing a heavily loaded wheelbarrow or cart, especially with sticky wheels
Holding things in the hands without any arm movement	Using one hand as a clamp in a sawing task
Holding the arms out or the elbows away from the body	Using a keyboard on a high desk and overreaching in many assembly tasks
Prolonged standing in one place	Doorman's task
Supporting the body weight by one leg while the other works a pedal	Operating a machine by pedaling in a standing posture
Bending the back forwards or sideways	Lifting or working on a low bench
Tilting the head forwards or backwards	Using a microscope
Raising the shoulders for long periods	Working above the elbow level

3.5.2 COMPONENTS OF PHYSICAL FITNESS

Physical fitness is an index of the body's ability to meet the demands of physical activities, and generally consists of the following four components (Whitney *et al.*, 1990):

- **Flexibility**, which is the ability to bend without sustaining any injury. Flexibility depends on the elasticity of muscles, tendons, and ligaments, as well as on the condition of the joints. A stretching program should be included with any fitness program.
- **Strength**, which is the ability of muscles to work against resistance. Strength training can improve muscular strength significantly. Strength training should be specific to the muscle group being trained. The best training is performing the actual activity for which the individual is being trained (**specificity of training principle**). Once a training level has been attained, no further training will take place unless the system is overloaded (**overload principle**). Repetition maximum (RM) training protocols have been shown to be successful in strength training (Asfour, 1980). A subject establishes a 10 RM value (the maximum amount of weight that the individual can handle for 10 repetitions under the task conditions). The training protocol then becomes to lift 10 repetitions at 50% of the 10 RM value, repeat with 10 repetitions at 75% of the 10 RM value, and finally 10 repetitions at the 10 RM value. Once the subject can lift the 10 RM value more than 12 times, a new 10 RM value is established. A

training program of three sessions per week for 6–12 weeks should show strength increases of 25% or more. Strength can be maintained by repeating the training protocol as little as once a week.

- **Muscle endurance**, which is the ability of a muscle to sustain repeated contractions over a period of time without becoming exhausted. Muscle endurance depends on the level of the exertion, whether the contraction is static or dynamic, and the motivation of the subject.
- **Cardiovascular endurance**, which is ability of the cardiovascular system to sustain an effort over a period of time. To improve cardiovascular endurance, a minimum of 20 min of aerobic exercise is necessary and must be repeated at least three times per week (Whitney *et al.*, 1990). The exercise must use most of the large muscle groups of the body.

3.5.3 MUSCULAR INCAPACITATION

Muscular incapacitation, whether mild or severe, can be induced by two primary situational causes (Murrell, 1965): a sudden exertion and a prolonged, continuous exertion of the muscle.

Under the first situation, the muscle is caught in the **state of shock** as an individual suddenly lifts a heavy object, especially in an awkward posture, or in a slipping incident tries to regain balance. Strains are likely to occur in many situations similar to these cases. Slipping incidents can often be foreseen and prevented by adequate design. Awkward lifting can be prevented by a proper job design and training individuals in proper lifting.

The second type of incapacitation, which develops by overexercising the muscle, is usually caused by the nature of job design. Knowledge of functional anatomy helps the job designer avoid allocating tasks that require an operator to use a few muscle groups continuously.

3.6 SKELETAL MUSCLES AND THEIR FUNCTIONS

Muscles are attached to bones by tendons. A tendon is a band of tough, inelastic fibrous tissue whose only function is to transmit the forces generated by the muscle to the bone to which it is attached. Each tendon is attached to an origin bone and an insertion bone (Fig. 3.3). Generally, the insertion bone moves when the muscle contracts.

As shown in Figure 3.3, the biceps at the front of the upper arm is attached to the radius (insertion bone) in the forearm and the scapula (origin bone) in the shoulder (Anthony and Kolthoff, 1975). Its contraction flexes the elbow joint. A muscle crosses one or more joints. The biceps muscle crosses both the elbow and shoulder joints.

To identify the muscle's function, physiologists (Crouch, 1972;

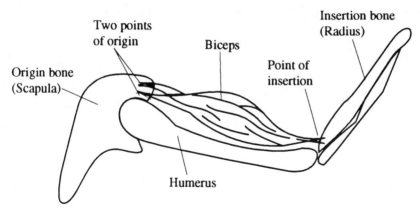

Fig. 3.3 Biceps muscle at the front of the upper arm is attached to radius (insertion) and scapula (origin) bones. Its contraction flexes the elbow joint.

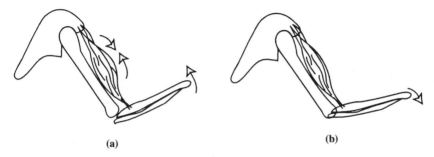

Fig. 3.4 Biceps as (a) a prime mover in elbow flexion and (b) an antagonist muscle during elbow extension.

Anthony and Kolthoff, 1975) have classified the skeletal muscles into prime movers, antagonists, and synergists:

- **Prime movers** – muscle or muscles whose contraction is the primary agent in producing a desired movement.
- **Antagonists** – muscles that relax or counteract to slow the action of a prime mover.
- **Synergists** – muscles that contract at the same time as the prime mover. By eliminating undesirable movements of a joint or holding a part steady, the synergist muscle assists the prime mover in producing a more effective movement.

As illustrated in Figure 3.4, the biceps brachii in the upper arm serves as a prime mover in flexing the elbow joint (Fig. 3.4a) and as an antagonist to the triceps brachii or gravity to slow the extension of the elbow (Fig. 3.4b).

3.7 MUSCULOSKELETAL SYSTEM AS A LEVER SYSTEM

When a muscle produces movement by pulling on the involved insertion bone across the joint, the insertion bone acts as a lever and the joint as a fulcrum of the lever. Similar to the mechanical situations, there are three classes of lever systems depending on the location of the joint (fulcrum) in relation to the points of force and resistance (Crouch, 1972):

- **Class I**: The joint (fulcrum) lies between the force point (where the muscle pulls on the lever) and resistance point (i.e., the weight to be lifted). The neck (Fig. 3.5a) is an example in the human body and a seesaw is a mechanical example of this class of lever).
- **Class II**: The resistance is between the joint (fulcrum) and the force (insertion) point. The foot (Fig. 3.5b) in walking is an example in the human body and a wheelbarrow is a mechanical example of this class of lever.
- **Class III**: The force point lies between the joint (fulcrum) and the resistance. The forearm (Fig. 3.5c) is an example in the human body and a fishing rod is a mechanical example of this class of lever.

The resistance can be the weight of the lever and its surrounding tissues, the counter force, other internal resistance (function of the joint, antagonistic muscles, etc.), and the external load.

3.7.1 MECHANICAL ADVANTAGE

The segment of the lever between the point of force exertion and the fulcrum is called the **force arm** and, likewise, the segment between the resistance and the fulcrum is called the **resistance arm**.

The mechanical advantage (MA) is the measure of the efficiency of the movements of parts of the body, which is computed as the ratio of the resistance overcome (R) to the force exerted (F):

$$\text{MA} = \frac{\text{Resistance } (R) \text{ overcome}}{\text{Force } (F) \text{ exerted}} = \frac{\text{Force arm}}{\text{Resistance arm}}$$

Example 3.1

Given the information in Figure 3.6, calculate the mechanical advantage and the muscular force required to move the weight upward.

MA = Force arm/resistance arm = 3/15 = 0.2
Muscle force required = R/MA = 10/0.2 = 50 lb (44.5 N)

Thus, a muscular force of greater than 50 lb would be required to move the weight upward.

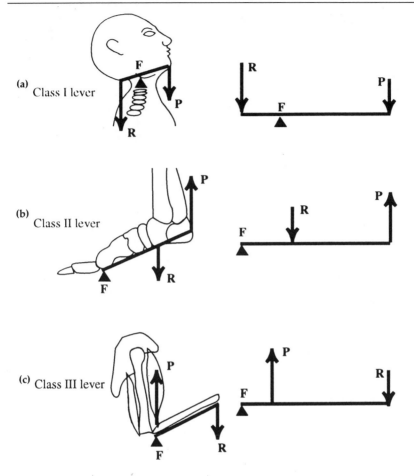

(a) Class I lever

(b) Class II lever

(c) Class III lever

Fig. 3.5 Lever classes in the human musculoskeletal system.

3.7.2 FORCE COMPONENTS

It should be noted that the muscle's angle of exertion continuously changes during the motion. The size of the angle affects the required muscular force in moving the insertion bone. Some part of the pull force exerted by the muscle is not contributing to the lever bone's movement, except when the pull force is at a right angle to the lever. The force is then divided into its two components (Fig. 3.7):

- **rotary** component, perpendicular to the lever and contributing to the lever's movement;
- **non-rotary** component, directed toward the fulcrum, parallel to the lever, and not contributing to the lever movement.

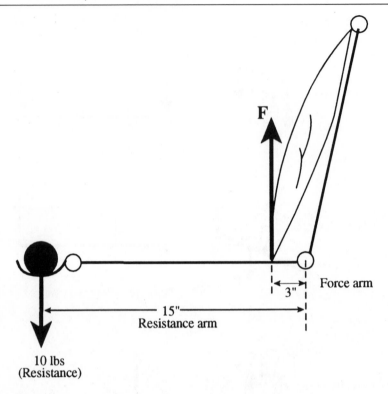

Fig. 3.6 An example of the mechanical advantage of the arm.

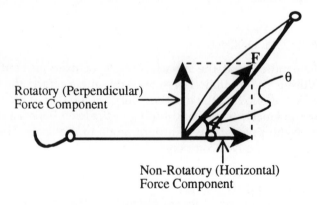

Fig. 3.7 Rotary and non-rotary components of a force.

If we let θ be the angle between the direction of the force and the lever axis, then:

Rotary force $= F \cdot Sin\ \theta°$
Non-rotary force $= F \cdot Cos\ \theta°$

It can be seen that as θ° gets smaller, the Sin θ and cons........ the rotary force component gets smaller and the non-rotary component becomes larger. This means that, as the angle at which the muscle exerts a pull force moves from right angle (to either side), there would be less mechanical advantage.

3.8 GUIDELINES FOR MUSCLE USE

If a static effort is repeated every day over a long period, permanent aches can appear in the limbs and may involve not only the muscles but also the joints, tendons, ligaments, and other tissues. Repeated static effort can therefore lead to a musculoskeletal disorder, that is, a damage to joints, ligaments, and tendons, such as (Grandjean, 1988):

- arthritis of the joints due to mechanical stress;
- inflammation of tendons and tendon sheaths (tendinitis and teno-synovitis);
- chronic degeneration of the joints (symptoms of arthrosis);
- muscle spasms;
- intervertebral disk troubles.

Consideration of the guidelines provided below in designing jobs and workplaces helps significantly reduce the risk of developing musculo-skeletal disorders and muscular inefficiency.

3.8.1 GENERAL GUIDELINES

The optimal use of muscle strength can be achieved by adapting the following guidelines:

- To maximize the muscle efficiency and skillfulness in performing a heavy workload, appropriate movements must be made to allow muscles to develop as much power as possible.
- When work involves a static load, the worker must assume an appropriate posture that employs as many strong muscles as possible in handling the load. This reduces the load on the involved muscle significantly.
- Since a muscle typically generates its greatest tension at or near its resting length, appropriate postures (resting lengths of muscles) should be assumed.

- If several muscles are involved in handling a load, the joint force exerted is maximized when as many muscles as possible contract simultaneously.

3.8.2 GUIDELINES FOR MUSCLE USE IN SITTING POSTURE

The following guidelines are for muscle use when sitting with the back supported against a backrest (Caldwell, 1959):

- The hand is significantly stronger in pronation (turning inwards) than during supination (turning outwards).
- The hand is significantly stronger when it is pulling downwards than when it is pulling upwards.
- The hand is more powerful when pushing than when pulling.
- Pulling strength is greatest at a grasping distance of 70 cm (28 in).
- Pushing strength is greatest at a grasping distance of 50 cm (20 in) in front of the axis of the body.

3.8.3 GUIDELINES FOR MUSCLE USE IN STANDING POSTURE

The following guidelines are for muscle use during a standing posture (Grandjean, 1988):

- The pushing force is greater than the pulling force.
- Both pushing and pulling forces are greater in the vertical plane than in the horizontal plane.
- The pushing and pulling forces are of the same order of magnitude whether the arms are held out sideways or forwards.

3.9 APPLICATIONS AND DISCUSSION

As in the previous chapter on the skeletal system, the muscular system is introduced to the ergonomist and basic concepts are presented. For a more thorough treatment of muscle contraction, texts on anatomy and physiology and kinesiology should be consulted. However, in understanding the physical demands of work, the ergonomist should also understand the basics of muscle contraction and how the worker generates the forces necessary to meet job demands.

REVIEW QUESTIONS

1. What is motor unit?
2. Name the three types of muscles and give an example of each. Which types are automatic and which voluntary?
3. What type of muscle fibers controls the size of blood vessels?
4. What type of muscle fibers has the property of being self-exciting?

5. What type of muscle fibers is fastest in contraction?
6. What type of muscles is of greatest interest to the ergonomist? Why?
7. There are several sources of ATP available for supporting muscle activities (contractions). List the four main sources, along with the general forms of the corresponding chemical reactions.
8. What are the fuel sources normally used for aerobic glycolysis? What are the amount of energy potentials and capacities for each fuel source?
9. What does a person's ability to perform at maximal or near maximal level depend on?
10. In performing what type of activities are slow-twitch and fast-twitch fibers utilized?
11. Describe the all-or-none law of physiology.
12. Describe the two basic strategies that are utilized to achieve increased muscle strength for performing a task at hand.
13. Describe the size principle of recruiting motor units for increasing muscle strength to perform a task at hand.
14. List and briefly explain the five functional characteristics of muscle tissues.
15. What is the effect of fatigue on the muscle contraction?
16. Explain why, in general, women can exert on average about 30% less force than men.
17. There are several different types of muscle contractions. Of the six common types of contractions listed in the text, which four are of concern in ergonomics and biomechanics?
18. Describe the dynamic and static muscle contractions. Also, explain why dynamic efforts are less fatiguing than static efforts, and why we cannot sustain a static muscular effort for very long.
19. Describe the physiological disadvantages of static effort, as compared to dynamic effort.
20. Give three examples of static components of work with corresponding situational work activities.
21. List the four components of physical fitness.
22. List the two primary situational causes of muscular incapacitation, whether mild or severe.
23. Describe the three classes of lever systems. Give examples for human skeletal and mechanical systems.
24. Briefly describe the functions of these classes of skeletal muscles: prime movers, antagonists, and synergists.
25. Describe the terms force arm and resistance arm, and mechanical advantage.
26. Describe the rotary and non-rotary components of force. If θ is the angle between the direction of the force and the lever axis, how are these components calculated?

27. What guidelines can you provide for the optimal use of muscle strength?
28. What guidelines can you provide for muscle use in sitting posture with the back against a backrest?
29. What guidelines can you provide for muscle use during standing posture?

REFERENCES

Anthony, C.P. and Kolthoff, N.J. (1975) *Textbook of Anatomy and Physiology*, 9th edn. C.V. Mosby, St Louis.

Asfour, S. S. (1980) *Energy Cost Prediction Models for Manual Lifting and Lowering Tasks*. PhD Dissertation. Texas Tech University, Lubbock, Texas.

Caldwell, L.S. (1959) *The Effect of the Special Position of a Control on the Strength of Six Linear Hand Movements*. Report no. 411. US Army Medical Research Laboratory, Fort Knox, Kentucky.

Crouch, J.E. (1972) *Functional Human Anatomy*, 2nd edn. Lea & Febiger, Philadelphia.

Grandjean, E. (1988) *Fitting the Task to the Man: A Textbook of Occupational Ergonomics*, 4th edn., Taylor & Francis, London.

Murrell, K.F.H. (1965) *Human Performance in Industry*. Reinhold, New York.

Whitney, E.N., Hamilton, E.M.N. and Rolfes, S.R. (1990) *Understanding Nutrition*, 5th edn. West, St Paul, MN.

Engineering anthropometry

4

4.1 INTRODUCTION

Anthropometry is the study of the dimensions and certain other physical characteristics of the human body such as weights (or masses), volumes, centers of gravity, inertial properties of body segments, and strengths of various muscle groups. Anthropometric measurements are a critical element in equipment and workplace/work-space design. Utilization of anthropometric data will enable designers to accommodate a desired portion of the potential user population in their designs.

Large volumes of tables and figures of anthropometric data have been compiled and published (Van Cott and Kinkade, 1972; NASA, 1978b) for people of different sexes and ages, and from different geographical regions of the world. These data have been also classified for various percentiles of populations (e.g., 5th, 10th, 50th, 90th, 95th percentiles).

In anthropometric tables, it is important to understand the population (a group of people) being described. The ergonomist might use data for "adult males" that was obtained from a military adult male population and may or may not be representative of the population of concern to the ergonomist. It may be as broad as the 18–55-year-old male population of the world or as narrow as the few people working on an assembly line in a specific production facility. Obviously, if the ergonomist is designing locker rooms for elementary schools, the population would be school-age children in the elementary grades, while if a locker room were being designed for a professional basketball team, the design population would be very different.

The basic application of anthropometry in design is finding appropriate dimensions to be incorporated in the design. There are two types of dimensions that determine what the design dimensions should be – clearance and reach dimensions.

Clearance dimensions determine the minimum space required for a human being to perform work activities in the workplace, such as operating and maintaining machines and so on. Clearances are established by the larger people from the expected user population (e.g., the size of a door frame is determined by the size of the largest expected user).

Reach dimensions determine the maximum space allowable for the human being who operates equipment and are established by the smaller people in the expected user population (e.g., control height is determined by accommodating shorter people).

4.2 BASIC ERGONOMIC DESIGN PHILOSOPHIES

There are three basic design philosophies utilized by ergonomists as they apply anthropometric data to design for their specific population. The philosophies are:

- **Design for the average.** The problem of designing for the average is that the design may end up fitting no one, because no one is average in all dimensions. However, the design-for-the-average philosophy is utilized for designs involving public facilities, such as park benches, bus seats, and other facilities used by a large variety of people.
- **Design for the extremes.** The problem of designing for extremes is the cost associated with such a design philosophy. Assuming that an automobile seat is designed to accommodate the smallest person, would it be feasible for the largest person to use the vehicle? Task criticality and understanding the extremes of the design population are obviously critical to such a design philosophy.
- **Design for a range.** The most common design philosophy of the ergonomist is to design for a range of the population. A typical range of the 5th to 95th percentile of the population is used. Such a design would be expected to accommodate 90% of the design population. Design ranges can be wider or narrower and are typically determined by task criticality and cost.

4.3 STATISTICAL BASIS OF ANTHROPOMETRY

Assume that the random variable X represents a specific anthropometric measure (e.g., seating height) of a certain population. Assume also that x_p is the pth percentile of X. Then we define

$$P(X \leq x_p) = p \qquad (4.1)$$

as the proportion of the population whose seating heights are equal to or less than x_p. That is, "$100*p$" percent of that population are of seating heights equal to or less than x_p.

Percentiles provide a basis for estimating the proportion of a population accommodated by a design. They can be used to determine the design cut-off points for rejecting those individuals whose body dimensions do not fit the design. Percentiles, thus, can be applied to assess size limitations imposed by a design. This is used in the selection of individuals who fit an existing design, and the proportion of the population inconvenienced by the design is found.

It is usually assumed that human anthropometric characteristics are normally distributed. That is, we can state that the anthropometric variable X has a normal distribution with mean μ and standard deviation σ. Then, $Z = (X - \mu)/\sigma$ has a standard normal distribution and we can rewrite equation (4.1) as:

$$P(X \leqslant x_p) = P[(X - \mu)/\sigma \leqslant (x_p - \bar{x})/s] = p \text{ or } P(Z \leqslant z_p) = p$$

where $z_p = (x_p - \bar{x})/s$ values are given in standard normal probability (often called the Z-distribution) tables for values of p (note that the size of the sample for which $\bar{x}$ is calculated should be $n \geqslant 30$; otherwise, t_p and the t-distribution table must be used instead of z_p). We can use $z_p = (x_p - \bar{x})/s$ to write the following equation:

$$x_p = \bar{x} + z_p \cdot s \tag{4.2}$$

where: $\bar{x}$ = sample mean;

s = sample standard deviation

x_p = pth percentile value of the variable X;

z_p = standard normal value corresponding to the pth percentile value of X.

The values of 0.005, 0.01, 0.025, 0.05, 0.10, 0.165, 0.25, 0.75, 0.835, 0.90, 0.95, 0.975, 0.99 and 0.995 are commonly used for p to indicate respectively the 0.5th, 1st, 2.5th, 5th, 10th, 16.5th, 25th, 75th, 83.5th, 90th, 95th, 97.5th, 99th and 99.5th percentiles. The z_p values are given in Table 4.1. (A standard normal distribution table is provided in Appendix C.)

Example 4.1

Based on an anthropometric measurement taken on a randomly selected sample from a given population, the mean seat-to-eye height was 31.8 in (80.8 cm) with a standard deviation of 1.3 in (3.3 cm). We

Table 4.1 Standard normal values for computing the commonly used percentiles

Percentile	0.5th	1st	2.5th	5th	10th	16.5th	25th
(p)	(0.005)	(0.01)	(0.025)	(0.05)	(0.10)	(0.165)	(0.25)
Z_p	−2.575	−2.327	−1.960	−1.645	−1.282	−0.974	−0.675

Percentile	75th	83.5th	90th	95th	97.5th	99th	99.5th
(p)	(0.75)	(0.835)	(0.90)	(0.95)	(0.975)	(0.99)	(0.995)
Z_p	0.675	0.974	1.282	1.645	1.960	2.327	2.575

calculate the seat-to-eye height of the 95th percentile; the 50th percentile; and the 10th percentile as follows:

- $x_{0.95} = 31.8 + 1.64\ (1.3) = 33.9$ in (86.1 cm)
- $x_{0.50} = 31.8$ in (80.8 cm)
- $x_{0.10} = 31.8 - 1.28\ (1.3) = 30.1$ in (76.5 cm)

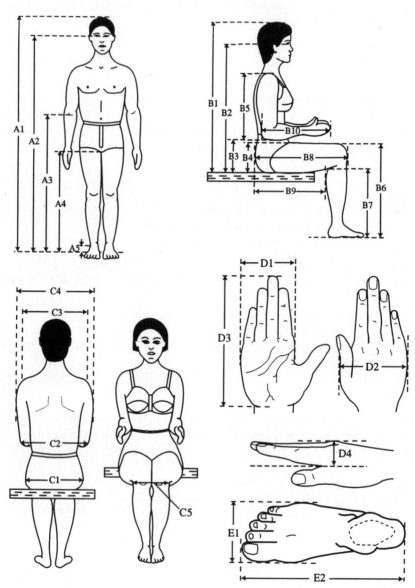

Fig. 4.1 Illustration of anthropometric measurements listed in Table 4.2.

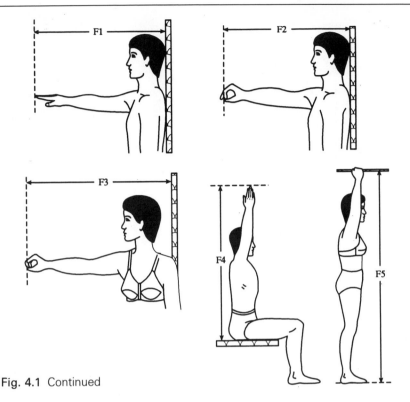

Fig. 4.1 Continued

4.4 ANTHROPOMETRIC DATA

There are two types of anthropometric data: structural and functional measurements.

- **Structural (static) measurements**: These measurements are concerned with the dimensions of the body segments at stationary positions (not in motion). They include the body's contour dimensions (e.g., stature, sitting height, length of upper arm, hip breadth) and skeletal dimensions (e.g., the distance between the centers of joints, such as between the hip and the knee).
- **Functional (dynamic) measurements**: These measurements are concerned with the dimensions of the body segments during physical activities. Examples of such measurements are reach envelopes, crawling length and height, kneeling height, and range of angular movement of joints.

Engineering anthropometry deals with the application of both static and dynamic measurements of people to derive recommendations for workplaces, tools, and products.

Table 4.2 Anthropometric data for the US adult population

Ill. no.	Measurement	Male population Mean	SD	Ref. pop.	Female population Mean	SD	Ref. pop	50–50 mixed male–female Mean	SD
Standing measurements									
A1	Stature (standing height)	173.2	6.9	[1]	160.3	6.6	[1]	166.8	9.3
A2	Eye height, standing	164.5	6.8	[6]	151.6	6.3	[2]	158.1	9.2
A3	Elbow height, standing	109.6	4.9	[3]	104.3	3.8	[5]	106.9	5.1
A4	Crotch height	83.4	4.5	[3]	74.6	3.9	[7]	79.0	6.1
A5	Ankle height	11.9	1.4	[3]	11.2	1.4	[7]	11.5	1.4
Seated measurements									
B1	Sitting height (erect)	90.5	3.7	[1]	84.7	3.6	[1]	87.6	4.7
—	Sitting height, relaxed (not pictured)	86.4	3.7	[1]	81.8	3.8	[1]	84.1	4.4
B2	Eye height, sitting	79.8	3.3	[3]	73.4	3.1	[7]	76.6	4.5
B3	Elbow rest height, sitting (from seat)	24.1	3.0	[1]	23.1	2.9	[1]	23.6	3.0
B4	Thigh clearance (thickness), sitting	14.3	1.7	[1]	13.7	1.8	[1]	14.0	1.8
B5	Shoulder-to-elbow length	36.5	1.8	[3]	33.7	1.5	[5]	35.1	2.2
B6	Knee height, sitting	54.2	2.9	[1]	49.7	2.7	[1]	52.0	3.6
B7	Popliteal height, sitting	44.0	2.7	[1]	39.7	2.6	[1]	41.9	3.4
B8	Buttock-to-knee length, sitting	59.1	2.9	[1]	56.8	3.1	[1]	58.0	3.2
B9	Buttock-to-popliteal length, sitting	49.4	3.1	[1]	48.0	3.1	[1]	48.7	3.2
B10	Forearm–hand length	47.9	2.7	[6]	43.5	1.7	[5]	45.7	3.1
Breadth measurements									
C1	Hip breadth, sitting	35.4	2.8	[1]	36.7	3.7	[1]	36.1	3.3
C2	Elbow-to-elbow breadth	42.0	4.7	[1]	39.0	5.4	[1]	40.5	5.3
C3	Shoulder (biacromial) breadth	39.6	2.1	[1]	35.4	1.9	[1]	37.5	2.9
C4	Shoulder (bideltoid) breadth	45.8	2.5	[3]	40.6	1.6	[5]	43.2	3.3
C5	Knee-to-knee breadth	21.4	1.8	[3]	17.7	0.9	[5]	19.5	2.3

Hand measurements

		Men			Women			Mixed	
D1	Hand breadth	8.9	0.5	[3]	7.5	0.5	[7]	8.2	0.8
D2	Hand breadth across thumb	10.5	0.6	[3]	9.2	0.6	[4]	9.8	0.9
D3	Hand length	19.7	1.0	[3]	17.2	0.9	[4]	18.4	1.6
D4	Hand thickness (at metacarpal III)	3.0	0.2	[3]	2.5	0.2	[4]	2.7	0.3

Foot measurements

		Men			Women			Mixed	
E1	Foot breadth	9.8	0.6	[3]	8.9	0.6	[7]	9.4	0.7
E2	Foot length	26.8	1.3	[3]	24.0	1.1	[7]	25.4	1.8

Other measurements

		Men			Women			Mixed	
F1	Arm reach (from wall)	87.1	4.6	[6]	78.8	3.3	[5]	83.0	5.6
F2	Thumb-tip reach (from wall)	79.2	4.2	[3]	74.1	3.8	[7]	76.6	4.8
F3	Thumb-tip reach, extended	89.6	4.5	[10]	83.7	4.8	[7]	86.7	5.5
F4	Vertical reach, sitting	139.8	5.7	[9]	132.9	4.7	[5]	136.4	6.3
F5	Vertical grip reach (standing)	209.6	8.5	[8]	199.2	8.4	[7]	204.4	9.9
F6	Buttock depth	24.0	2.2	[9]	20.4	1.0	[5]	22.2	2.5

Body weight

		Men			Women			Mixed	
	Weight (kg) (not pictured)	74.9	12.6	[1]	63.7	13.8	[1]	69.3	14.4

Notes

- Ill. No. = Illustration number, a cross-reference to Figure 4.1 and Table 4.3.
- The 50–50 mixed male–female population means and standard deviations (SD) were calculated according to Tayyari (1993).
- All data were extracted from NASA (1978b), unless otherwise stated in the following note.
- Ref. pop. means the referenced population, listed as:

[1] = US civilians of age 18–72 years;
[2] = US adult civilians (Kroemer et al., 1994);
[3] = US Air Force basic trainees (men);
[4] = US Air Force basic trainees (women);
[5] = Stewardesses;
[6] = Army drivers (men; Damon et al., 1966);
[7] = US Air Forces enlisted women;
[8] = US Military personnel (men; Damon et al., 1966);
[9] = Air traffic controllers (men);
[10] = US Air Force flying personnel.

Table 4.3 Recommended clothing allowances and the definitions of the anthropometric measurements

Ill. no.	Dimension	Clothing allowance cm (in)		Definition of the dimension	Other considerations
A1	Stature (standing height)			The height of the top of the head; measured from floor; the subject stands erect, looking straight ahead	The average slump in stature from erect to the normal standing position is about 2 cm (0.75 in). At the end of the day, stature decreases by 2.4 cm (0.95 in), due to compression of the inter-vertebral disks
	Men's shoes	2.5	(1.0)		
	Women's shoes*	7.5	(3.0)		
	Work boots	3.25	(1.3)		
	Civilian headgear	2.5	(1.0)		
	Hard hat	7.5	(3.0)		
A2	Eye height, standing			The height of the inner corner of the eye (from floor); the subject stands erect, looking straight ahead	Consider the 2-cm (0.75-in) average slump in stature during normal standing
	Men's shoes	2.5	(1.0)		
	Women's shoes*	7.5	(3.0)		
	Work boots	3.25	(1.3)		
A3	Elbow height, standing			The height of the radial (the depression at the elbow between humerus and radius) measured from floor	Consider the slumping effect. The preferred work-surface height depends on the kind of task and the tools used
	Men's shoes	2.5	(1.0)		
	Women's shoes*	7.5	(3.0)		
	Work boots	3.25	(1.3)		
A4	Crotch height			The height of the midpoint of the crotch	
	Men's shoes	2.5	(1.0)		
	Women's shoes*	7.5	(3.0)		
	Work boots	3.25	(1.3)		
A5	Ankle height			The height of the level of minimum circumference of the leg	
	Men's shoes	2.5	(1.0)		
	Women's shoes*	7.5	(3.0)		
	Work boots	3.25	(1.3)		

B1	*Sitting height* Heavy clothing (under buttocks) Civilian headgear Hard hat	0.6 (0.25) 2.5 (1.0) 7.5 (3.0)	The height of the top of the head, measured from the sitting surface; the subject sits erect	The average slump from the erect to the comfortable sitting position is 3.75 cm (1.5 in) for men and 2.75 cm (1.1 in) for women
—	*Sitting height, relaxed* Heavy clothing (under buttocks) Civilian headgear Hard hat	0.6 (0.25) 2.5 (1.0) 7.5 (3.0)	The height of the top of the head, measured from the sitting surface; the subject sits relaxed (normal)	The same average slump from the erect to the comfortable sitting position is 3.75 cm (1.5 in) for men and 2.75 cm (1.1 in) for women
B2	*Eye height, sitting* Heavy clothing (under buttocks)	0.6 (0.25)	The height of the inner corner of the eye above the sitting surface; the subject sits erect, looking straight ahead	The same average slump as for the sitting height
B3	*Elbow-rest height*	No difference	The height of the bottom of the tip of the elbow above the sitting surface; the subject sits erect, the upper arm vertical at one side, the forearm at a right angle to the upper arm	That under the buttocks is balanced by that under the elbow
B4	*Thigh clearance* Light clothing Heavy clothing	0.5 (0.2) 3.5 (1.4)	The height of the highest point of the thigh above the sitting surface; the subject sits erect, knees and ankles at right angles	
B5	*Shoulder–elbow length* Light clothing Heavy clothing	0.5 (0.2) 2.5 (1.0)	The vertical distance from the acromion (at the uppermost point of the largest edge of the shoulder) to the bottom of the elbow, measured with the elbow bent 90° and the forearm held horizontal	Hunching the shoulder (bowing the back) increases the needed shoulder–elbow space

Table 4.3 Continued

Ill. no.	Dimension	Clothing allowance cm (in)		Definition of the dimension	Other considerations
B6	*Knee height, sitting*			The height of the uppermost point on the knee (not the kneecap); the subject sits erect, knees and ankles at right angles	Depending on their locations, pedals and footrests may raise the foot and the knee
	Men's shoes and light clothing	2.5	(1.0)		
	Women's shoes* and light clothing	7.5	(3.0)		
	Work boots and heavy clothing	3.75	(1.5)		
B7	*Popliteal height, sitting*			The height of the underside of the thigh immediately behind the knee; the subject sits erect, knees and ankles at right angles	Depending on their locations, pedals and footrests may raise or lower the popliteal region
	Men's shoes	2.5	(1.0)		
	Women's shoes*	7.5	(3.0)		
	Work boots	3.25	(1.3)		
B8	*Buttock-to-knee length*			The horizontal distance from the rearmost surface of the buttock to the front of the kneecap; the subject sits erect, knees and ankles at right angles	Shifting the buttocks forward on the seat requires additional clearance
	Light clothing	0.5	(0.2)		
	Heavy clothing	1.75	(0.7)		
B9	*Buttock-to-popliteal length*			The horizontal distance from the rearmost surface of the buttock to the back of the lower leg; the subject sits erect, knees and ankles at right angles	When seat length is fixed, coats and trousers push the popliteal region forward about 0.5 cm (0.2 in) by light clothing and 1.25 cm (0.5 in) by heavy clothing
	Light clothing	No difference			
	Heavy clothing				

Code	Dimension	cm (in)	Description	Notes
B10	*Forearm–hand length*		The distance from the tip of the elbow to the tip of the middle finger; the subject sits erect, the upper arm vertical at side, forearm, hand, and fingers extended horizontally	For manipulation by fingers, subtract 1.25 cm (0.5 in) for flip, 2.5 cm (1.0 in) for push; for manipulation by thumb and forefinger, subtract 7.5 cm (3.0 in); for grasp by whole hand, subtract 12.75 cm (5.0 in)
	Light clothing without gloves	0.5 (0.2)		
	Gloves	0.5 (0.2)		
	Medium clothing and gloves	1.25 (0.5)		
	Heavy clothing and gloves	2.5 (1.0)		
C1	*Hip breadth, sitting*		The breadth of the body, across the widest portion of the hips	Additional space should be added for relaxation and changing position
	Light clothing	1.25 (0.5)		
	Medium clothing	2.5 (1.0)		
	Heavy clothing	5.0 (2.0)		
C2	*Elbow-to-elbow breadth*		The distance across the lateral surfaces of the elbows; the subject sits erect, upper arms vertical and touching the sides, forearms held horizontally	The space needed across the elbows for relaxation is about 7 cm (2.75 in). Depending on the types of work space and controls, additional space is needed for the trunk and arm movements to operate controls
	Light clothing	1.25 (0.5)		
	Medium clothing	5.0 (2.0)		
	Heavy clothing	11.25 (4.5)		
C3	*Shoulder (biacromial) breadth*	No difference	The horizontal distance across the shoulders from right to left acromion	
C4	*Shoulder (bideltoid) breadth*		The horizontal distance across the maximum lateral protrusions of the right and left deltoid muscles; the subject sits erect, upper arms vertical and touching the sides, forearms held horizontally	Arm, trunk and shoulder movements require additional space across the shoulders
	Light clothing	0.75 (0.3)		
	Heavy clothing	3.75 (1.5)		

Table 4.3 Continued

Ill. no.	Dimension	Clothing allowance cm (in)		Definition of the dimension	Other considerations
C5	*Knee-to-knee breadth, sitting* Light clothing Medium clothing Heavy clothing	1.25 2.5 5.0	(0.5) (1.0) (2.0)	The maximum horizontal distance across the lateral surfaces of the knees, measured with knees at right angles, gently touching each other	Additional lateral space must be provided for both knees to allow relaxation
D1	*Hand breadth (at metacarpal)* Woolen or leather gloves Arctic handwear	0.75 2.5	(0.3) (1.0)	The maximum breadth of the hand, across the distal ends of the meta-carpal bones of the index and little fingers; the right hand held straight and stiff, fingers together	Functional hand breadth is larger than anthropometric hand breadth since in practice the hand is not held straight all the time
D2	*Hand breadth across thumb* Woolen or leather gloves Arctic handwear	0.75 2.5	(0.3) (1.0)	The breadth of the hand, measured at the level of the distal end of the first metacarpal of the thumb	
D3	*Hand length* Woolen or leather gloves Arctic handwear	0.6 1.25	(0.25) (0.5)	The distance from the base of the hand to the tip of the middle finger; the hand held straight and stiff	Gloves do not effectively increase functional hand reach
D4	*Hand thickness* Woolen or leather gloves Arctic handwear	0.5 3.75	(0.2) (1.5)	The thickness (distance between the dorsal and palmar surfaces) of the knuckle of the middle finger, fingers extended	The bent resting or working hand is thicker in the palm than the hand held flat

E1	Foot breadth			The maximum breadth of the foot, measured at right angles to its long axis; the subject stands with weight equally distributed on both feet
	Shoes	0.75	(0.3)	
	Work boots	1.25	(0.5)	
E2	Foot length			The length of the foot measured parallel to its long axis, from the back of the heel to the tip of the longest toe; the subject stands with weight equally distributed on both feet
	Shoes	3.0	(1.2)	
	Work boots	4.0	(1.6)	
F1	Arm reach (from wall)			The distance from the wall to the tip of the middle finger; the subject stands erect, with heels, buttocks, and shoulders pressed against a wall, the right arm and hand extended horizontally and maximally
	Light clothing	0.75	(0.3)	
	Light gloves	0.5	(0.2)	
	Medium and heavy clothing and gloves	1.25	(0.5)	Heavy clothing pushes the subject forward. However, it may hamper maximum reach. For fingertip manipulation, subtract 1.3 cm (0.5 in) for flip, 2.5 cm (1.0 in) for push; for manipulation by thumb and forefinger, subtract 7.6 cm (3.0 in); for grasp by whole hand, subtract 12.7 cm (5.0 in). Kyphosis (humpback) in old people increases their forward arm reach
F2	Thumb-tip reach (from wall)			The distance from the wall to the tip of the thumb measured while keeping both shoulders against the wall, the right arm extended forward, and the index finger touching the tip of the thumb
	Light clothing	0.75	(0.3)	
	Heavy clothing	1.25	(0.5)	

Table 4.3 *Continued*

Ill. no.	Dimension	Clothing allowance cm (in)		Definition of the dimension	Other considerations
F3	*Thumb-tip reach, extended*			Similar to thumb-tip reach, except that the right shoulder is extended as far as possible while keeping the left shoulder firmly against the wall	
	Light clothing	0.75	(0.3)		
	Heavy clothing	1.25	(0.5)		
F4	*Vertical reach, sitting*			The height of the tip of the middle finger above the sitting surface when the arm, hand, and fingers are extended vertically	The same considerations as noted for sitting height
	Woolen or leather gloves	0.6	(0.25)		
	Arctic handwear	1.25	(0.5)		
	Heavy clothing (under buttocks)	0.6	(0.25)		
F5	*Vertical grip reach (standing)*			The height of a pointer held horizontal in the subject's fist when the arm is maximally extended upward	The same considerations as noted for stature
	Men's shoes	2.5	(1.0)		
	Women's shoes*	7.5	(3.0)		
	Work boots	3.25	(1.3)		
F6	*Buttock depth*			The depth of the torso at the level of the maximum posterior protrusion of the buttock. The subject stands erect	
	Light clothing	0.5	(0.2)		
	Heavy clothing	1.75	(0.7)		

Sources: Damon *et al.* (1966), NASA (1978a), and Panero and Zelnik (1979).
*The adjustment for women's shoes should be based on heel height. Typically the adjustment should fall between 2.5 and 7.5 cm. The designer may choose a liberal or conservative estimate depending on the dimension and its application.

Table 4.2 provides the human body dimensions (for the US population) for use in design. The designer should be cautious about the fact that the body dimensions are based on the nude body, and that clothing adds to these dimensions. The work space needed for a 95th-percentile male with heavy clothing is larger than the work space for a 95th-percentile male with light clothing and much larger than the work space necessary for a 5th-percentile female with light clothing. Footwear and gloves are most often the center of attention in designing workplace and equipment. Therefore, it is necessary to make allowances for the operator's clothing and work gear. Table 4.3 contains recommended clothing allowances as well as the definitions of the anthropometric measurements listed in Table 4.2.

Figure 4.2 shows two extreme cases, comparing a 5th-percentile civilian in street clothing with a 95th-percentile soldier in arctic clothing (Hertzberg, 1972).

4.5 BODY SURFACE AREA

A close estimation of the whole body surface area (BSA), and its segmental surface areas, is important to the ergonomist, engineer, and many other professionals in such studies as:

- human response to thermal conditions (since the heat exchange takes place at the body surface);
- metabolic energy expenditure (e.g., the basal metabolic rate is usually

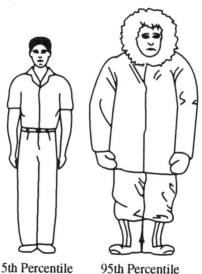

5th Percentile   95th Percentile

Fig. 4.2 Comparison of a 5th-percentile civilian in street cothing with a 95th-percentile soldier in Arctic clothing. Adapted from Hertzberg (1972).

expressed in kilocalories or oxygen consumption per square meter of the BSA);

- exposure to ionizing radiation.

The most commonly used formula for estimating the BSA is the DuBois equation (DuBois and DuBois, 1916):

$$BSA = 0.007184 * Wt^{0.425} * Ht^{0.725} \tag{4.3}$$

where: BSA = body surface area (m^2), Wt = body weight (kg), and Ht = body height (cm).

4.6 BODY SEGMENTS

Determination of the center of mass (or center of gravity) and weight distribution of the human body and its segments is needed in human biomechanical assessments. Such data, along with understanding basic mechanics, are useful in:

- design of equipment operated or occupied by humans in various postures;

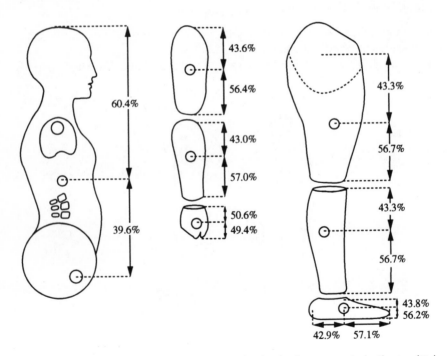

Fig. 4.3 The locations of centers of mass in the body segments in the sagittal plane indicated by percentages of the body segments. Adapted from Dempster (1955).

- design of workstations (e.g., seats for industrial workers, vehicle operators and aircraft pilots);
- biomechanical applications.

The locations of the centers of mass (c.m.) of body segments as a percentage of the length of their corresponding body segments are graphically shown in Figure 4.3.

Since the mass of each individual body segment increases with the total body mass increase, the mass of each segment can be expressed in terms of a percentage of the total body mass (Winter, 1990). Data on the masses of body segments (Table 4.4) as a percentage of the total body mass were obtained from the published literature (NASA, 1978a). The data are also presented graphically in Figure 4.4.

4.7 USE OF ANTHROPOMETRIC DATA IN DESIGN

To achieve an appropriate fit between the operator and the design (of equipment, workplace, etc.), the following procedure for the use of anthropometric data should be utilized:

1. Define the equipment's potential user population (e.g., worldwide population, US civilian).
2. Choose the proportion of the population to be accommodated by the design (i.e., 90% and 95% are most common).
3. Determine the body dimensions important in the design (i.e., seating eye height, knuckle height, etc.). Then, in turn, determine the type of accommodation, whether it is a reach or clearance situation, or

Table 4.4 Masses of body segments as a percentage of the whole body mass

Group body segments as a percentage of		Individual body segment mass as a percentage of		
	Total body mass (%)		Group segment mass (%)	Total body mass (%)
Head and neck	8.4	Head	73.8	6.2
		Neck	26.2	2.2
Torso (trunk)	50.0	Thorax	43.8	21.9
		Lumbar	29.4	14.7
		Pelvis	26.8	13.4
Each arm (total)	5.1	Upper arm	54.9	2.8
		Forearm	33.3	1.7
		Hand	11.8	0.6
Each leg (total)	15.7	Thigh	63.7	10.0
		Lower leg (shank)	27.4	4.3
		Foot	8.9	1.4

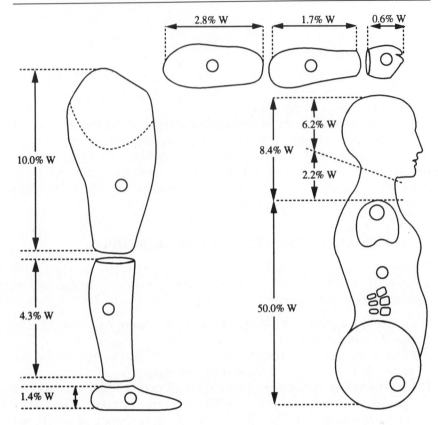

Fig. 4.4 The segmental body masses as a percentage of the whole body mass.

whether it is to accommodate the middle band of the user population.

4. Determine the percentile values of the dimensions (determined in step 3) for the chosen proportion of the population (determined in step 2) from anthropometric tables. Where the percentile values are not readily provided, use the mean and standard deviation of the dimension from anthropometric data, and equation 4.2.

5. Determine the type of clothing and personal equipment worn by the users (i.e., light summer clothing, heavy winter clothing, gloves, etc.) and make the relevant clothing allowances (see Table 4.3).

When designing for a mixed population of males and females, and the anthropometric data for the combined population are not readily available, the larger population (i.e., males) determines the clearance dimensions, and the smaller population (i.e., females) determines the reach dimensions.

Example 4.2

A designer is concerned with the layout of a push-button control that has to be located on a vertical panel such that about 95% of the US male population of industrial workers can reach the control while standing. The problem is how high the control should be installed.

The designer needs to know the 5th percentile of vertical arm-reach for males. From standard normal tables, $z_{0.05} = -1.645$; and from Table 4.2 the mean and standard deviation are found to be $\bar{x} = 209.6$ cm, $s = 8.5$ cm. Noting that the data for vertical reach are not given in the table, the values for the vertical grip reach, which are fairly reasonable, may be chosen. The vertical grip reach value for the 5th percentile (of the male population) is calculated as follows:

$$x_p = \bar{x} + z_p \cdot s$$
$$x_{0.05} = 209.6 - 1.645\ (8.5) = 195.6 \text{ cm (77 in)}$$

Thus, the control button should be located no higher than about 196 cm (77 in). However, a height increment of about 2.5 cm (1 in) may be applied to account for shoes worn by operators.

Example 4.3

In Example 4.2, assume that the designer is to accommodate at least 95% of both the male and female populations. Here the designer needs to know the 5th percentile of vertical arm-reach for the female population (i.e., the shortest expected user). The vertical reach for the 5th percentile of the female population is calculated as follows:

$$x_p = \bar{x} + z_p \cdot s$$
$$x_{0.05} = 199.1 - 1.645\ (8.6) = 185 \text{ cm (72.8 in)}$$

In this case the control button should be located at 185 cm (72.8 in) above the floor. A slight height increment of 2.5 cm (1 in) may be added to account for the shoes worn by operators.

Example 4.4

What should be the height between the seat and the inside of the roof of automobiles to fit the lower 90% of the US population?

$$p = 0.90$$
$$z_p = z_{0.90} = 1.282 \text{ (from standard normal tables, or Table 4.1)}$$

The seat-to-roof height may be calculated based on the sitting height for the 50–50 mixed male–female population. The average sitting height and its standard deviation, respectively, are $\bar{x} = 87.6$ cm and $s = 4.7$ cm. Now, the seat-to-roof height can be found as:

$$x_{0.90} = 87.6 + 1.282 \, (4.7) = 93.6 \text{ cm } (36.9 \text{ in})$$

This height may be set at 96 cm (37.8 in) or 97 cm (38.2 in) to account for the use of civilian headgear and heavy clothing.

4.8 APPLICATIONS AND DISCUSSION

Anthropometry, or the measurement of people, provides the foundation for many ergonomics analyses. There are numerous examples of the application of anthropometric data. Seat adjustments, access opening sizes, control locations, hand grip sizes, and workplace dimensions are a few examples of where anthropometric data can be utilized to "design for the user population." Tables of anthropometric data are readily available from many sources. However, care must be taken to insure the use of appropriate data. The population of concern is the group of people involved in the ergonomics analysis. A population could range in size from one individual to 10 or 20 people working in a department, to the population of adult drivers in the world. The population that might be appropriate for one study might not be appropriate for another. If standard data are to be used, the ergonomist should provide the rationale or justification for why that specific data set is being used.

Often the ergonomist is faced with compromises when deciding on the most appropriate design. For example, in workplace design for a seated operator, the design specifications based on popliteal height, thigh clearance, resting elbow height, and the thickness of the work surface may result in conflicts in recommendations. For example, suppose that the task is to design a workstation that would accommodate 90% of the US adult female population. Such a design would examine data from the 5th percentile to the 95th percentile. For simplicity, we will assume that a foot rest will be designed to accommodate the smaller workers, so only the 95th percentile data will be examined. To convert average (mean) data to the 95th percentile we use:

$$X_{0.95} = \mu + 1.645\sigma$$

Therefore, using Tables 4.2 and 4.3, we can calculate the following dimensions of the 95th percentile:

Popliteal height $= 39.7$ cm $+ 1.645 \, (2.6$ cm$) = 44.0$ cm
Elbow-rest height (sitting) $= 23.1$ cm $+ 1.645 \, (2.9$ cm$) = 27.9$ cm
Thigh clearance $= 13.7$ cm $+ 1.645 \, (1.8$ cm$) = 16.7$ cm
Thigh clearance adjustment $= 0.5$ cm (assuming light clothing)
Shoe height adjustment $= 2.5$ cm (we assume that in an industrial

workplace, women would not be wearing high heels, and that their shoe adjustment would be similar to that for male workers)

Given a work surface thickness of 15 cm, we must now calculate the minimum work surface height. If the work is to be performed at elbow-rest height, we can find the recommended height in two ways:

- Minimum work surface height = popliteal height + shoe height adjustment + elbow height (sitting)
  = 44.0 + 2.5 + 27.9 cm
  = 74.4 cm
- Minimum work surface height = popliteal height + shoe height adjustment + thigh clearance + thigh clearance clothing adjustment + table thickness
  = 44.0 + 2.5 + 16.7 + 0.5 + 15.0
  = 78.7 cm

In this case, a conflict exists. What can be compromised? Can the table thickness be reduced to 10.7 cm or less? If table thickness cannot be reduced, we might compromise thigh clearance by incorporating a chair with a downward-sloping seat pan, or we might raise the work surface height and provide arm rests to alleviate static postural stress. Input from the worker would be valuable in determining the feasible alternative for workplace modifications. The nature of the work and the worker will determine the appropriate compromises.

REVIEW QUESTIONS

1. Define anthropometry.
2. What is the basic application of anthropometric data? What are the two types of dimensions that determine the design dimensions?
3. Describe the applications of clearance and reach dimensions in workstation designs.
4. List the three basic design philosophies utilized by ergonomists as they apply anthropometric data to design for their specific population. Which philosophy is most commonly used?
5. Briefly state the problem associated with designing for the average. Explain where the "design-for-the-average" philosophy is applied.
6. Briefly state the problem associated with designing for the extremes.
7. Based on an anthropometric measurement taken on a randomly selected sample ($n \geqslant 30$) from a given population, the mean eye height was 160 cm with a standard deviation of 12 cm. Calculate the

eye height of: (a) the 90th percentile; (b) the 50th percentile; and (c) the 5th percentile of the population.

8. Explain the differences between the two types of anthropometric measurements, structural and functional measurements, and give two examples of each.

9. Assuming a shoe-sole thickness of 2.5 cm, what range of seat height adjustability can accommodate 90% of the US 50–50 mixed male–female civilian population?

10. Assume that the statures of the 5th and 90th percentiles of a given population are, respectively, 149.35 cm and 178.60 cm. Find the average and standard deviation of this measurement.

11. Calculate the body surface area for an individual of 82 kg body weight and 180 cm height.

12. Calculate the weights of the head and neck, trunk, and each upper arm, forearm, and hand of a person weighing 80 kg.

13. Describe the procedure for the use of anthropometric data that should be used to achieve an appropriate fit between the operator and the design (of equipment, workplace, etc.).

REFERENCES

Damon, A., Stoudt, H.W. and McFarland, R.A. (1966) *The Human Body in Equipment Design*. Harvard University Press, Cambridge, MT.

Dempster, W.T. (1955) *Space Requirements of the Seated Operator*. Wright-Patterson Air Force Base, Ohio, WADC-TR 55–159.

DuBois, D. and DuBois, E.F. (1916) A formula to estimate the approximate surface area if height and weight be known. *Archive of Internal Medicine* **17**, 863–871.

Hertzberg, H.T.E. (1972) Engineering anthropology. In: *Human Engineering Guide to Equipment Design* (Van Cott, H.P. and Kinkade, R.G., eds.) US Government Printing Office, Washington, DC.

Kroemer, K.H.E., Kroemer, H.B. and Kroemer-Elbert, K.E. (1994) *Ergonomics: How to Design for Ease and Efficiency*. Prentice Hall, Englewood Cliffs, N.J.

NASA (1978a) *Anthropometric Source Book, Volume I: Anthropometry for Design*. NASA reference publication 1024 (NASA RP-1024). National Aeronautics and Space Administration, Scientific and Technical Information Office, Lyndon B. Johnson Space Center, Houston, Texas. (For sale by the National Technical Information Service, Springfield, Virginia 22161.)

NASA (1978b) *Anthropometric Source Book, Volume II: A Handbook of Anthropometric Data*. NASA reference publication 1024 (NASA RP-1024). National Aeronautics and Space Administration, Scientific and Technical Information Office, Lyndon B. Johnson Space Center, Houston, Texas. (For sale by the National Technical Information Service, Springfield, Virginia 22161.)

Panero, J. and Zelnik, M. (1979) *Human Dimensions and Interior Space*. Whitney Library of Design, Watson-Guptil Publications, New York.

Tayyari, F. (1993) Mathematical models for combining anthropometric data. In: *The Proceedings of the M.M. Ayoub Occupational Ergonomics Symposium* (Smith,

J.L., ed). Institute for Ergonomics Research, Department of Industrial Engineering, Texas Tech University, Lubbock TX 79409, pp. 15–17.

Van Cott, H.P. and Kinkade, R.G. (eds) (1972) *Human Engineering Guide to Equipment Design*. US Government Printing Office, Washington, DC.

Winter, D.A. (1990) *Biomechanics and Motor Control of Human Movement*, 2nd edn. John Wiley, New York, NY.

Biomechanical bases of ergonomics

5

5.1 INTRODUCTION

Biomechanics combines engineering physics (mechanics), anthropometry and basic medical sciences (biology and physiology), through mathematical relationships. It utilizes the laws of physics to describe biological phenomena in the human body. Biomechanics principles are used to study the responses of the human body to loads and stresses placed on the body in the workplace. Biomechanical models are often utilized to analyze the forces and torques on segments of the body and to compare those forces to muscle strength limitations of people, to predict stressful work postures and conditions.

Biomechanics is a very powerful tool available to ergonomists. A biomechanical analysis is typically utilized for conditions involving large forces (push, pull, lifting, holding, etc.) or work postures that impose stress on the body.

5.2 DEFINITIONS AND BASIC MECHANICS

To understand biomechanics principles better, it is necessary for the reader to become acquainted with descriptive terminology used in this field of study. A simplified relationship among subdivisions of mechanics is depicted in Figure 5.1. However, only those divisions of mechanics that are most directly concerned with occupational biomechanics are subdivided into corresponding branches.

In a biomechanical analysis, the body segments are assumed to be rigid links that rotate about joint centers. **Rigid body mechanics** are based on Newton's laws and deal with the interrelationships among the forces acting upon rigid bodies. Both static and dynamic analyses can be conducted using rigid body mechanics.

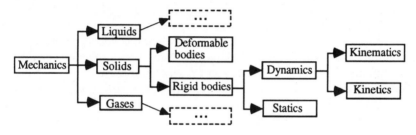

Fig. 5.1 A simplified subdivision of mechanics.

Statics is the study of the body at rest. Static analysis involves the calculation of composition and resolution of forces, moments, and torques, such that the body remains in static equilibrium. A variety of static conditions exist in industry. The obvious example is holding an object. But there are numerous less obvious examples of static activities. Standing can be a static activity, as can be a variety of postures assumed during seated work. Some researchers have argued that "slow" dynamic activities can be analyzed with static models.

Dynamics is the study of the body in motion. With upper- and lower-extremity movements, as frequently occur in industry, dynamic analyses become very important. Examples of dynamic activities in industry include walking, pushing, pulling, manual materials handling, hand work, foot work, and many others. Dynamics can encompass both kinematics and kinetics components.

Kinematics is that area of mechanics which describes the motion of a body without considering the forces causing the motion. Kinematics variables include linear and angular displacement, velocity, and acceleration. **Velocity** is the time rate of change in displacement. It is expressed in units of length per unit of time, such as meters per second ($m \cdot s^{-1}$), kilometers per hour ($km \cdot h^{-1}$), and miles per hour (mph). **Acceleration** is the time rate of change in velocity. It is expressed in terms of units of velocity per unit of time, such as meters per second per second ($m \cdot s^{-2}$). Kinematics profiles of joint centers as well as centers of segment masses are common in biomechanical analyses.

Kinetics deals with the forces acting on the body. The term kinetics refers to the forces that cause the movement. These forces include both internal and external forces. In biomechanics, internal forces are those forces generated by or acting on muscles, ligaments and joints. External forces come from ground (i.e., gravitational force) and other external sources (e.g., an object lifted, a cart pushed, wind resistance). Examples of kinetic analyses include evaluation of moments of external forces at a joint, estimation of muscular forces generated to overcome these moments, calculation of compressive and shear forces acting on the body joints, and determination of energy changes in the body due to muscular activities. The application of kinetic analyses in biomechanics is tremendous. Winter (1990) believes that an even larger part of future biomechanical studies will be based on kinetic analyses.

Force is defined as a physical quantity that can accelerate and/or deform a body. The international unit of measurement of force is the newton (N), where one newton is the amount of force required to accelerate a mass of one kilogram (kg) by one meter per second per second (m · s^{-2}). If an unbalanced force acts on a body, the body accelerates in the direction of the force. Conversely, if a body is accelerating, there must be an unbalanced force acting on it in the direction of the acceleration. The unbalanced force acting on a body is proportional to the product of the mass and acceleration produced by the unbalanced force ($F = m · a$). Since a force has both magnitude and direction, it is a vector quantity.

Work results when a resisting force acts on a body to produce motion in the body. The amount of work is calculated by the product of effective force and the distance through which the force acts. Here, effective force means the projection of the actual force in the direction of the displacement. Consider that a constant external force (F) acts on a body at an angle (θ) with the direction of motion and causes it to be displaced a distance (d). The work is calculated by the product of the force (F), Cos $\theta°$, and distance moved (d); that is:

$$\text{Work} = (F · \text{Cos } \theta°) · d \tag{5.1}$$

where F = actual force, and $F · \text{Cos } \theta°$ = effective force (the projection of the actual force in the direction of the displacement).

If F and d are in the same direction, then Cos $\theta°$ = Cos $0°$ = 1 and $W = F · d$. If F and d are in opposite directions, then Cos $\theta°$ = Cos $180°$ = −1 and the work is negative. This means that work is done by the body, not by the acting external force. For example, in lowering a heavy box, work is done by the body against the force of gravity.

From the above definition and equation 5.1, it becomes clear that no work is done by force if there is no displacement (i.e., the distance is zero). For example, a person can become tired by pushing against a wall for a whole afternoon without performing work. Therefore, the degree of tiredness cannot be used as a valid index of work performance.

The **energy** of a body is its ability to do work. Since the energy of a body is measured in terms of the work it can do, it has the same units as work. Energy can neither be created nor destroyed, but only transformed from one kind to another (law of conservation of energy).

The **potential energy** (E_p) of a body is its ability to do work because of its position or state.

$$E_p = m \cdot g \cdot h = W \cdot h \tag{5.2}$$

where E_p is the potential energy of a mass of m lifted through a vertical distance (h), g is the acceleration due to gravity, and $W = m \cdot g$ is the weight lifted through a vertical distance (h).

The **kinetic energy** (E_k) of a body is its ability to do work because of its motion.

$$E_k = \tfrac{1}{2} m \cdot v^2 \tag{5.3}$$

E_k is the kinetic energy of a mass (m) moving with a velocity (v).

Workload or **power** is the rate of performing work per unit time; that is:

Average power = (work done) ÷ (time taken to do the work) (5.4)

By replacing work by its equivalent (force · distance) in equation 5.4, that is:

Average power = (force) · (distance of displacement)
              ÷ (time the force applied)

we can write:

Average power = (force) · (velocity of body to which
                          force is applied) (5.5)

The velocity of the body is in the direction of the applied force.

The **principle of work** can be expressed in the following form:

Work input = useful work output
           + work done against friction (5.6)

The **actual mechanical advantage (AMA)** of a machine (e.g., a lever) is force ratio:

AMA = (force exerted by machine on load)
     ÷ (force used to operate machine) (5.7)

or

AMA = (resistance) ÷ (effort) (5.8)

The **ideal mechanical advantage (IMA)** of a machine is distance ratio:

IMA = (distance moved by the applied force)
     ÷ (distance moved by load) (5.9)

or

IMA = (force arm) ÷ (resistance arm) (5.10)

Since friction is usually present, AMA < IMA.

The **efficiency** of a machine can then be determined by:

$$\text{Efficiency of a machine} = \frac{\text{work output}}{\text{work input}} = \frac{\text{power output}}{\text{power input}} \qquad (5.11)$$

Momentum is a vector quantity whose direction is that of its velocity. It is calculated by the product of the body mass and velocity:

$$\text{Momentum of a body} = \text{(mass of body)} \cdot \text{(velocity of body)} = m \cdot v \qquad (5.12)$$

Impulse is a vector quantity whose direction is that of its force, and is calculated as follows:

$$\text{Impulse} = \text{(force)} \cdot \text{(length of time the force acts)} = F \cdot t \qquad (5.13)$$

The change of momentum produced by an impulse is numerically equal to the impulse. Thus, if an unbalanced force F acting for a time t on a body of mass m changes its velocity from an initial value v_o to a final value v_t,

$$\text{Impulse} = \text{change in momentum} \qquad (5.14)$$

$$F \cdot t = m \cdot (v_t - v_o) \qquad (5.15)$$

Friction is the force exerted tangentially by the surface of one object on another with which it is in contact, and resists motion of either surface relative to the other.

Torque (or moment of a force about an axis) is the effect of a force producing rotation about an axis. It is measured by the product of the force and the perpendicular distance from the axis of rotation to the line of action of the force.

$$\text{Torque (or moment)} = \text{(force)} \cdot \left[\begin{array}{l} \text{perpendicular distance from axis} \\ \text{to the line of action of the force} \end{array} \right]$$
$$(5.16)$$

Effort (in a rotation) may be defined as the product of torque (moment) times time duration in which the force is sustained:

$$e = M \cdot t = F \cdot d_v \cdot t \qquad (5.17)$$

where: e = effort;
M = moment;
t = time duration of sustaining (holding) the force (weight);
F = force (weight);
d_v = perpendicular distance from the axis of rotation to the line of action of the force.

Gravitational acceleration is the acceleration of a body produced by the earth's gravity, which is approximately 9.8 m $\cdot$ s^{-2}. Therefore, the gravitational force acting on a 1-kg mass is about 9.8 N.

Normal (compressive and **tensile) forces** are the forces directed perpendicularly toward (compressive) or away from (tensile) a cross-sectional surface of a body (Fig. 5.2).

Tangential (shear) forces are the forces acting parallel to any cross-sectional surface of the body (Fig. 5.3).

Stress is the force applied to one unit of surface area. Hence, stress is pressure and its common unit of measurement is the pascal (Pa), where 1 Pa $= 1$ N $\cdot$ m^{-2}.

5.3 NEWTON'S LAWS OF MOTION

- A body will maintain its state of rest or of uniform motion (at a constant speed) along a straight line unless compelled by some unbalanced force to change that state. In other words, a body accelerates only if any unbalanced force acts upon it.
- An unbalanced force (F) acting on a body produces in it an acceleration (a), which is in the direction of the force and directly

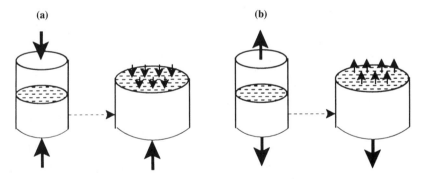

Fig. 5.2 Normal forces: (a) compressive forces; (b) tensile forces.

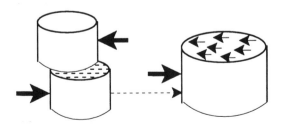

Fig. 5.3 Tangential (shear) force.

proportional to the force, and inversely proportional to the mass (m) of the body.

$$K \cdot a = F/m \text{ or } F = K \cdot m \cdot a \quad (5.18)$$

where K is a proportionality constant. If suitable units are chosen so that K = 1, then

$$F = m \cdot a \quad (5.19)$$

- To every action (or force), there is an equal and opposite reaction (or force).

Newton's laws of motion will be utilized in biomechanical model development. However, before discussing biomechanical models, it is appropriate to examine the units of mechanics. The units are presented in Table 5.1.

5.4 BIOMECHANICAL ANALYSIS AND MODELS

5.4.1 STATIC ANALYSIS AND MODELS

In many cases, the ergonomist can examine a "worst-case" scenario and determine the forces and stresses placed on the body as a result of the load on the body and the required body posture. Usually the activity can be represented as a two-dimensional task (e.g., twisting is not recommended for manual materials handling activities). Once the decision has been made to conduct a two-dimensional static analysis, the following information must be obtained: external forces acting on the body and their directions, body posture, and the body segment parameters (segment masses and locations of centers of mass) of the person being analyzed.

External forces are usually the weight of the object being lifted, lowered, or held, or the push/pull forces being generated as a result of

Table 5.1 Units of mechanics

Entity	MKS system*	cgs system**	English system
Force	N (kg $\cdot$ m $\cdot$ s^{-2})	dyn (g $\cdot$ cm $\cdot$ s^{-2})	lb (slug $\cdot$ ft $\cdot$ s^{-2})
Work (energy)	J (N $\cdot$ m)	erg (dyn $\cdot$ cm)	ft-lb (ft $\cdot$ lb)
Power	W (J $\cdot$ s^{-1})	W (10^7 $\cdot$ erg $\cdot$ s^{-1})	hp (55 ft-lb $\cdot$ s^{-1})
Momentum	kg $\cdot$ m $\cdot$ s^{-1}	g $\cdot$ cm $\cdot$ s^{-1}	slug $\cdot$ ft $\cdot$ s^{-1}
Impulse	N $\cdot$ s	dyn $\cdot$ s	lb $\cdot$ s
Torque	N $\cdot$ m	dyn $\cdot$ cm	ft $\cdot$ lb

*MKS system: meter–kilogram–second system of units
**cgs system: centimeter–gram–second system of units

the task. In cases such as postural analysis, the external forces might be zero and the internal forces caused by the body posture and segment masses might be of interest. Once the body posture and external forces have been identified, simple trigonometry can be utilized to resolve the external and internal forces into horizontal and vertical components.

Body segment parameters are obtained from standard data as discussed in Chapter 4 (Table 4.4 and Figures 4.3 and 4.4). Required data for a static analysis include measurement of the segment lengths and the body weight of the person being analyzed. Once such data have been obtained, segment masses and centers of mass can be estimated.

To calculate the forces and torques involved in a condition of static equilibrium, the following conditions must be met:

$$\Sigma F_x = 0 \text{ (the sum of forces in the } x\text{-direction} = 0)$$
$$\Sigma F_y = 0 \text{ (the sum of forces in the } y\text{-direction} = 0) \qquad (5.20)$$
$$\Sigma M = 0 \text{ (the sum of moments about a fixed point} = 0)$$

If the sums of horizontal and vertical forces and the moments are equal to zero, then there are no forces to result in motion and the system is in a state of static equilibrium.

Example 5.1 (static analysis)

The task to be analyzed requires that a male worker pick a container off a conveyor (located 35 cm above the floor) and lift the container to a cart (shelf located 65 cm above the floor). The container has a mass of 15 kg. This task is performed 360 times per shift. The task is illustrated in Figure 5.4.

The following measurements were taken from the male worker:

Distance from wrist to center of mass (c.m.) of hand (SL_1)	0.07 m
Distance from wrist to elbow (SL_2)	0.28 m
Distance from elbow to shoulder (SL_3)	0.30 m
Distance from shoulder to L5/S1 disk (SL_4)	0.36 m
Angle of hand from horizontal (θ_1)	30°
Angle of forearm from horizontal (θ_2)	30°
Angle of upper arm from horizontal (θ_3)	80°
Angle of trunk from horizontal (θ_4)	45°
Body weight (mass)	70 kg

Before calculating the forces and moments for a static analysis, we can easily determine how much external work is being done on this particular job.

$$\text{Work} = m \cdot g \cdot h \cdot f \qquad (5.21)$$

where: m = mass of the load in kilograms (kg);

Fig. 5.4 A 70-kg man lifts 15-kg part containers from a 35-cm height conveyor and places them onto a cart 65 cm above the floor.

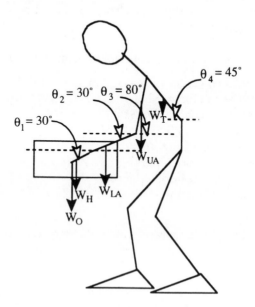

Fig. 5.5 A free body diagram showing the forces and body postures.

g = gravitational constant (9.8 m · s^{-2});
h = height of lift in meters (m);
f = frequency (number of lifts per shift).

Work = (15 kg/lift) · (9.8 m · s^{-2}) · (0.65 m – 0.35 m) · (360 lifts/shift)
Work = 15 876 J per shift

Now, we are ready to continue our static analysis. We can perform this with a segment-by-segment analysis. First, we need to state our assumptions and draw a free body diagram (Fig. 5.5) showing the forces and body postures.

Assumptions

• Body segment parameters from Table 4.4 and Figures 4.3 and 4.4 are appropriate. Mass distribution of the trunk is 0.45 M (which represents the sum of the masses of the head, neck, thorax, and lumbar segments).
• Centers of mass remain constant and can be represented by single points.
• The body is symmetric, with the external load evenly distributed between the right and left hands.

If we are interested, we can find the resulting forces and moments at each joint. In this case, we will calculate the forces and moments until we reach the L5/S1 disk, where we will find the compression and shear forces on the L5/S1 disk as well as the required muscle force of the erector spinae muscles to maintain static equilibrium.

For the hand segment (Fig. 5.6) where:

W_0 = Force due to the weight of the external load = $m · g$
= 15 kg · 9.8 m · s^{-2} = 147 N

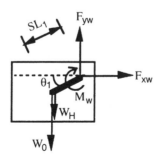

Fig. 5.6 The free diagram for the hand segment.

W_H = Force due to the weight of the hand = $m_H \cdot g$
= (0.006 · 70 kg) · (9.8 m · s^{-2}) = 4.1 N
M_w = Resultant moment at the wrist to maintain static equilibrium
F_{xw} = Resultant force in x-direction at the wrist to maintain static equilibrium
F_{yw} = Resultant force in y-direction at the wrist to maintain static equilibrium
θ_1 = Angle of the hand relative to horizontal
(θ_1 = 30° for this example)
SL_1 = Measured length from wrist to c.m. of hand (at handles of box) (SL_1 = 0.07 m for this example)
$\Sigma F_x = F_{xw} = 0$
$\Sigma F_y = F_{yw} - W_o - W_H = 0$
$\Sigma M_w = M_w - (W_0 + W_H) \cdot SL_1 \cdot Cos\ \theta_1$

Then, we find the following (for each wrist):

$F_{xw} = 0$
$F_{yw} = W_0 + W_H = (147\ N)/2 + 4.1\ N = 77.6\ N$
$M_w = (77.6\ N) \cdot (0.07\ m) \cdot (\cos 30°) = 4.7\ N \cdot m$

For the lower arm segment In drawing the free body diagram for the lower arm segment (Fig. 5.7), we include forces at the wrist that are equal in magnitude and opposite in direction from those just calculated for the hand segment (this is necessary to maintain static equilibrium at the wrist).

where: W_{LA} = Force due to the weight of the lower arm
= $m_{LA} \cdot g$ = (0.017 · 70 kg) · (9.8 m · s^{-2}) = 11.7 N

$M_w = 4.7\ N \cdot m$
$F_{xw} = 0$

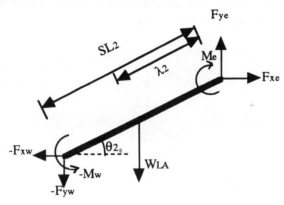

Fig. 5.7 The free body diagram for the lower arm segment.

$F_{yw} = 77.6$ N
θ_2 = Angle of the lower arm relative to horizontal
$\quad$ = 30° for this example
SL_2 = Measured length from wrist to elbow
$\quad$ (for this example $SL_2 = 0.28$ m)
λ_2 = Location of c.m. as a portion of SL from elbow
$\quad$ = 0.43 (or 43%)
M_e = Resultant moment at the elbow to maintain static
$\quad$ equilibrium
F_{xe} = Resultant force in x-direction at the elbow to maintain static
$\quad$ equilibrium
F_{ye} = Resultant force in y-direction at the elbow to maintain static
$\quad$ equilibrium
$\Sigma F_x = - F_{xw} + F_{xe} = 0$
$\Sigma F_y = - F_{yw} - W_{LA} + F_{ye} = 0$
$\Sigma M_e = M_e - M_w - W_{LA} \cdot \lambda_2 \cdot SL_2 \cdot \text{Cos } \theta_2 - F_{yw} \cdot SL_2 \cdot \text{Cos } \theta_2$
$\quad - F_{xw} \cdot SL_2 \cdot \text{Sin } \theta_2 = 0$

From these equations, we find the following (for each elbow):

$F_{xe} = 0$
$F_{ye} = F_{yw} + W_{LA} = 77.6$ N $+ 11.7$ N $= 89.3$ N
$M_e = 4.7$ N $\cdot$ m $+ 11.7$ N $\cdot 0.43 \cdot 0.28$ m $\cdot 0.866 + 77.6$ N $\cdot 0.28$ m
$\quad \cdot 0.866 = 24.7$ N $\cdot$ m

For the upper-arm segment (Fig. 5.8) Again, the resultant forces and moment at the elbow, calculated from the lower-arm segment, are included with equal magnitudes and opposite directions for the upper-arm segment.

where: W_{UA} = Force due to the weight of the upper arm
$\quad$ = $m_{UA} \cdot g = (0.028 \cdot 70$ kg$) \cdot (9.8$ m $\cdot$ s$^{-2}) = 19.2$ N

$F_{xe} = 0$
$F_{ye} = 89.3$ N
$M_e = 24.7$ N $\cdot$ m
θ_3 = Angle of the lower arm relative to horizontal
$\quad$ = 80° for this example
SL_3 = Measured length from the elbow to shoulder
$\quad$ (for this example $SL_3 = 0.30$ m)
λ_3 = Location of c.m. as a portion of SL from shoulder
$\quad$ = 0.436 (or 43.6%)
M_s = Resultant moment at the shoulder to maintain static
$\quad$ equilibrium
F_{xs} = Resultant force in x-direction at the shoulder to maintain
$\quad$ static equilibrium

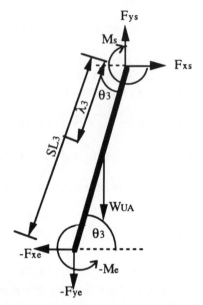

Fig. 5.8 The free body diagram for the upper arm segment.

F_{ys} = Resultant force in y-direction at the shoulder to maintain static equilibrium

$$\Sigma F_x = -F_{xe} + F_{xs} = 0$$
$$\Sigma F_y = -F_{ye} - W_{UA} + F_{ys} = 0$$
$$\Sigma M_e = M_s - M_e - W_{UA} \cdot \lambda_3 \cdot SL_3 \cdot \cos \theta_3 - F_{ye} \cdot SL_3 \cdot \cos \theta_3 - F_{xe} \cdot SL_3 \cdot \sin \theta_3 = 0$$

From these equations, we find the following (for each shoulder):

$$F_{xs} = 0$$
$$F_{ys} = F_{ye} + W_{UA} = 89.3 \text{ N} + 19.2 \text{ N} = 108.5 \text{ N}$$
$$M_s = 24.7 \text{ N} \cdot \text{m} + 19.2 \text{ N} \cdot 0.436 \cdot 0.30 \text{ m} \cdot 0.174 + 89.3 \text{ N} \cdot 0.30 \text{ m}$$
$$\cdot 0.174 = 29.8 \text{ N} \cdot \text{m}$$

For the trunk segment The free body diagram for the trunk segment is presented in Figure 5.9. Note that the trunk segments ends at L5/S1 (actually, the disk between the fifth lumbar and the first sacral vertebrae). In this case it was necessary to estimate λ_4, the location of the center of mass of the trunk segment.

where:

$$W_T = \text{Force due to the weight of the trunk}$$
$$= m_T \cdot g = (0.45 \cdot 70 \text{ kg}) \cdot (9.8 \text{ m} \cdot \text{s}^{-2}) = 308.7 \text{ N}$$

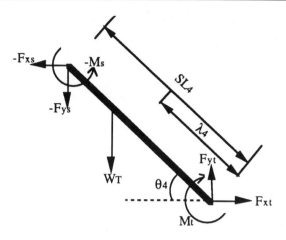

Fig. 5.9 The free body diagram for the trunk segment.

$F_{xs} = 0$
$F_{ys} = 108.5$ N for each shoulder $= 217.0$ N for both shoulders
$M_s = 29.8$ N $\cdot$ m for each shoulder $= 59.6$ N $\cdot$ m for both shoulders
$\theta_4 = $ Angle of the trunk relative to horizontal $= 45°$ for this example
$SL_4 = $ Measured length for L5/S1 to shoulder; for this example $SL_4 = 0.36$ m
$\lambda_4 = $ Location of c.m. as a portion of SL from L5/S1: $\lambda_4 = 0.67$ (estimated)
$M_t = $ Resultant moment at L5/S1 to maintain static equilibrium
$F_{xt} = $ Resultant force in x-direction at L5/S1 to maintain static equilibrium
$F_{yt} = $ Resultant force in y-direction at L5/S1 to maintain static equilibrium
$\Sigma F_x = - F_{xs} + F_{xt} = 0$
$\Sigma F_y = - F_{ys} - W_T + F_{yt} = 0$
$\Sigma M_e = M_t - M_s - W_T \cdot \lambda_4 \cdot SL_4 \cdot \text{Cos } \theta_4 - F_{ys} \cdot SL_4 \cdot \text{Cos } \theta_4 - F_{xs} \cdot SL_4 \cdot \text{Sin } \theta_4 = 0$

From these equations, we find the following:

$F_{xt} = 0$
$F_{yt} = F_{ys} + W_T = 217.0$ N $+ 308.7$ N $= 525.7$ N
$M_e = 59.6$ N $\cdot$ m $+ 308.7$ N $\cdot 0.67 \cdot 0.36$ m $\cdot 0.707 +$
    217.0 N $\cdot 0.36$ m $\cdot 0.707 = 167.5$ N $\cdot$ m

Examination of the calculations in the static analysis shows that a resultant moment of 167.5 N $\cdot$ m at L5/S1 must be countered if the body is to remain in static equilibrium. If we assume that the erector spinae muscle group is the only muscle group in the back active to counter the moment at L5/S1, and we know the moment arm of the erector spinae muscle group, we can estimate the muscle force necessary in the erector spinae muscle group to maintain static equilibrium. (Such assumptions allow for a simplification of the problem, but may not be appropriate for a more detailed analysis.)

If the moment arm of the erector spinae muscle group is 0.04 m (from L5/S1), we can determine the muscle force required by:

$$F \cdot d = 167.5 \text{ N} \cdot \text{m}$$

$$F = \frac{167.5 \text{ N} \cdot \text{m}}{0.04 \text{ m}} = 4187 \text{ N}$$

where: F = Muscle force required in erector spinae to maintain static equilibrium

d = Moment arm length of erector spinae muscle group (from L5/S1)

Now, if we wish to examine the compressive and shear forces acting on the disk between the fifth lumbar and first sacral vertebrae (L5/S1), we can use the calculations above. Since the trunk (L5/S1) is bent at a 45° angle, the vertical force can be resolved into equal compressive and shear components. The vertical force, other than that exerted by the erector spinae muscle group, is the sum of the weights of the load (box), the arms and trunk. Therefore:

$$
\begin{aligned}
F_V &= \text{Total vertical force acting upon L5/S1 disk} \\
&= W_0 + W_H + W_{LA} + W_{UA} + W_T \\
&= 147 + 2(4.1) + 2(11.7) + 2(19.2) + 308.7 = 525.7 \text{ N}
\end{aligned}
$$

Then, as shown in Figure 5.10, the vertical force due to weight of the

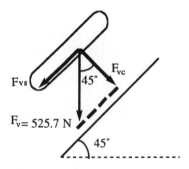

Fig. 5.10 A free body diagram for finding compressive and shear components of vertical forces acting upon L5/S1 disk.

box, arms and trunk, which is 525.7 N, is resolved into its compressive (F_{vc}) and shear (F_{vs}) components. The calculations are as follows:

$$F_{vc} = 525.7 \cdot \text{Cos } 45° = 371.7 \text{ N}$$
$$F_{vs} = 525.7 \cdot \text{Sin } 45° = 371.7 \text{ N}$$

Therefore, the total compressive (F_c) and shear (F_s) forces are found as follows:

$$F_c = 371.7 \text{ N} + 4187 \text{ N} = 4558.7 \text{ N}$$
$$F_s = 371.7 \text{ N}$$

5.4.2 DYNAMIC ANALYSIS AND MODELS

Time-displacement data are often obtained for joint center movement and/or center of segment mass movement. Through a process of inverse dynamics (velocity is obtained by taking the derivative of displacement and acceleration is found by taking the derivative of velocity), the kinematics of a body in motion can be determined. Techniques exist (e.g., video-based systems, and a variety of sensor-based systems) to collect time-displacement data automatically. Many of these systems also incorporate computer software to smooth the data and calculate the velocity and acceleration profiles of the motion.

In cases where automated data collection and analysis are not available, the velocity and acceleration can be found through finite difference equations. To find the velocity at time i, a two-time unit interval is used:

$$v_i = \frac{d_{i+1} - d_{i-1}}{2\Delta t} \tag{5.22}$$

where: v_i = velocity at time i
d_{i+1} = displacement at time $i + 1$
d_{i-1} = displacement at time $i - 1$
Δt = sampling interval (time between samples)

The acceleration at time i can be found similarly as:

$$a_i = \frac{v_{i+1} - v_{i-1}}{2\Delta t} \tag{5.23}$$

or, substituting velocities with their displacement derivatives, the acceleration equation can be rewritten as:

$$a_i = \frac{d_{i+2} - 2d_i + d_{i-2}}{4\Delta t^2} \quad \text{or} \quad a_i = \frac{d_{i+1} - 2d_i + d_{i-1}}{\Delta t^2} \tag{5.24}$$

In most cases, the data sampling system has some inherent errors that result in "rough raw data". Such data should be smoothed before applying the finite differences to obtain velocity and acceleration profiles. It is easy to examine a graph of acceleration data to determine if the data need smoothing. Most human movement is relatively smooth (very little jerking). If very erratic acceleration patterns exist, more likely than not, the data are in need of smoothing.

A variety of smoothing routines are available for biomechanical data. Polynomial curve fits have been used successfully in some cases to smooth data (although higher-ordered polynomials, such as ninth-order, may be necessary not to over-smooth the data). Perhaps the most popular data-smoothing methods are the digital filters and the spline routines. Digital filters utilize a cut-off frequency to smooth the noise from the data. Cubic and quinitic splines have gained popularity as smoothing routines. A numerical methods text can provide additional information on digital filters and spline-smoothing routines.

Data smoothing is an art as well as a science. Data can contain too much noise (unsmoothed data) or data can be over-smoothed with important information being lost due to the over-smoothing. Data should be smoothed enough to represent the movement being observed. Usually such movement is smooth (acceleration profiles are smooth, not jerky).

Once a smooth data set has been obtained, the analysis can begin. For those situations involving no (or very little) movement, the body can be considered to be in static equilibrium, and a static analysis can be conducted. In such cases a series of static analyses can be conducted for each set of data sampling points. A graphical interpretation of the data can reveal the maximum forces and moments and the point during the activity where they occurred. In static equilibrium, the sums of the forces and moments acting about a point must be equal to zero (see equation 5.20).

For dynamic data, Newton's second law is utilized to develop the equations of motion. In general, the forces can be determined from:

$\Sigma F_x = m \cdot a_x$ (where m is segment mass and a_x is acceleration in the x-direction)

$\Sigma F_y = m \cdot a_y$ (where a_y is acceleration of the segment mass in the y-direction)

$\Sigma M_{cm} = I_{cm} \cdot \ddot{\theta}$ (where M_{cm} are the moments about the center of mass and I_{cm} is the moment of inertia of the segment about its c.m., and $\ddot{\theta}$ is the angular acceleration of the segment)   (5.25)

The values for a_x, a_y, and $\ddot{\theta}$ are obtained from the kinematic analysis (displacement–time relationships). The segmental masses can be approximated by using the data in Table 4.4.

The moment of inertia of a rigid body is a measure of the body's resistance to rotational kinematics. In the dynamic equations of motion, the sum of the moments about the center of mass was equal to the product of the moment of inertia times the angular acceleration of the segment. The angular acceleration is obtained from the kinematic data, but the moment of inertia is usually estimated from the body segment parameters.

If the body segment were considered a solid, uniform density circular cylinder of length l and radius r, the moment of inertia about the center of mass is:

$$I_{cm} = \frac{1}{4} mr^2 + \frac{1}{12} ml^2 \qquad (5.26)$$

However, since body segments rotate about their ends of the segment (joint centers) rather than about their centers of mass, the moments of inertia are often given about the proximal or distal end of the segment. We can use the **parallel axis theorem** to find the moment about the center of mass. The parallel axis theorem states that:

$$I_{end} = I_{cm} + mk^2 \qquad (5.27)$$

where k = distance from center of mass to the end of the segment.

For the solid cylinder:

$$I_{end} = I_{cm} + m \left(\frac{1}{2} l \right)^2$$

$$I_{end} = \frac{1}{4} mr^2 + \frac{1}{12} ml^2 + \frac{1}{4} ml^2 \qquad (5.28)$$

$$I_{end} = \frac{1}{4} mr^2 + \frac{1}{3} ml^2$$

In a biomechanical analysis, we are usually not willing to assume that body segments are appropriately represented by uniform-density solid circular cylinders. Moment of inertia data can be found in Dempster (1985) or McConville *et al.* (1980). Often the data are presented as radius of gyration data as a function of segment length. The radius of gyration (k) is defined as:

$$k = \sqrt{\frac{I_{cm}}{m}} \quad \text{or} \quad I_{cm} = mk^2$$

If k is given as a percentage of segment length, then:

$$I_{cm} = m(k \cdot SL)^2 \qquad (5.29)$$

where SL = segment length

Radius of gyration values for body segments can be obtained from Table 5.2. Note that the radii of gyration are represented as a percentage of segment length, and must be multiplied by the segment length and the product squared to determine moment of inertia. For an individual subject, segment length can be easily measured. For population data, segment length can be represented as a percentile of the population (e.g., 50th or 95th percentile).

Note that other body segment parameters such as segment mass and location of the center of mass (as a percentage of segment length) are presented in Table 4.4 and Figures 4.3 and 4.4.

5.5 MUSCULAR STRENGTH

Both static and dynamic biomechanical analyses can provide information about the forces and moments generated during various activities. As previously illustrated, we can estimate muscle forces necessary to perform an activity or to maintain static equilibrium. In order to know if that activity can be safely performed by a variety of people, we would like to know the strength of those people in the muscle groups being used for the activity. An ergonomically designed job should insure that

Table 5.2 Radius of gyration as a percentage of segment length (from Dempster, 1955)

Segment	Anatomical reference point (proximal end/distal end)	Radius of gyration (as % of segment length)	
		From proximal end	From distal end
Hand	Wrist axis/knuckle II middle finger	0.587	0.577
Forearm	Elbow axis/wrist axis	0.526	0.647
Hand and forearm	Elbow axis/ulnar styloid	0.827	0.565
Upper arm	Glenohumeral axis/elbow axis	0.542	0.645
Trunk, head, and neck	Greater trochanter/glenohumeral axis	0.830	0.607
Upper leg	Greater trochanter/femoral condyles	0.540	0.653
Lower leg	Knee axis/ankle axis	0.528	0.643
Foot	Lateral malleolus/metatarsal II (head)	0.690	0.690
Lower leg and foot	Knee axis/medial malleolus	0.735	0.572

the capabilities and limitations of the working population not be exceeded by the demands of the job. In the case of biomechanically limiting jobs, this means that the forces and moments required by the job should not exceed the strength capabilities of the workers assigned to those jobs.

The term muscular strength means the maximum force that a muscle can generate under particular conditions (Chaffin and Andersson, 1991). Since the muscular force is generated voluntarily by the subject, strength is often referred to as **maximum voluntary exertion**. For this reason, the measured muscular strength is expected to be below the physiological tolerance of the musculoskeletal (or muscle–tendon–bone) system. However, workers performing heavy tasks that require exertions greater than the person's maximum voluntary strength can be at a greater risk of injury (Chaffin *et al.*, 1978; Keyserling *et al.*, 1980).

Before embarking on any strength-testing program, the National Institute For Occupational Safety and Health (NIOSH, 1981) has provided a set of guidelines that should be considered. The guidelines apply equally to static or dynamic strength testing. The strength-testing guidelines are (adapted from NIOSH):

- Is the test safe to administer? A test that includes a risk of injury to those taking the test is obviously a problem. Suggestions for improving the safety of the test are: reiterating to subjects that the test is a measure of their maximum voluntary strength, isolation of testing from other subjects, and no provision for feedback to the subject (these suggestions are to eliminate overstress from competition).
- Does the strength test give reliable, quantitative values? Will the test provide the same conclusions if conducted by another individual or if conducted on another day? A quantitative score, such as weight lifted or push/pull force generated, is necessary rather than a subjective classification such as strong or weak.
- Is the strength test related to specific job requirements? Many legal avenues are opened to individuals who fail a strength test if that test is not related to job performance. If the job requires muscle exertions of the upper body, then those muscle groups should be a part of the strength test.
- Is the strength test practical? Practicality can vary somewhat. For a large company a strength-test device that costs several thousands of dollars might be practical, whereas for a small company, the same device might be cost-prohibitive. Practicality should explore cost, time to conduct the test, time to train test administrators, and physical space requirements.
- Does the strength test predict the risk of future injury or illness? It is difficult to evaluate the predictive capabilities of a strength test until the test has been implemented for some time. The important issue to

remember here is that a strength-testing program should be dynamic. It should be continually assessed along with the demands of the jobs. A well-designed and monitored program should be able to build up a set of historical data to support the predictive capabilities of the test.

Static (isometric) strength is defined as the maximum voluntary muscular exertion (contraction) of a body part (e.g., the arms, legs, or the back) in a restrained position without movement. Static muscle strength evaluation should be performed according to a standardized procedure such as that proposed by Chaffin (1975) in the American Industrial Hygiene Association Ergonomics Guide for static strength testing. The subject should be asked to build up to maximum exertion in a couple of seconds and then attempt to hold that maximum for 2 or 3 s. The test administrator should watch subjects and ask them to relax if muscle tremors occur or if the force begins to deteriorate. Patients should be asked to generate only what they feel is a safe exertion against a fixed resistance without producing additional symptoms (Johns, 1987).

Dynamic strength is defined as the maximum voluntary muscular exertion (contraction) of a body part (e.g., the arms, legs, or the back) while in motion. Dynamic muscle strength evaluation should be performed according to the psychophysical guidelines proposed by Snook (1978), which allow subjects to self-select increasing weights until they reach a safe maximum exertion for the test being conducted without overexertion that could lead to injury. Since most strength tests are self-limiting, it becomes a fine line between what is a maximum acceptable exertion and what is a potentially hazardous exertion. A consistent set of instructions to the subject will help minimize variations in testing.

There are several popular types of dynamic strength tests. **Isotonic strength tests** refer to those tests in which a constant level of tension is exerted throughout the range of motion. There are commercially available machines that vary the tension through a series of cams and gears to produce an isotonic exercise. **Isokinetic strength tests** are those tests that require a constant velocity of movement. Again, there are commercially available machines that restrict movement to a fixed velocity. Subjects are to move as fast as possible and exert their maximum force throughout the range of motion. The device limits the speed of movement and typically produces a graph of the torque curve generated during the test. **Isoinertial strength tests** are those tests that utilize a constant load. Free weights and restricted travel weight machines are representative of such tests.

In all strength testing, the safety of the subject must be emphasized. A well-documented testing program is essential. Ideally, a strength-testing program will involve medical personnel as well as ergonomists. The test administrators should be well-trained and understand the demands of the jobs in the work environment. The strength-testing program must be continuously updated and monitored to reflect the changes in the workplace. Another valuable consideration of a strength-testing program is to provide baseline data for employees regarding their strength. If employees are injured, the data will be invaluable in determining when a rehabilitation program has returned them to their pre-injury strength. A program of periodic monitoring, such as that utilized for a hearing conservation program, would be very helpful.

5.6 APPLICATIONS AND DISCUSSION

Biomechanical analyses can range from a relatively simple static analysis of one or two segments to a very detailed three-dimensional dynamic analysis of multiple segments. The level of depth of a biomechanical analysis is determined by the nature of the problem being examined. If the task is a static or quasi-static activity, then a static model would be appropriate. The ergonomist should make every effort to insure that the proper assumptions and analysis techniques are used for each biomechanical analysis. If the ergonomist is using a biomechanical analysis to compare two or more alternatives – rather than arriving at a value that might be compared to values obtained by other researchers – then the assumptions and analysis type might be of less interest, since the same would be used for all analyses. For example, if two alternative arm positions were being evaluated, the relative differences in forces and torques at the wrist, elbow, and shoulder joints might be adequate to determine the preferred design. However, if absolute values are obtained for forces and torques and they are to be compared to data in design guidelines or other existing data, much more care is required. The assumptions of the analysis and the analysis technique itself must be consistent with the comparative data. It would not be appropriate to conduct a static analysis and compare the results with a dynamic analysis done elsewhere.

One of the primary benefits of a biomechanical analysis is the development of a biomechanical model of the activity. If the ergonomist can develop such a model, then modifications can be made to the model parameters and their effects determined through the model rather than by trial and error with the worker. For example, the ergonomist could evaluate changes in the workplace, changes in external loads and forces, and changes in worker characteristics by analyzing data in the model and then using the analysis to develop recommended changes to create a more ergonomically suitable workplace.

REVIEW QUESTIONS

1. Define the following areas of mechanics: statics, dynamics, kinematics, and kinetics.
2. What types of calculations does static analysis involve?
3. Give examples of static and dynamic activities in industry.
4. What are kinematics variables?
5. Define the following terms: velocity, acceleration, force, and work.
6. What are the sources of internal and external force of concern in biomechanics?
7. What types of evaluation are performed using kinetic analyses?
8. Describe the characteristics of the factor causing acceleration in a body.
9. Explain how the amount of work is calculated.
10. Define the term energy, and state the law of conservation of energy.
11. Define the term potential energy, and express its mathematical equation.
12. Define the term kinetic energy, and express its mathematical equation.
13. Define the terms workload and power, and show how to calculate them.
14. Write an expression for the principle of work.
15. Write the expressions for calculating the actual mechanical advantage (AMA) and ideal mechanical advantage (IMA). Which one is usually greater? Why?
16. Write an expression for the efficiency of a machine.
17. Write an expression for momentum.
18. Write an expression for impulse.
19. What is friction?
20. Define the term torque (moment). Explain how it is calculated.
21. Define the term effort.
22. Define compressive and shear forces.
23. Define the term stress.
24. State Newton's laws of motion.
25. Define the term muscular strength.
26. List the questions given in the NIOSH guidelines that must be answered before initiating any muscular strength testing.
27. Define the terms isometric strength and dynamic strength (muscular).
28. Describe isotonic, isokinetic, and isoinertial types of dynamic strength tests.
29. A worker is loading a truck with 33-lb boxes at a rate of 300 boxes per hour. The boxes are lifted at their bottom from the floor and the truck bed is 3 ft above the floor ($g = 9.8 \text{ m} \cdot \text{s}^{-2}$, 1 lb = 0.454 kg, 1 in = 2.54 cm, $1 \text{ N} \cdot \text{m s}^{-1} = 1 \text{ J} \cdot \text{s}^{-1} = 1 \text{ W}$):

(a) Calculate the amount of work performed in 1 hour.

(b) Calculate the level of workload (in watts) performed by the worker.

30. A man is carrying a 25-kg box with both hands in front of his body. The load is carried equally with both arms, while the elbows are kept at 120° (i.e., the angle between the forearm and the biceps is 120°) and the forearm horizontally (Fig. 5.11). Assume that the force produced by the biceps muscle acts parallel to the long axis of the humerus. The following information is also available:

- The weight of each hand/forearm (W_A) is 1.5 kg.
- The center of mass of the hand/forearm is $B = 0.18$ m from the elbow.
- The point of lift is $A = 0.35$ m from the elbow.
- The elbow flexor (biceps) pulls on the forearm at a point approximately $C = 0.06$ m from the elbow.
- Gravitational acceleration $= g = 9.8$ m · s⁻.

(a) Calculate the moment about each elbow produced by the weight of the box and arm.

(b) Calculate the muscular force required through each forearm flexor (biceps).

(c) Calculate the reactive forces at each elbow joint (i.e., R_x and R_y).

31. Repeat Question 30, assuming that the load is carried while the upper arms are kept vertical and the elbows at 120° (Fig. 5.12). Assume also that the force produced by the biceps muscle acts parallel to the long axis of the humerus.

32. As shown in Figure 5.13, an individual weighing 70 kg is performing a leg-lift exercise (one leg at the time) with the leg fully extended, creating an angle of 19° with the horizontal plane. Table 5.3 shows

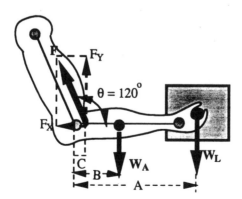

Fig. 5.11 A load carried with both arms: the shoulder joint is flexed 30°, the forearms are kept horizontally and thus the elbow is flexed 120°.

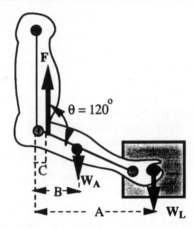

Fig. 5.12 A load carried with both arms: the upper arms are kept vertically and the elbow is flexed 120°.

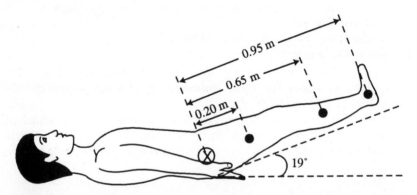

Fig. 5.13 An individual performing a leg-lift exercise with fully extended knees. The center of rotation is the hip joint which is marked with a cross. The hip joint is flexed 19°.

Table 5.3 The subject's anthropometric data

Body segment	Distance from the center of gravity to the center of rotation (m)	Percentage of total body weight
Foot (each)	0.95	1.4
Lower leg (each)	0.65	4.5
Thigh (each)	0.20	9.7

the relevant anthropometric data for the individual. (The gravitational acceleration is 9.8 m · s⁻².)

(a) Calculate the total mass of the leg flexed 19°, flexed at hip.

(b) Calculate the torque on the hip produced by the weight of the flexed leg.

(c) Calculate the moment arm (horizontal distance) from the center of rotation at the hip joint to the center of gravity of the whole leg (i.e., the common center of gravity of the three body segments involved).

(d) Calculate the actual distance between the hip joint and the center of gravity of the whole leg.

33. A 75-kg man lifts a 15-kg box from a 30-cm high conveyor belt and places it on to a 120-cm high shelf (Fig. 5.14). The combined center of gravity of the trunk, head, and arms is 1.5 cm dorsal to the transverse axis of motion in the L5/S1 disk. The man's upper-body weight (including the torso, head, and arms) is 68.6% of his body

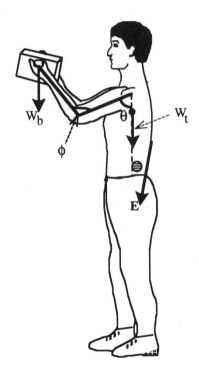

Fig. 5.14 The subject lifts a 15-kg box from a 30-cm high conveyor and places it onto a 120-cm high shelf. The center of rotation is the L5/S1 disk which is marked with a ⊜ sign.

weight. The center of gravity of the box is 40 cm from the vertical plane passing through the L5/S1 disk. The force (E) generated by the spinae erector muscles, which have a combined moment arm of 5 cm from the L5/S1 disk, are assumed to counteract the forward-bending moment. The line of application of the spinae erector muscles forms a 15° angle with the vertical plane at the L5/S1 disk. The following information is also available for the situation:

The angle between the upper arm and the vertical plane: $\theta = 56°$

The angle between the upper arm and the forearm: $\phi = 100°$

Body weight (W)	165 lb (75 kg)
Weight of the head and neck	8.0% of W
Weight of the torso (trunk)	50.0% of W
Weight of the arms and hands	10.0% of W
Weight of the legs and feet	32.0% of W

Moment arm of the weight of the body excluding the weight of the legs	0.06 m
Moment arm of the force generated by the erector spinae muscles	0.05 m
Moment arm of the weight of the box in the hands	0.40 m

(a) Calculate the magnitude of the force generated by the spinae erector muscles.
(b) Calculate the reaction force in the L5/S1 disk.
(c) Calculate the compressive force and shear force on the L5/S1 disk.

34. Figure 5.15 has been provided to study the effect of load distribution on the hip joint of an individual standing on one leg (the right leg), while carrying a 55-lb (26-kg) suitcase in the hand on the opposite side of the standing leg, i.e., his left hand (Fig. 5.15a) and a 55-lb suitcase in each hand (Fig. 5.15b). The center of rotation of the hip joint is marked with an X. Use the following data in answering the questions:

The line of application of the force generated by the abductor muscles (A), as shown in the free body diagram presented in Figure 5.15c, forms a 70° angle to the transverse (horizontal) plane. It is assumed that the abductor muscles absorb the entire torque on the hip joint.

(a) Calculate the torque (T_A) that must be counteracted by the abductor muscles of the right hip joint when the man carries a suitcase in his left hand (Fig. 5.15a) to balance his body in the frontal plane.
(b) Calculate the torque (T_B) that must be counteracted by the abductor muscles of the right hip joint when the individual

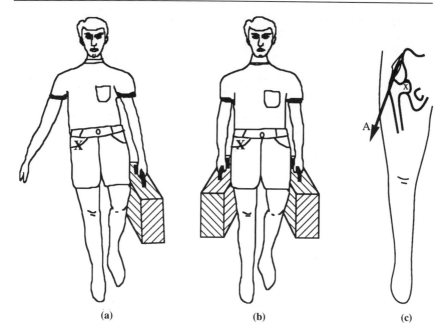

Fig. 5.15 Figure for question 34.

carries a suitcase in each hand (Fig. 5.15b) to balance his body in the frontal plane.

(c) Calculate the reaction force (R_A) on the right hip of the individual in Figure 5.15a.

(d) Calculate the reaction force (R_B) on the right hip of the individual in Figure 5.15b.

REFERENCES

Chaffin, D.B. (1975) Ergonomics guide for the assessment of human static strength. *American Industrial Hygiene Association Journal* **36**: 505–510.

Chaffin, D.B. and Andersson, G.B.G. (1991) *Occupational Biomechanics*, 2nd ed. John Wiley, New York, NY.

Chaffin, D.B., Herrin, G.D. and Keyserling, W.M. (1978) Preemployment strength testing. *Journal of Occupational Medicine* **20**: 403–408.

Dempster, W.T. (1955) *Space Requirements of the Seated Operator*. WADC TR-55-159. Wright Patterson Air Force Base, OH.

Johns, R.E. Jr. (1987) Quantifying patient capabilities to return to work. In: *Ergonomic Interventions to Prevent Musculoskeletal Injuries in Industry*. American Conference of Governmental Industrial Hygienists, Lewis Publishers, Chelsea, MI.

Keyserling, W.M., Herrin, G.D., Chaffin, D.B., Armstrong, T.J. and Foss, M.L. (1980) Establishing an industrial strength testing program. *American Industrial Hygiene Association Journal* **41**: 730–736.

McConville, J., Churchill, T., Kaleps, I., Clauser, C. and Cuzzi, J. (1980) *Anthropmetric Relationships of Body and Body Segment Moments of Inertia.* AFAMRL-TR-80-119. Aerospace Medical Research Laboratory, Wright-Patterson Air Force Base, OH.

NIOSH (1981) *Work Practices Guide for Manual Lifting.* DHHS publication number 81–122. Cincinnati, OH.

Snook, S.H. (1978) The design of manual handling tasks. *Ergonomics* **21**: 963–985.

Winter, D.A. (1990) *Biomechanics and Motor Control of Human Movement,* 2nd ed. John Wiley, New York, NY.

Work physiology 6

6.1 INTRODUCTION

Work physiology is the study of the functions of the human organisms affected by muscular work stresses. **Work physiologists** devise work systems that allow individuals to accomplish their work without developing excessive fatigue, so that at the end of the work day not only can they recover from the work-induced fatigue for returning to work on the next day, but also they will be able to enjoy their off-the-job leisure time.

6.2 PHYSIOLOGICAL RESPONSES

The body's physiological responses to physical workload involve the musculoskeletal and cardiovascular systems. Muscular forces are required to perform the physical work, that is, to hold and move the load from one point to another. Muscular activities (muscle contraction and extension) during physical work require energy. Supplying the demanded energy creates loads on the cardiovascular system (heart and blood vessels) and respiratory (lungs) system. The heart must pump faster to deliver the increased oxygen demand through blood vessels to the involved muscles. The rate of ventilation (inhalation and exhalation) must increase to supply the additional oxygen requirements. These physiological responses are directly related to the work intensity (workload). They are assessed in terms of such parameters as heart rate, body temperature (e.g., oral, rectal, and/or skin temperature), blood pressure, respiration (ventilation) rate, oxygen consumption rate, and concentration of metabolites in saliva, blood, and/or urine.

Ergonomists typically examine energy expenditure rates to assess physiological demand on workers. In theory, if workers are required to exert less than 50% of their energy expenditure capacity during the work day, then they should not become physiologically strained during the work day. In order to provide a margin of safety, many companies and ergonomists recommend that the workload should not exceed 33% of the worker's capacity for the work day.

The 33% and 50% values are derived from the estimates of the individual's anaerobic threshold (AT). The AT is defined as that point at which lactic acid (a product of anaerobic glycolysis) begins to accumulate in the blood at levels above the basal rate of lactic acid production. For untrained individuals, their AT occurs at approximately 50% of their aerobic capacity ($\dot{V}O_{2max}$). For highly trained athletes, their AT may be as high at 85% of their $\dot{V}O_{2max}$. If individuals work at a rate that is below their AT (i.e., no excess lactic acid is accumulating in the blood), then their aerobic system should be able to handle the workload, and they should not be physiologically strained from the work.

6.3 ENERGY AND ENERGY SOURCES OF THE BODY

6.3.1 AEROBIC AND ANAEROBIC PROCESS

The body obtains the energy demanded by the muscles involved in physical activities in two ways – aerobically (i.e., in the presence of oxygen) and anaerobically (in the absence of oxygen). When the person performs a normal-paced work activity, which can be sustained for a relatively long period of time, the oxygen entering into the body through the lungs is sufficient to meet the energy requirement. Most industrial tasks are of this nature.

However, the energy necessary to perform tasks of a very high intensity is supplied anaerobically. The energy demanded at the start of a lower-intensity activity is also supplied anaerobically to the level at which the cardiovascular system has made all required adjustments (to reach a steady state) to supply the demanded energy aerobically. Under the anaerobic conditions the body develops an oxygen deficit that must be repaid during the recovery period upon ceasing the physical activity.

6.3.2 ENERGY SUPPLY DURING REST (BASAL METABOLISM)

Under resting conditions (basal metabolism), approximately two-thirds of the energy consumed by the body is supplied by fat and the rest by carbohydrates. Under such conditions the body's oxygen transport system is capable of supplying sufficient oxygen to the muscle cells, hence only the aerobic process is in operation. For most people basal metabolism requires an oxygen consumption level of between one-fourth and one-third of a liter of oxygen per minute.

6.3.3 ENERGY SUPPLY DURING WORK

Figure 6.1 illustrates several interesting aspects of the body's physiological response to work, in terms of oxygen uptake during a work bout.

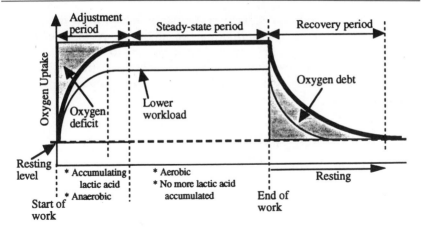

Fig. 6.1 Oxygen consumption profile for a prolonged work bout.

Note that the oxygen uptake does not start at zero, but rather starts at the resting (or basal) rate of the individual, and will return to the resting rate after recovery from work. The initial demands to suppy oxygen to the muscles are not met aerobically by the body (we do not have an instant response to increased oxygen demand). As a result, an oxygen deficit occurs, which is due to the anaerobic processes contributing to the increased work demands. In a short period of time (typically 3–5 min) a steady state will be reached in which the aerobic system is supplying the required oxygen to the body. During actual oxygen measurements, the steady-state level may show minor fluctuations due to sampling differences from one sampling period to another. When the work bout is completed, the body does not immediately return to its basal conditions. Due to elevated respiration rates, increased metabolic rates and the need to repay the oxygen debt, the oxygen uptake declines over time until it finally returns to its basal rate. Recovery time is generally much longer than the time it took to build up to a steady-state oxygen consumption level. Depending on the intensity of the activity and the fitness of the worker, total recovery can take from a few minutes to an hour. The recovery curve shows that initial recovery is very rapid, but total recovery takes much greater time. Typically 80–90% of the recovery takes place in the first few minutes of the recovery period.

As the work intensity changes, the oxygen consumption curve changes. As illustrated in Figure 6.1, an increase in the intensity of work will lead to a faster rise in early oxygen consumption (steeper slope of the initial oxygen consumption curve during the period of oxygen deficit) as well as a higher level of oxygen consumption at steady state. Under steadily increasing workloads, the oxygen consumption curve will continue to increase to new steady-state values until the aerobic capacity ($\dot{V}O_{2max}$) of the individual is attained. At supramaximal workloads, the slope of the oxygen consumption curve will be greater, but the steady-state oxygen consumption will be at the individual's capacity. The recovery period for a higher work intensity also increases, as expected.

6.4 CATEGORIES OF WORK

Work may be categorized into prolonged or intermittent work, as described below.

6.4.1 PROLONGED WORK — anaerobic threshold.

Prolonged or continuous work typically involves submaximal efforts (below the AT) for a relatively long period of time (maintained for a few minutes to several hours). During such type of work the oxygen consumption reaches a steady-state level, as shown in Figure 6.1. The following general observations can be made during prolonged work:

- The major food fuels in prolonged work are carbohydrates and fats. After approximately 1 h of activity the glycogen stored in the muscles and liver may be depleted if the metabolic demand is sufficiently high. At that point, the major foodstuff supplying energy will be fats.
- The majority of adenosine triphosphate is supplied through aerobic processes.
- For submaximal work activities (at less than the AT), lactic acid accumulation would occur only at the beginning of work, before oxygen consumption reaches a steady-state level. With very low workloads lactic acid can be resynthesized during work and should not accumulate above the resting level.
- Anaerobic glycolysis is not needed once steady-state oxygen consumption is reached and lactic acid accumulation does not reach a higher level.

6.4.2 INTERMITTENT WORK

Intermittent work usually consists of sustained effort for short periods of time followed by periods of rests. The change in oxygen consumption during typical high-effort intermittent work is illustrated in Figure 6.2. During this type of work the level of oxygen consumption does not reach a steady-state condition (does not level off); that is, the level of exhaustion is achieved before the cardiovascular system reaches a steady-state condition. The following observations are usually made during highly demanding intermittent work:

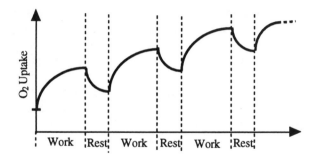

Fig. 6.2 Oxygen consumption profile for an intermittent work bout.

- The major food fuel during intermittent work is carbohydrates, with fat usage minor, and proteins negligible.
- The supplied energy is mainly through anaerobic processes; the aerobic system is not capable of supplying the energy.
- With higher workloads more accumulation of lactic acid results.
- The oxygen consumption level continues to increase until a maximum is reached or exhaustion occurs.

6.5 RESPIRATION

The process of exchanging carbon dioxide and oxygen between the body's organisms and the environment is called **respiration**. In humans, the gas exchange takes place in the lungs, at the alveolar–capillary and tissue–capillary membranes, by diffusion. This diffusion is effected primarily by the level of the partial pressure gradient of gases in the air. The partial pressure of a gas is the pressure exerted by the individual gases in a gas mixture. At higher altitudes the barometric pressure and, hence, the partial pressure of oxygen is lower than at sea level, therefore oxygen taken up by the blood is reduced. It should be noted that the percentage of oxygen concentration at any altitude (any level above or below sea level) is the same as that at sea level, only barometric pressures are different. In the body, the partial pressure of oxygen is at its highest level in the lungs and lowest in the tissues. In contrast, the partial pressure of carbon dioxide is highest in the tissues and lowest in the lungs. The partial pressure differences facilitate the gas exchange diffusion process.

At sea level the barometric pressure of dry air is assumed to be 760 mmHg. To determine the partial pressure of oxygen, the percentage of air that is oxygen is multiplied by the air pressure of 760 mmHg. If we assume that air consists of 20.93% oxygen, then the partial pressure of oxygen would be 159 mmHg ($0.2093 \cdot 760$ mmHg). Likewise, the partial pressure of nitrogen would be 600 mmHg (0.79 · 760 mmHg), and the partial pressure of carbon dioxide would be 0.3 mmHg (0.0004 · 760 mmHg). For expired air (which has been saturated with water vapor, heated to body temperature, and has completed the gas exchange in the lungs), the partial pressures are: P_{O_2} = 116 mmHg, P_{N_2} = 565 mmHg, P_{CO_2} = 32 mmHg, and P_{H_2O} = 47 mmHg. It is obvious that significant differences can result due to differences in water vapor and its resulting partial pressure. To provide some measure of consistency, gas volumes are reported under a few standard conditions:

- **STPD**: standard temperature pressure dry is perhaps the most common method of reporting gas volumes. The gas volume is converted to an equivalent volume of dry gas at sea level (P_B = 760 mmHg and T = 0°C). All gas volumes should be reported in STPD to allow uniform comparisons with values in the literature.
- **BTPS**: body temperature pressure saturated is popular because it is the value actually measured. It is generally assumed that the expired gas is at body temperature (37°C) and saturated at 100% relative humidity (P_{H_2O} = 47 mmHg). The danger of reporting gas volumes in BTPS is that, unless barometric pressure is also reported, comparisons with data at STPD are not possible.
- **ATPS**: ambient temperature pressure saturated is occasionally used to report gas volumes. ATPS conditions arise when air samples have been stored and allowed to reach ambient conditions before being analyzed. Both ambient temperature and pressure must be reported to allow comparisons to STPD volumes.

6.5.1 FACTORS AFFECTING GAS EXCHANGE

The factors, other than partial pressure, affecting gas exchange include:

- the length of diffusion path;
- the number of red blood cells and/or their hemoglobin concentration;
- surface area available for diffusion, e.g., further opening of alveoli and capillaries during exercise, thereby increasing diffusion capacity.

6.5.2 GAS TRANSPORT

Oxygen that diffuses from the alveoli (in the lungs) into the alveoli-capillary blood is transported by the blood to the tissues, where it is diffused into the tissue cells. Carbon dioxide removed from tissues is diffused into the tissue-capillary blood and transported to the alveoli (in the lungs), also by the blood. The following comments can be made for gas transportation:

- Gas exchange takes place due to pressure differences in the system.
- Oxygen is transported mainly in red blood cells and in small amounts in the plasma.
- Carbon dioxide is transported mainly by red blood cells and with water in the blood.
- Carbon monoxide from cigarette smoke can significantly reduce the oxygen carrying capacity of the blood.

6.6 THE CIRCULATORY SYSTEM

The circulatory system is the plumbing system of the body. Oxygenated blood is delivered to the capillary beds where gas exchange can take place in the tissues. Oxygen is diffused into the cells, and carbon dioxide and waste products are diffused into the blood stream. In the lungs, the partial pressure differences between the venous blood and the air in the lungs will result in carbon dioxide being removed from the blood and eliminated through exhalation, and oxygen being extracted from the inhaled air and transported by the hemoglobin in the blood for delivery to the body.

The circulatory system consists of arteries (which carry oxygenated blood away from the heart), capillaries (which allow gas transfer to occur), and veins (which return venous blood to the heart). These three vessels are very different in structure and represent very different functions. An important function of the arteries is pressure regulation to insure adequate pressure in the capillary beds for gas exchange to occur. As a result, the arteries contain more smooth muscle than capillaries or veins. The arteries regulate pressure by constricting the smooth-muscle walls and reducing the cross-sectional area of the artery. The capillaries are thin-walled vessels to allow easy diffusion of gases. The veins are a passive return system and, as such, need to be expandable to allow for accumulation of blood from the capillaries. The veins contain less smooth-muscle tissue than the arteries to allow them to expand. The veins also contain a series of one-way valves that allow blood to flow only toward the heart, and not to return toward the capillaries. The venous return system is non-pressurized and relies on the contraction of skeletal muscles to move blood through the system of valves and back to the heart.

6.6.1 THE HEART AS A PUMP

The heart is the pump that circulates the blood to allow the various gas exchanges to take place. The heart consists of four chambers (two upper chambers – the atria – and two lower chambers – the ventricles). The atria (also called auriculae or auriculas) serve as collection chambers to refill the ventricles. The ventricles contain the blood that will be pressurized and pumped out to the body. The heart is also divided into a left and right side (each having an atrium or auricle and a ventricle). The left side of the heart pumps blood to the body to supply the tissues, while the right side of the heart pumps venous blood to the lungs for gas exchange to take place with inhaled air in the lungs.

Blood flow is turbulent as it is pumped out of the heart in pulses of 60–80 times per minute. The two values obtained from a typical blood pressure measurement (systolic pressure and diastolic pressure) represent flow during pumping action of the heart and pressure during the "quiet period" of heart activity. A blood pressure reading of 120/80 could be averaged to a single pressure value of 100 mmHg, which is given as a typical adult value.

6.6.2 CARDIAC OUTPUT

Cardiac output is the volume of blood ejected into the main artery by the left ventricle per unit time (typically expressed in liters per minute or $l \cdot min^{-1}$). It is a product of **stroke volume** times **heart rate**. Stroke volume is the volume of blood ejected from the left ventricle during a ventricular contraction (usually expressed in milliliters per heartbeat or $ml \cdot beat^{-1}$). Stroke volume can change with both level of activity and level of training. Heart rate is defined as the number of ventricular beats per minute. Heart rate is generally assumed to be the same as pulse rate, which can be easily measured at the superficial palmar arch artery at the wrist or the carotid artery in the neck. We can express cardiac output by the following equation:

$$Cardiac\ output = SV \cdot HR \qquad (6.1)$$

If stroke volume (SV) in this equation is in $ml \cdot beat^{-1}$ and heart rate (HR) in beats per minute (bpm), the result must be divided by $1000\ ml \cdot l^{-1}$ to measure cardiac output in $l \cdot min^{-1}$.

At rest the cardiac output for an adult male is approximately $5–6\ l \cdot min^{-1}$. Cardiac output can increase to $25–30\ l \cdot min^{-1}$ during heavy work. During increased metabolic activity, blood flow is distributed to the working muscles. A constant volume of blood is required for the brain, and a constant percentage of cardiac output is required for the nourishment of the cardiac muscle. Blood flow to internal organs (such as the digestive system) may be reduced during heavy metabolic activity.

There is a correlation between resting cardiac output and body surface area, called **cardiac index** (Ganong, 1987). This index is cardiac output per minute per square meter of body surface area, which averages

$3.2 \, 1 \cdot m^{-2}$. Hence, to estimate cardiac index, the subject's body surface area may be multiplied by 3.2; that is:

$$\text{Cardiac index (cardiac output at rest)} = 3.2 \cdot \text{BSA} \qquad (6.2)$$

where BSA is body surface area in square meters and cardiac output is measured in $1 \cdot min^{-1}$.

Åstrand and Rodahl (1986) suggested the use of **oxygen pulse** as a relative measure of the stroke volume. This is the ratio of oxygen uptake at rest divided by heart rate.

6.7 METABOLISM

Metabolism is a series of biomechanical processes that take place in the body. It may be defined as the collective chemical process of the conversion of foodstuffs and oxygen into mechanical work (both internal and external) and heat.

6.7.1 BASAL METABOLISM

The rate of metabolic activity of a subject after 12 h fasting and 8 h resting (Consolazio *et al.*, 1963) is called the basal metabolic rate (BMR). The basal rate is generally considered the level of energy expenditure (or oxygen consumption) required to keep an individual alive and functioning, but not performing any external work. Basal metabolism generally falls in the range of one-fifth to one-third of a liter of oxygen consumed per minute (or 1.0 to 1.5 kcal of energy expenditure per minute). The following BMR predictive models were originally presented from Carpenter (1939):

For male subjects

$$\begin{aligned}
\text{BMR (kcal} \cdot 24 \text{ h}^{-1}) &= 66.473 + 13.7516 \cdot W_b \\
&\quad + 500.33 \cdot H_b - 6.755 \cdot A
\end{aligned}$$

or $\qquad\qquad\qquad\qquad\qquad\qquad\qquad\qquad\qquad\qquad\qquad$ (6.3)

$$\begin{aligned}
\text{BMR (kcal} \cdot \text{h}^{-1}) &= 2.7697 + 0.5729 \cdot W_b + 20.8471 \cdot H_b \\
&\quad - 0.2815 \cdot A
\end{aligned}$$

For female subjects

$$\begin{aligned}
\text{BMR (kcal} \cdot 24 \text{ h}^{-1}) &= 655.0955 + 9.5634 \cdot W_b + 184.96 \cdot H_b \\
&\quad - 4.6756 \cdot A
\end{aligned}$$

or $\qquad\qquad\qquad\qquad\qquad\qquad\qquad\qquad\qquad\qquad\qquad$ (6.4)

$$\begin{aligned}
\text{BMR (kcal} \cdot \text{h}^{-1}) &= 27.2956 + 0.3985 \cdot W_b + 7.7067 \cdot H_b \\
&\quad - 0.1948 \cdot A
\end{aligned}$$

where: BMR = basal metabolic rate;
        W_b = the subject's body weight (kg);
        H_b = the subject's height (m);
        A = the subject's age (years).

The International Organization for Standardization (ISO, 1987) presented these equations in the international standard units (SI units) in the following forms:

For male subjects

$$BMR = 0.04833 [66.473 + 13.7516 \cdot W_b + 500.33 \cdot H_b - 6.755 \cdot A] \div BSA \qquad (6.5)$$

For female subjects

$$BMR = 0.04833 [655.0955 + 9.5634 \cdot W_b + 184.96 \cdot H_b - 4.6756 \cdot A] \div BSA \qquad (6.6)$$

where: BMR = the basal metabolic rate ($W \cdot m^{-2}$ BSA)
        [1 W = 0.85985 kcal $\cdot$ h^{-1}];
        W_b = the subject's body weight (kg);
        H_b = the subject's height (m);
        A = the subject's age (years);
       BSA = the subject's body surface area (m^2; Chapter 4);
$$0.04833 = \frac{1.16 \, W / kcal \cdot h^{-1}}{24 \, h}$$

The **Met** is also occasionally used as a unit of metabolism, where 1 Met represents the metabolic energy produced at rest (basal rate). Sometimes work rates are presented in terms of Mets. A rate of 3 Mets would indicate that the individual's metabolic rate was three times the resting metabolic rate.

6.7.2 CALORIMETRY

Calorimetry is the process of measuring the heat generated by an energy source. It refers to the measurement of the metabolic energy expenditure by the human body. There are two general methods of calorimetry – direct and indirect calorimetry.

6.7.2.1 Direct calorimetry

Direct calorimetry is obtained through the measurement of heat (energy) generated by a person in a calorimeter. A calorimeter is basically a box-like chamber (Fig. 6.3) containing multiple-layer walls. These walls are made of a series of cold-water pipes for cooling, electrical elements for heating, and a thermometer for measuring the temperature. The chamber is maintained such that the temperature across the walls remains constant and any heat loss is prevented.

The heat generated by the subject in the chamber is absorbed by the circulating water in the pipes through the inside walls of the chamber. The increase in the water temperature is recorded and, then, the equivalent (energy) is calculated. Any changes in the heat content of the

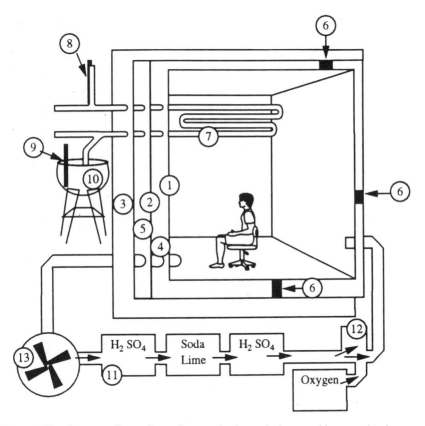

Fig. 6.3 The Atwater–Rosa–Benedict respiration calorimeter. Heat production as well as gas exchange of the subject can be measured. The air-sealed chamber has copper walls (1 and 2) surrounded by a cork insulation wall (3). Dead air spaces are shown by 4 and 5. Thermocouples (6) detect temperature differences between copper walls. Cold water flows through tubings (7), past thermometers (8 and 9), into collecting apparatus (10). Respirator gases are pulled through the fan (13) into sulfuric acid absorber (11), followed by soda lime absorber and second acid absorber, then through spirometer (12). Adapted from Consolazio *et al.* (1963).

body are also estimated from the changes in the subject's body temperature.

However, the high cost of construction, difficulty of operation, and restriction of physical activities due to the size are usually associated with this apparatus which limits its applications. These problems encouraged investigators to develop other methods of calorimetry (e.g., indirect calorimetry) to measure metabolic heat production.

6.7.2.2 Indirect calorimetry

The usual method of measuring a metabolic rate is indirect calorimetry, by which the metabolic rate is calculated from the measured oxygen consumption rate. It can be performed through either an **open-circuit method** or a **closed-circuit method**.

Open-circuit method

In the open-circuit method, test subjects inhale atmospheric air from the outside, while their expired air is collected in either a respiratory gasometer (e.g., Max Plank gasometer or Douglas bag) or metabolimeter (e.g., Metabolic Measurement Cart – MMC – or Oxylog) to measure the volume of the expired air and/or oxygen consumption. The expired air volume must be corrected for the standard conditions (i.e., STPD) and is analyzed for its oxygen and carbon dioxide content. Then oxygen consumption and carbon dioxide production are calculated. The following is the STPD correction factor:

$$\text{STPD}_{\text{correction}} = \left[\frac{P_B \text{ (mmHg)} - P_{H_2O} \text{ (mmHg)}}{760}\right]\left[\frac{273}{273 + T \text{ (°C)}}\right] \quad (6.7)$$

where: P_B = ambient barometric pressure (mmHg);
P_{H_2O} = vapor tension of water (mmHg) at the temperature of the gasometer, which is obtained from empirical tables (Table 6.1);
T = temperature of the gasometer or collected gas sample (°C).

The energy expenditure, with a practically acceptable accuracy level, can be estimated using Weir's formula (Weir, 1949):

$$E(\text{kcal} \cdot \text{min}^{-1}) = 4.92 \, (F_{IO_2} - F_{EO_2}) \cdot \dot{V}_E \quad (6.8)$$

where: $\dot{V}_E$ = Volume of expired air in $1 \cdot \text{min}^{-1}$ at STPD;
F_{IO_2} = Fraction of oxygen in the inspired air (usually assumed to be 0.2093);

Table 6.1 The vapour tension of water (mmHg) between 0 and 49°C shown in steps of 1°C

Temperature (°C)	0	1	2	3	4	5	6	7	8	9
0	4.6	4.9	5.3	5.7	6.1	6.5	7.0	7.5	8.0	8.6
10	9.2	9.8	10.5	11.2	12.0	12.8	13.6	14.5	15.5	16.5
20	17.5	18.7	19.8	21.1	22.4	23.8	25.2	26.7	28.3	30.0
30	31.8	33.7	35.7	37.7	39.9	42.2	44.6	47.1	49.7	52.4
40	55.3	58.3	61.5	64.8	68.3	71.9	75.7	79.6	83.7	88.0

Adapted from Ricci (1967).

F_{EO_2} = Fraction of oxygen in the expired air (a fraction of the volume of expired air);

4.92 = Kilocalories of energy equivalent to 1 l of oxygen consumed.

If we use the approximation of 1 l of oxygen = 5 kcal of energy expenditure, then equation 6.8 above can be rewritten as:

$$E(\text{kcal} \cdot \text{min}^{-1}) = 5.0(\dot{V}_E) (F_{IO_2} - F_{EO_2}).$$

Closed-circuit method

In the closed-circuit method of indirect calorimetry, the subject is completely cut off from the ambient air and breathes through a closed respiratory system. The respiratory system initially contains pure oxygen. The carbon dioxide content of the subject's expired air is continuously removed as it passes through soda lime. The rate of oxygen consumption is measured from the measured decrease in the gas volume, in the closed system.

Example 6.1 (calculations for work metabolic rate)

The following data were collected in a physical activity performed by a male subject using a Max Plank pulmonary gasometer and an oxygen analyzer:

Pre-test measurements

Subject's body weight (nude): 75 kg
Subject's height: 178 cm
Weight of subject's clothes: 4 kg
Weight of pulmonary gas-meter: 4 kg

Experimental measurements

Gasometer reading started at: 10:08:00 (h:m:s)
Gasometer reading ended at: 10:18:30 (h:m:s)
Initial gasometer reading: 31205.5 l
Final gasometer reading: 31541.5 l
Gasometer temperature: 25°C
Barometric pressure: 755 mmHg
Oxygen content of the expired air: 16.5%
Oxygen content of the inspired (ambient) air: 20.93%
Energy equivalent to 1 l of oxygen consumed: 5 kcal

The expired air volume over the observed time, corrected for standard conditions (i.e., STPD), is calculated as follows:

$$V_E = 541.5 - 205.5 = 336 \text{ l}$$
$$\text{(volume of air expired in 10.5 min)}$$

$$\text{STPD}_{correction} = \left[\frac{P_B \text{ (mmHg)} - P_{H_2O} \text{ (mmHg)}}{760} \right] \left[\frac{273}{273 + T \text{ (°C)}} \right]$$

$$= \left[\frac{755 - 23.8}{760} \right] \left[\frac{273}{273 + 25} \right] = 0.88139$$

$$V_E \text{ (STPD)} = 336 \times 0.88139 = 296.1 \text{ l per 10.5 min}$$

The pulmonary ventilation rate ($l \cdot min^{-1}$), oxygen consumption rate ($l \cdot min^{-1}$; corrected for standard conditions; i.e., STPD), and rate of energy expenditure ($kcal \cdot h^{-1}$) during the activity are found as follows:

$$\dot{V}_E = 296.1/10.5 = 28.2 \text{ l} \cdot min^{-1}$$
$$\dot{V}_{O_2} = 28.2 (20.93 - 16.5)/100 = 1.25 \text{ l} \cdot min^{-1}$$
$$E = (1.25 \text{ l} \cdot min^{-1}) \times (60 \text{ min} \cdot h^{-1}) \times (5 \text{ kcal} \cdot l^{-1})$$
$$= 375 \text{ kcal} \cdot h^{-1}$$

The time-weighted average rate of energy expenditure during the work day, if the subject works for 50 min and rests for 10 min (with a resting energy expenditure of 100 $kcal \cdot h^{-1}$) is:

$$E(\text{average}) = 375(50/60) + 100(10/60) = 329 \text{ kcal} \cdot h^{-1}$$

Indirect calorimetry using the oxylog and gasometers (gas volume-measuring devices) have proven to be reasonably accurate for field measurement of metabolic rates of work (Harrison *et al.*, 1982; Louhevaara *et al.*, 1985). The levels of error of estimate for these devices have been reported to be within 10% of the Douglas bag method.

However, the major drawbacks associated with the indirect methods of measuring oxygen consumption are the requirement of special equipment and expertise, interference with the worker's job performance, and requirement of a number of workers to perform the job tasks to obtain sufficient information about the job energy demands. Some investigators have tried to resolve these problems by developing methods for predicting the metabolic rates of the job tasks in question. The next section describes prediction methods for metabolic rates of work.

6.7.3 PREDICTION OF METABOLIC RATE

Predictive methods for assessing metabolic rate of work have appealed to practitioners as an invaluable tool. A major advantage of these methods is that they do not interfere with the worker's job performance. Another major advantage of predictive methods, that other methods do not have, is that they can be used in the design stage of a job.

These methods are based on some sort of task analysis. The oxygen demands of the individual job tasks are estimated from previously established relationships between the tasks and their corresponding metabolic rate. The task analysis should be conducted in such a manner that every task component of the job is well-defined. The analyst must be knowledgeable of the types of data that are required by the predictive method selected. They present much greater potential for estimation error than do the calorimetry methods. The accuracy of an estimation depends on the ability of the analyst and the refinement and calibration of the predictors in the methods. The ability of the analyst can be enhanced with a detailed checklist and appropriate training. The calibration of the predictors in the method should be made by applying the methods to several simulated or typical tasks prior to the assessment of the actual tasks, while measuring the metabolic rate of the tasks by an indirect calorimetry method. A comparison of the predicted and measured data should evoke ways for improvements.

Two most convenient methods of predicting metabolic rates are **table look-up** and **schematic** approaches. Because of the convenience, table look-up has become a popular approach for assessment of metabolic rates. They have been recommended by various investigators and practitioners, especially for use in field studies. A large number of tables, in which occupational tasks are grouped together according to their metabolic demands, can be found in the literature (Passmore and Durnin, 1955; Eastman Kodak Company, 1983; NIOSH, 1986; ACGIH, 1994).

Using the schematic approach, the analyst conducts estimation of metabolic rates for job tasks by coding the tasks according to schema and then the codes are converted into their caloric values. A method for systematic workload estimation (SWE) was developed (Burford *et al.*, 1984) for assessment of the metabolic cost of work performed in underground mines (Ramsey *et al.*, 1986). Appendix D presents a step-by-step guide (Tayyari *et al.*, 1989) for training observers and analysts in conducting studies in which an estimation of the time-weighted average metabolic rate of workers is necessary.

6.8 PHYSICAL WORK CAPACITY

Physical work capacity (PWC) refers to the maximum capabilities of the physiological systems to produce energy for muscular work. In a normal healthy person, the PWC is directly related to the capability of the cardiovascular system to provide oxygen to the working muscles and to remove waste products of metabolism. PWC can be defined in terms of specific muscle group activities (e.g., PWC for arm lifting tasks) or for whole-body activities (e.g., inclined treadmill walking, ladder climbing, or other activities involving the major muscle groups of the body). Standard tests for PWC normally utilize a treadmill or cycle ergometer to provide a constant workload for the subject. For untrained subjects, PWCs obtained on a cycle ergometer may be 3–5% less than those obtained on an inclined treadmill.

6.8.1 AEROBIC CAPACITY

Aerobic capacity (also called maximal aerobic capacity or power, and maximal oxygen uptake) is the maximum level of oxygen uptake. It is denoted by $\dot{V}_{O_2\,max}$ and usually expressed in liters per minute $(l \cdot min^{-1})$ or milliliters of oxygen per kilogram body weight per minute $(ml \cdot kg^{-1} \cdot min^{-1})$. Synonyms for aerobic capacity include PWC, maximal oxygen uptake, and $\dot{V}_{O_2\,max}$.

6.8.2 FACTORS AFFECTING AEROBIC CAPACITY

Aerobic capacity varies significantly among individuals. It is affected by many factors, such as:

- somatic factors: body size, age, sex;
- psychic factors: attitude, motivation;
- environment: altitude, temperature, humidity, etc.;
- nature of work: workload (or intensity), duration, rhythm, technique;
- physiological characteristics of the individual which are genetically determined (inherited at birth).

$\dot{V}_{O_{2max}}$ of females is about 70–75% of that of comparable males (i.e., about 25–30% lower than males). $\dot{V}_{O_{2max}}$ reaches its peak at about 18–25 years of age and deteriorates with age.

6.8.3 ASSESSMENT OF AEROBIC CAPACITY

Aerobic capacity can be determined either directly or indirectly. Direct assessment of aerobic capacity involves maximal testing and is usually performed on young, well-trained subjects (athletes and physically active college students). Submaximal testing (indirect assessment), which is much less stressful, is more appropriate for industrial populations.

6.8.3.1 Direct method (maximal exercise test)

- Workloads are administered that will tax the cardiovascular system and the oxygen uptake to their limits (HR_{max} and $\dot{V}O_{2\ max}$).
- The workload is started at a safe low level and is then increased either continuously (progressively) or discontinuously (intermittently) until the maximum oxygen intake level is reached. Maximum $\dot{V}O_2$ is reached when an increase in workload is not accompanied by a corresponding increase in oxygen consumption. Maximum $\dot{V}O_2$ should occur at maximal heart rate (approximately 220 − age).
- The disadvantage of the direct method of measuring $\dot{V}O_{2max}$ is the risk associated with stressing the physiological systems to the limits of their capacity.
- The advantage of this method is its accuracy.

6.8.3.2 Indirect methods (submaximal exercise test)

There are several indirect methods of estimation of $\dot{V}O_{2max}$ by measurement of submaximal $\dot{V}O_2$ and heart rate. The two most widely used submaximal tests for estimation of $\dot{V}O_{2max}$ are the regression method and a method based on the Åstrand–Åstrand nomogram (Åstrand and Rodahl, 1986). Investigators have developed other conventional methods for such estimation (Siconolfi *et al.*, 1985; Tayyari, 1995). Some of these estimation methods are described below:

Regression method This indirect method is based on the following two factors: the linear relationship between HR and $\dot{V}O_2$ at submaximal workloads (Fig. 6.4), and an age-dependent expected maximal heart rate.

$$HR_{max} = 220 - age \qquad (6.9)$$

In measuring oxygen consumption, an oxygen analyzer is needed. The oxygen consumption is determined by the difference in the fraction

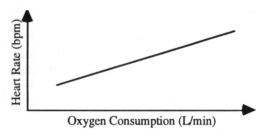

Fig. 6.4 Illustration of the relationship between heart rate and the rate of oxygen consumption.

of oxygen between the inspired and expired air, multiplied by the volume of air breathed (equation 6.10).

$$V_{O_2} = (F_{IO_2} - F_{EO_2}) \cdot V_E \tag{6.10}$$

where: V_{O_2} = Volume of oxygen consumed (usually in liters per minute, i.e., $\dot{V}_{O_2}$);

F_{IO_2} = Fraction of oxygen in inspired air (usually assumed to be 0.2093);

F_{EO_2} = Fraction of oxygen in expired air (usually ranges from 0.15 to 0.18);

V_E = Volume of expired air (usually in liters per minute, i.e., $\dot{V}_E$).

Heart rate is normally measured with a heart rate monitor, or through palpation of the carotid artery. Heart rate is reported in beats per minute (bpm).

At least three steady-state submaximal data sets of $\dot{V}_{O_2}$ and HR should be obtained. Steady state is usually achieved in 3–5 min at each workload. Data obtained using heart rates below 110–120 bpm should not be used to predict $\dot{V}_{O_{2max}}$ since non-linear changes occur in $\dot{V}_{O_2}$ versus HR at lower heart rates. During early responses to work, the stroke volume of the heart changes, as does the fraction of oxygen extracted in pulmonary ventilation. After the heart rate reaches the 110–120 bpm range, a reasonably linear relationship usually exists.

Once the three or more pairs of $\dot{V}_{O_2}$ and HR have been obtained, linear regression can be used to predict max $\dot{V}_{O_2}$. A linear relationship is established between heart rate and $\dot{V}_{O_2}$.

$$\dot{V}_{O_2} = b + m \cdot HR \tag{6.11}$$

where: b = the Y intercept of the regression equation;

m = the slope of the regression equation

If a suitable calculator or computer program is not available to calculate the linear regression equations, the intercept and slope can be determined from the normal equations:

$$b = \frac{\Sigma HR^2 \Sigma \dot{V}O_2 - \Sigma HR \Sigma HR \cdot \dot{V}O_2}{n\Sigma HR^2 - (\Sigma HR)^2}$$

and   (6.12)

$$m = \frac{n\Sigma HR \cdot \dot{V}O_2 - \Sigma HR \Sigma \dot{V}O_2}{n\Sigma HR^2 - (\Sigma HR)^2}$$

Once the regression equation has been established, $\dot{V}O_{2max}$ can be estimated by using the predicted HR_{max} (i.e., $220 - age$) in the regression equation and solving for $\dot{V}O_{2max}$.

Example 6.2

The following data were collected on a 35-year-old male subject riding a cycle ergometer:

HR	$\dot{V}O_2$
120	1.05
134	1.35
153	1.80

The regression equation is calculated as $\dot{V}O_2 = -1.69 + 0.0228 \cdot HR$. The subject's maximum HR is estimated to be 185 bpm ($HR_{max} = 220 - 35$). The subject's $\dot{V}O_{2max}$ can now be predicted:

$$\dot{V}O_{2max} = -1.69 + 0.0228 \cdot (185)$$
$$\dot{V}O_{2max} = 2.52 \; l \cdot min^{-1}$$

The disadvantage of this method is the variation of maximum heart rate among individuals, even in the same age group. For example, of two 30-year-old individuals, one may have a HR_{max} of less than 180 bpm while the other has a maximum of 210 bpm, rather than the expected HR_{max} of 190 bpm each.

Åstrand–Åstrand nomogram This method (discussed in detail in Åstrand, 1960) is also based on submaximal measurements of oxygen uptake and heart rate. The nomogram is used along with an age-correction factor (also see Åstrand and Rodahl, 1986, and Rodahl, 1989 for more explanation). The disadvantage of this method is the possibility of errors in extracting data from the nomogram, especially by those not well-trained in using it.

Tayyari and Ramsey (1985) presented a convenient model based on the Åstrand–Åstrand nomogram that prevents errors in using the nomogram and enhances the ability of practitioners in estimation of the maximum oxygen uptake (aerobic capacity), while they still can benefit from the accuracy of the nomogram. The general form of the model is:

$$\dot{V}O_{2max} = AG \cdot (131.5 \cdot \dot{V}O_2) \div (HR + GF - 72) \qquad (6.13)$$

where:　　$\dot{V}O_2$ = the subject's submaximal oxygen uptake (in $l \cdot min^{-1}$);

　　　　　　HR = the subject's heart rate corresponding to the measured submaximal $\dot{V}O_2$ (in bpm);

　　　　　　GF = a gender factor (GF = 10 for males, and GF = 0 for females);

　　　　　　AG = an age correction factor that is calculated by the following equation:

$$AG = 1.12 - 0.0073 \cdot age \qquad (6.14)$$

However, by replacing the values of the gender factor in equation 6.13, the following equations can be stated for the two genders:

For men:　　$\dot{V}O_{2max} = AG \cdot (131.5 \cdot \dot{V}O_2) \div (HR - 62)$　(6.15)

For women: $\dot{V}O_{2max} = AG \cdot (131.5 \cdot \dot{V}O_2) \div (HR - 72)$　(6.16)

Example 6.3

The aerobic capacity of a 40-year-old male subject with a heart rate of 138 bpm at a workload requiring a $\dot{V}O_2$ of $2.5\ l \cdot min^{-1}$ is calculated as follows:

AG = 1.12 − 0.0073 · (40) = 0.828
$\dot{V}O_2$ = 0.828 · (131.5 · 2.5) ÷ (138 − 62) = 3.58 $l \cdot min^{-1}$ (corrected for age)

Therefore, the predicted aerobic capacity of the subject is about 3.58 $l \cdot min^{-1}$.

Both HR_{max} and $\dot{V}O_{2max}$ decline with age (Fig. 6.5). Maximum aerobic capacity is reached in the early 20s and decreases as an individual ages. The rate of decline can vary from individual to individual and is influenced by the individual's level of physical activity and fitness.

Step test In many facilities oxygen analysis equipment is not available for determining FEO_2 values. In such cases, alternative methods of estimating $\dot{V}O_{2max}$ are required. Step tests have become a popular alternative for estimation of $\dot{V}O_{2max}$ due to their simplicity, ease of conducting the test, use of heart rate as the measured parameter,

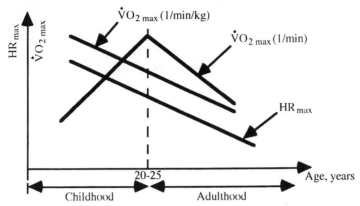

Fig. 6.5 The relationships between age and maximum heart rate (HR$_{max}$) and $\dot{V}O_{2\ max}$.

portability, and low cost. A large variety of step tests have been developed (examples are found in most exercise physiology textbooks and in Åstrand and Rodahl, 1986, and Kroemer *et al.*, 1990).

A step test protocol developed by Siconolfi *et al.* (1985) appears to be well-suited for use in industry. Siconolfi's test is a relatively low-impact test designed for individuals aged 19–70 years. A 10-in (25-cm)-high bench is utilized in the protocol. Siconolfi recommended that the test be conducted in three stages of three-minute work bouts separated by one minute rest periods. Siconolfi's test was based on predicted $\dot{V}O_2$ value obtained from the ACSM guidelines (American College of Sports Medicine, 1991) and then using that $\dot{V}O_2$ value in the Åstrand–Åstrand Nomogram to obtain a predicted $\dot{V}O_{2max}$ that would then be modified to account for age. The ACSM prediction is based on step height and frequency of stepping and can be written as:

$$\dot{V}O_2 = 0.35 \cdot f + 2.395 \cdot f \cdot h \qquad (6.17)$$

where: f = frequency of stepping (steps $\cdot$ min^{-1});
h = height of step (meters)
$\dot{V}O_2$ = steady-state oxygen consumption
(ml $\cdot$ kg^{-1} $\cdot$ min^{-1})

Thus, for a 10-in (25.4-cm) step and a stepping frequency of 30 steps $\cdot$ min^{-1} the steady-state for that activity would be estimated as:

$$\dot{V}O_2 = 0.35 \cdot 30 + 2.395 \cdot 30 \cdot 0.0254$$
$$\dot{V}O_2 = 28.7 \text{ ml} \cdot \text{kg}^{-1} \cdot \text{min}^{-1}$$

Using the steady-state heart rate for a step test (collected after 3 minutes of performing the step test), $\dot{V}O_{2max}$ can now be estimated from equation 6.13, which is modified to account for age by equation 6.14.

Example 6.4

The following data were collected on a 42-year-old woman (body weight = 61 kg) performing a step test on a 12-in (30.5-cm) step at a frequency of 30 steps · min^{-1}. She reached a heart rate of 133 bpm after 3 minutes of stepping.

$$\dot{V}O_2 = 0.35 \cdot 30 + 2.395 \cdot 30 \cdot 0.305$$
$$\dot{V}O_2 = 32.4 \text{ ml} \cdot \text{kg}^{-1} \cdot \text{min}^{-1}$$

Converting $\dot{V}O_2$ into l · min^{-1},

$$\dot{V}O_2 = 32.4 \cdot 61 \text{ kg} \cdot 1000^{-1} = 2.0 \text{ l} \cdot \text{min}^{-1}$$

Using equation 6.13 (or equation 6.16 could be used directly) and equation 6.14, $\dot{V}O_{2max}$ can now be estimated as:

$$\dot{V}O_{2max} = AG \cdot (131.5 \cdot \dot{V}O_2) \div (HR + GF - 72)$$
$$\dot{V}O_{2max} = (1.12 - 0.0073 \cdot 42) \cdot (131.5 \cdot 2.0) \div (133 + 0 - 72)$$
$$\dot{V}O_{2max} = 3.5 \text{ l} \cdot \text{min}^{-1}$$

Tayyari's conventional method A conventional method has recently been introduced (Tayyari, 1995) which estimates $\dot{V}O_{2max}$ based on the person's body weight and measured heart rate during walking on a treadmill. This method uses the following equation:

$$\dot{V}O_{2max} = \frac{0.263(W_b + 10)V + 13.15}{HR + G - 72} \qquad (6.18)$$

subject to the age-correction multiplier for individuals over 30 years of age given by equation 6.14 (i.e., AG = 1.12 − 0.0073 · age).

where: $\dot{V}O_{2max}$ = the subject's maximal oxygen uptake (in l · min^{-1});
W_b = the subject's body weight (kg);
V = the subject's speed of walking on a treadmill (km · h^{-1});
HR = the subject's heart rate (bpm) during walking on the treadmill;
G = a gender factor (GF = 10 for males, and GF = 0 for females).

Example 6.5

A 32-year-old male subject weighing 82 kg records a heart rate of 143 bpm while walking on a treadmill at 7.5 km · h^{-1}. Using Tayyari's method, predict the subject's $\dot{V}O_{2max}$.

$$\dot{V}_{O_2max} = \frac{0.263(W_b + 10)V + 13.15}{HR + G - 72}$$

$$= \frac{0.263(82 + 10)(7.5) + 13.15}{143 + 10 - 72}$$

$$\dot{V}_{O_2max} = 2.40 \; l \cdot min^{-1}$$

Correcting for age: $AG = 1.12 - 0.0073 \cdot age = 1.12 - 0.0073(32)$
$= 0.8864$
$\dot{V}_{O_2max} = 2.40 \cdot 0.8864 = 2.13 \; l \cdot min^{-1}$

6.9 FATIGUE AND ITS EVALUATION

The distinction between muscular (physical) fatigue and mental fatigue is widely accepted. Muscular fatigue manifests itself by aching and pain signals in the fatigued muscles, and by weakness and reduced ability of muscular movement. As Edholm (1967) stated, all the factors involved in fatigue development are not understood; the accumulation of lactic acid and swelling in the muscle certainly play a part. High concentration of lactic acid completely disables muscular contraction, and lower concentrations impair it. Edholm related muscular swelling to the increased blood flow that leads to an increased fluid volume in the muscle tissues. Training can improve the person's ability to sustain a physically demanding activity for a longer period of time without developing the aching sensation that may be elicited in untrained individuals.

Brouha (1967) introduced a methodology of fatigue evaluation that is based upon the recovery of heart rate curves, which are measured during the first 3 min immediately after work stops. It is determined in three phases as follows:

- Phase 1: P_1 = heart rate (in bpm) during the second 30 s of the first minute (pulse between 30 s and 60 s is counted and multiplied by 2);
- Phase 2: P_2 = heart rate (in bpm) during the second 30 s of the second minute (pulse between 1.5 and 2 min is counted and multiplied by 2);
- Phase 3: P_3 = heart rate (in bpm) during the second 30 s of the third minute (pulse between 2.5 and 3 min is counted and multiplied by 2)

where P_1, P_2, P_3 are, respectively, called the first, second and third recovery pulse rate.

Brouha stated that when fatigue occurs and cumulatively builds up, the initial recovery pulse rate (P_1) after successive bouts of work becomes progessively higher, and the heart rate during the recovery period remains at an elevated level for a progressively longer time (Fig. 6.6). Such conditions indicate that satisfactory recovery does not take place between the successive work bouts. A slow return of heart

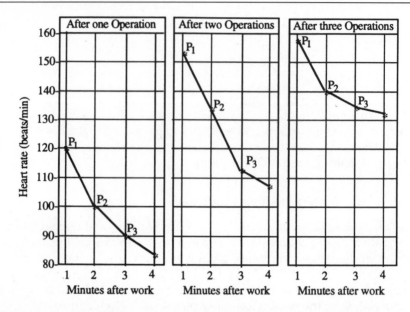

Fig. 6.6 Increasing level of the recovery pulse-rate curves and decreasing speed of recovery after successive operations. Adapted from Brouha (1967).

Table 6.2 Summary of Brouha's fatigue assessment techniques

Condition	Criteria
Normal curves*	$P_1-P_3 \geqslant 10$ bpm, and $P_1, P_2, P_3 \leqslant 90$ bpm
No-recovery curves*	$P_1-P_3 < 10$ bpm, and $P_3 > 90$ bpm
Inverse recovery curves*	$P_3 > 90$ bpm and $P_1-P_3 \leqslant -10$ bpm (or $P_3-P_1 \geqslant 10$ bpm)‡
No increasing cardiac strain during the day†	$P_1 \leqslant 110$ bpm, and $P_1-P_3 \geqslant 10$ bpm

*Recovery pulse rate curves.
†Such stress can be sustained throughout the shift in a physiological steady state.
‡The recovery heart rate is progressively increasing.

rate toward the resting level is a clear evidence of **physiological fatigue** in the worker.

Brouha uses these recovery pulse rates to diagnose the suitability of workloads. As summarized in Table 6.2, **normal curves** include all cases in which the third recovery pulse rate (P_3) is at least 10 bpm slower than the first (P_1); and those in which all three recovery pulse rates (P_1, P_2, and P_3) are below 90 bpm. **No-recovery curves** are those in which the difference between the first (P_1) and third (P_3) recovery pulse rates is less than 10 bpm and in which the third reading (P_3) remains above 90 bpm. **Inverse recovery curves** are those in which the third pulse reading (P_3) is above 90 bpm and paradoxically greater than the first reading (P_1) by 10 bpm or more.

Brouha also offers a technique to distinguish work levels that can be sustained for the work-shift from those that cannot. He argued that when the average value of the first recovery pulse (P_1) is maintained at about 110 bpm or less, and when the deceleration between the first (P_1) and the third (P_3) recovery pulse rates is at least 10 bpm, no increasing cardiac strain occurs as the work-shift progresses. Whether this level of strain is produced by the workload alone or by a combination of workload and heat exposure, it appears that such stress can be sustained throughout the work-shift in a physiological steady state, given that the sequence of work and rest periods is adequately organized.

6.10 CLASSIFICATION OF WORKLOAD

Physical work is usually classified based on the energy requirements and expressed in kcal · h^{-1} (kilocalories per hour) or W (watts). However, the factors used to classify the workload are somewhat arbitrary and hence factors such as oxygen uptake, heart rate, pulmonary ventilation rate, and, sometimes, rectal temperature are used in workload classification. An example of such classifications is given in Table 6.3.

6.11 ENDURANCE IN PHYSICAL WORK

Endurance is the ability to sustain an activity. It is measured in units of time (e.g., seconds, minutes, hours). It is related to PWC; that is, an individual with a higher PWC is able to endure a given physical task much longer than an individual with a lower PWC.

Table 6.3 Classification of physical workload using oxygen consumption and energy expenditure

Workload class	Oxygen consumption $\dot{V}_{O_2}$ (l · min^{-1})	Energy expenditure (kcal · h^{-1})	Heart rate (bpm)
Sedentary	0.25–0.33	75–100	60–80
Light	0.33–0.5	100–150	70–90
Moderate	0.5–1.0	150–300	80–110
Heavy	1.0–1.5	300–450	100–130
Very heavy	1.5–2.0	450–600	120–150

The time duration the person is able to sustain the physical work until forced to give up due to fatigue is called the **endurance time limit**. For light to moderate levels of work, employees might not reach their endurance time limit (the limit is beyond the 8-h working shift). In such cases mental fatigue could still be a limiting factor in the ability of the worker to perform satisfactorily. However, if workers reach their endurance time limit during the shift, some type of work–rest schedule should be implemented.

It is important to note at this time that the term rest is often misunderstood. In the context of work–rest scheduling, rest represents a less demanding activity. Rest does not necessarily mean going to a break area and sitting down to visit with coworkers. Such breaks (in the USA typically given at one-fourth and three-fourths of the way through a shift) may be of social significance and do provide for physiological recovery. However, a rest break can also involve non-physically demanding activities such as filling out paperwork, entering data into a computer, housekeeping, delivering a report to the supervisor's office, or any other activity that does not require extensive use of the muscles being used on the physically demanding task.

6.12 WORK–REST SCHEDULING

Energy expenditure may be used to determine the amount of work and rest required to establish a "fair-day" workload. For this purpose, the energy expenditures during all job tasks must be determined (or closely estimated) to find the average energy expenditure during a typical work cycle.

The generally accepted standard workload in the UK and the USA is equivalent to the workload generated by a man of average physique walking without carrying a load in a straight line on level ground at a speed of 4 mph (6.4 km · h^{-1}) (International Labor Office (ILO), 1979). An average (standard) man refers to a healthy 25-year-old male of 65 kg body weight and 1.75 m height (Durnin and Passmore, 1967). This workload, however, requires regular rest pauses. The energy expenditure equivalent of this standard workload is about 370 W or 5.3 kcal · min^{-1} for a 65-kg man (Tayyari, 1987).

The duration of work bouts and rest pauses should be regulated to account for the worker's age. The work endurance, like the maximum aerobic (or work) capacity, may be negatively age-dependent.

Edholm (1967) recommended a total energy expenditure of 2000 kcal during work as a maximum that can be expected for work to continue for months or years. This workload requires an average energy expenditure of 4.2 kcal · min^{-1} during an 8-h workshift. He also suggested the measurement of heart rate as an indication of whether the workload is too heavy for the worker since the figure may be too high as it tends to produce exhaustion by the end of the shift. The following equation can be used to determine the rest duration required to achieve this level of work energy expenditure recommended by Edholm:

$$T_R = \frac{480\,M - 2000}{M - 1.5} \tag{6.19}$$

where T_R is the required rest duration in minutes and M is the metabolic energy cost of work, while the resting level of energy expenditure is assumed to be 1.5 kcal · min^{-1}.

Murrel (1965) suggested the following equations for calculation of the total recovery time required during the work-shift:

$$T_R = T_S \left(\frac{M - S}{M - 1.5} \right) \tag{6.20}$$

where: T_R = Total recovery (resting) time in minutes;
T_S = Total work-shift length in minutes (e.g., 480 min for an 8-h shift);
M = Average energy expenditure (kcal · min^{-1});
S = The level of energy expenditure (kcal · min^{-1}) adopted as a standard metabolic rate for the entire work-shift;
1.5 = Resting level of energy expenditure in kcal · min^{-1} for an "average man."

Equations 6.19 and 6.20 calculate only the total time during the work-shift that should be taken as rest. They do not determine the optimal arrangement of work–rest cycles. In an optimal work–rest regimen, the worker stops at the time when lactic acid starts to accumulate in the body; that is, at the first feeling of fatigue which is prior to exhaustion. Rest duration required at this time will be minimal (Murrel, 1965). Furthermore, the work-bout duration should not be left to workers to determine based on their sensation of fatigue. Work may go on for too long and, therefore, a longer rest period will be required to recover from the fatigue developed. This reduces the worker's efficiency and productivity.

Müller (1953) reported that the energy reserve of the so-called average man is about 25 kcal, and it will not be drawn upon as long as the work demands an energy expenditure of less than 5 kcal · min^{-1}. The time for depletion of this energy reserve at a workload requiring M kcal · min^{-1} can be calculated by the following equation (Tayyari, 1994):

$$T_{max} = \frac{25}{M - 5} \tag{6.21}$$

T_{max} in equation 6.21 is the maximum work-bout duration (in minutes), beyond which the work should stop to prevent the development of excessive fatigue. Thus, the work-bout duration (T_W) should be:

$$T_W \leq T_{max} \tag{6.22}$$

or

$$T_W \leq \frac{25}{M - 5} \tag{6.23}$$

A workload requiring M kcal · min^{-1} metabolic energy expenditure will remove a total of $T_W \cdot (M - 5)$ kcal from the 25 kcal energy reserve in T_W minutes. The removed energy can be restored at a rate equal to $5 - 1.5 = 3.5$ kcal · min^{-1} (where 1.5 kcal · min^{-1} is assumed to be the resting energy expenditure). Then, the rest duration required for complete restoration of the energy reserve is determined as follows:

$$T_R = \frac{T_W (M - 5)}{5 - 1.5} = \frac{T_W (M - 5)}{3.5} \tag{6.24}$$

For example, the maximum work-bout duration for a workload requiring 6 kcal · min^{-1} is 25 min. This is calculated using equation 6.21 as follows:

$$T_{max} = 25/(6 - 5) = 25 \text{ min}$$

If the work-bout duration is, for example, 14 min, then according to equation 6.24 subjects need only 4 min to rest to restore their energy reserve, as calculated below:

$$T_R = 14(6 - 5)/3.5 = 4 \text{ min}$$

If the work-bout duration is extended to the recommended maximum of 25 min, the required rest for recovery must be approximately 7 min.

6.13 APPLICATIONS AND DISCUSSION

Physiological assessment of work is most appropriate for those jobs performed at moderate to high frequency which utilize large muscle masses. Good examples of such tasks include many manual materials handling activities, tasks involving a great deal of walking, and jobs requiring a great deal of body movement. The level of sophistication of the analysis determines the equipment requirements and methods utilized.

At the very least sophisticated level of physiological assessment, the ergonomist may take only pulse rate data, either through palpation or by utilizing one of several pulse meter devices commercially available. For example, there are finger pulse meters available which would allow the ergonomist briefly to interrupt workers and have them place a finger in the meter to obtain a working pulse rate. More sophisticated telemetry systems are also available. In such cases, the worker might be asked to wear a chest band (with dry electrodes), and then periodically during the day, the ergonomist could sample the worker's pulse rate with an appropriate receiver. The primary advantage to such a telemetry system is that it would not interfere with the worker and productivity. Once pulse rate data are obtained, oxygen consumption or energy expenditure data could then be estimated from Table 6.3 or from an individual's $HR/\dot{V}O_2$ curve. For more precise estimates, individual $HR/\dot{V}O_2$ responses could be established under controlled conditions (in a laboratory) simulating the work conditions of concern.

At the most sophisticated level, physiological assessment would be direct measurement of oxygen consumption. This is typically accomplished under controlled laboratory conditions where the job is simulated, or in the actual work environment. The advantages to a laboratory simulation are the control of the work parameters, the availability of equipment that might not be suitable for the work environment (because of power requirements, space requirements, etc.), and the non-interference with production. The obvious disadvantage is that the data are collected in a simulated environment rather than the actual work environment.

Portable systems are available for oxygen consumption measurement. Typically such systems have small oxygen meters that can be worn on the back or over a shoulder. The level of interference of the equipment should be determined before using a portable system. Also, oxygen analysis systems require the collection of expired air, which means that the worker must wear a face mask or use a mouth piece and nose clip. Such masks and hoses might interfere with workers' ability to perform their job safely.

Once heart rate, oxygen consumption, and/or energy expenditure data are collected, they have very little meaning until the capacity of the worker (or work population) has been compared. The most commonly recommended guideline is that a worker should be able to work continuously (8 h) at a work rate that requires one-third or less of that worker's physiological capacity. For example, a worker who has an aerobic capacity of 3.5 l min^{-1} could be expected to work at a job that required 1.2 l min^{-1} of oxygen, or less, without requiring any rest breaks other than those normally used (15-min breaks, lunch, personal breaks) during the work-shift.

REVIEW QUESTIONS

1. Define work physiology and the role of work physiologists.
2. Explain how muscular activities load the cardiovascular and respiratory systems.
3. Name four parameters used in assessment of physiological responses to physical workload.
4. Theoretically, what workload, as assessed in terms of energy expenditure, does not physiologically strain the worker during the work day? What workload is usually recommended by companies and ergonomists to provide a margin of safety?
5. Define anaerobic threshold (AT). What is the AT for untrained individuals and highly trained athletes?
6. What system of energy supply (aerobically or anaerobically) will be in effect for performing tasks of a normal-paced work? How about performing a very-high-intensity task?
7. What proportion of the energy consumed under resting conditions is supplied by fat and what proportion by carbohydrates? How much of the oxygen consumption, under such conditions, is supplied by the aerobic process, and how much by the anaerobic process?
8. Explain the changes in the oxygen supply during the initial stage of a submaximal physical work bout, steady state, and during the recovery period after the work bout is completed. Also, describe when and why oxygen deficit and oxygen debt occur.
9. Explain the differences between prolonged work and intermittent work.
10. Define the respiration process, and describe how it takes place in the human body.
11. Describe the differences in the partial pressures of oxygen and carbon dioxide between the lungs and other body tissues. Also, state what facilitates the gas exchange between the body and the environment.
12. Define the acronyms STPD, BTPS, and ATPS, and state their corresponding conditions.
13. What factors, other than partial pressure, can affect the gas exchange between the body and the environment?
14. Describe the gas transportation process in the body.
15. Describe the functions of the arteries, capillaries, and veins in the circulatory system.
16. Describe the cardiac output, stroke volume, and heart rate, and the relationship among them.
17. What is the cardiac index? How can it be approximated?
18. Define metabolism.

19. What is the rate of basal metabolism?
20. What is Met?
21. What is calorimetry?
22. Briefly describe direct calorimetry.
23. Briefly describe the open-circuit and closed-circuit methods of calorimetry.
24. The following data were collected for a task performed by a female subject using a pulmonary gasometer and an oxygen analyzer:

    Pre-test measurements:
      Subject's body weight (nude): 70 kg
      Subject's height: 165 cm
      Weight of subject's clothes: 3 kg
      Weight of pulmonary gasometer: 4 kg

    Experimental measurements:
      Gasometer reading started at: 10:13:00 (h:m:s)
      Gasometer reading ended at: 10:20:30 (h:m:s)
      Initial gasometer reading: 51005.5 l
      Final gasometer reading: 51268.0 l
      Gasometer temperature: 26°C
      Barometric pressure: 752 mmHg
      Oxygen content of the expired air: 16.68%
      Oxygen content of the inspired (ambient) air: 20.93%
      Energy equivalent to 1 l of oxygen consumed: 5 kcal

    Determine the energy expenditure during performing the task.
25. What is physical work capacity?
26. What is aerobic capacity?
27. List the factors that affect aerobic capacity.
28. Describe the direct method of measuring maximum oxygen uptake. What are the advantages and disadvantages of this method?
29. What is the basis for using the regression method to estimate maximum oxygen uptake?
30. The following data were collected on a 38-year-old male subject, riding a cycle ergometer:

HR	$\dot{V}O_2$
120	1.20
135	1.50
154	1.90

    Using the regression method, calculate the subject's $\dot{V}O_{2max}$.
31. Using the data given in the previous question and Tayyari and Ramsey's method, estimate the subject's maximum oxygen consumption.
32. A 35-year-old man, weighing 79 kg, performed a step test on a 15-in

(38.1-cm) high step at a frequency of 25 steps $\cdot$ min^{-1}. He reached a heart rate of 138 bpm after 3 minutes of stepping. Estimate the subject's maximum aerobic capacity.

33. Assume that a 32-year-old woman of 70 kg body weight, walking on a treadmill at speed 4.8 km $\cdot$ h^{-1} (3 mph) has a heart rate of 120 bpm. Using Tayyari's conventional method, estimate the subject's $\dot{V}O_{2max}$.

34. What conclusion can be made when a worker's initial pulse rate after successive bouts of work becomes progressively higher?

35. How would muscular (physical) fatigue manifest itself? What two factors are involved in fatigue development?

36. How do high and low concentrations of lactic acid affect muscular contraction?

37. What is the probable cause of muscular swelling?

38. How does Brouha define normal curves, no-recovery curves, and inverse recovery curves?

39. Define endurance. What is the unit for measuring endurance? How is endurance related to physical work capacity?

40. Define the endurance time limit. For light to moderate levels of work, how soon may an employee reach this limit? What should be done when the workers reach their endurance time limit?

41. In the context of work–rest scheduling, what is meant by rest?

42. Define the generally accepted standard workload in the UK and the USA and its equivalent energy expenditure rate.

43. The energy expenditure for a certain job, performed by a 65-kg man, is 6.25 kcal $\cdot$ min^{-1}. Calculate the following:

    (a) the maximum work-bout duration;
    (b) the length of required rest period for the man to restore his energy reserve if he works for the full allowed maximum work-bout duration;
    (c) the length of required rest period for the man to restore his energy reserve if each work-bout is 15 min long.

REFERENCES

ACGIH (1994) *1994–1995 Threshold Limit Values for Chemical Substances and Physical Agents and Biological Exposure Indices*. American Conference of Governmental Industrial Hygienists, Cincinnati, OH.

American College of Sports Medicine (1991) *Guidelines for Exercise Testing and Prescription*. Lea & Febiger, London.

Åstrand, I. (1960) Aerobic work capacity in men and women with special reference to age. *Acta Physiol. Scand.* **49**(Suppl. 169).

Åstrand, P.O. and Rodahl, K. (1986) *Textbook of Work Physiology: Physiological Bases of Exercise*, 3rd edn. McGraw-Hill, New York.

Brouha, L. (1967) *Physiology in Industry*, 2nd edn. Pergamon Press, Oxford.

Burford, C.L., Ramsey, J.D., Tayyari, F., Lee, C.H. and Stepp, R.G. (1984) A method for systematic workload estimation (SWE). In: *The Proceedings of the Human Factors Society – 28th Annual Meeting*. Human Factors Society, Santa Monica, CA, pp. 997–999.

Carpenter, T.M. (1939) *Tables, Factors, and Formulas for Computing Respiratory Exchange and Biological Transformations of Energy*, 3rd edn. Carnegie Institution of Washington, Washington.

Consolazio, C.F., Johnson, R.E. and Pecora, L.J. (1963) *Physiological Measurements of Metabolic Functions in Man*. McGraw-Hill, New York.

Durnin, J.V.G.A. and Passmore, R. (1967) *Energy, Work and Leisure*. Heinemann Educational, London.

Eastman Kodak Company, The Ergonomics Group (1983) *Ergonomic Design for People at Work*, vol. 2. Van Nostrand Reinhold, New York.

Edholm, O.G. (1967) *The Biology of Work*. World University Library, McGraw-Hill, New York.

Ganong, W.F. (1987) *Review of Medical Physiology*, 13th edn. Appleton & Lange, Norwalk, CT.

Harrison, M.H., Brown, G.A. and Belyavin, A.J. (1982) The "Oxylog": an evaluation. *Ergonomics* **25**: 809–820.

ILO (1979) *Introduction to Work Study*, 3rd edn. International Labor Office, Geneva.

ISO (1987) *Ergonomics–Determination of Metabolic Rate*. ISO/DIS 8996 (draft). International Organization for Standardization (ISO), Geneva.

Kroemer, K.H.E., Kroemer, H.J. and Kroemer-Elbert, K.E. (1990) *Engineering Physiology: Bases of Human Factors/Ergonomics*, 2nd edn. Van Nostrand Reinhold, New York.

Louhevaara, V., Ilmarinen, J. and Oja, P. (1985) Comparison of three field methods for measuring oxygen consumption. *Ergonomics* **28**: 463–470.

Müller, E.A. (1953) The physiological bases of rest pauses in heavy work. *Quarterly Journal of Experimental Physiology* **384**: 205–215.

Murrell, K.F.H. (1965) *Human Performance in Industry*, Reinhold Publishing Corporation, New York.

NIOSH (1986) *Criteria for a Recommended Standard: Occupational Exposure to Hot Environments*. USDHHS (NIOSH) publication no. 86–113. National Institute for Occupational Safety and Health, US Department of Health and Human Services, Cincinnati, OH.

Passmore, R. and Durnin, J.V.G.A. (1995) Human energy expenditure. *Physiological Reviews* **35**: 801–840.

Ramsey, J.D., Burford, C.L., Dukes-Dobos, F.N., Tayyari, F. and Lee, C.H. (1986) Thermal environment of an underground mine and its effect upon miners. *Annals of American Conference of Governmental Industrial Hygienists*, **14**: 209–223.

Ricci, B. (1967) *Physiological Basis of Human Performance*. Lea & Febiger, Philadelphia.

Rodahl, K. (1989) *The Physiology of Work*, Taylor & Francis, London.

Siconolfi, S.F., Garber, C.E., Lasater, T.M. and Carleton, R.A. (1985) A simple, valid step test for estimating maximal oxygen uptake in epidemiologic studies. *American Journal of Epidemiology* **121**: 382–390.

Tayyari, F. (1987) Estimating energy expenditure in level walking. In: *Trends in*

Ergonomics/Human Factors IV (Asfour, S.S., ed.). Elsevier Science, Amsterdam, pp. 285–290.

Tayyari, F. (1994) Objective determination of work–rest regimen. In: *Advances in Industrial Ergonomics and Safety VI* (Aghazadeh, F., ed.). Taylor & Francis, London, pp. 213–215.

Tayyari, F. (1995) Estimation of maximum oxygen uptake: a convenient method for industry. In: *Advances in Industrial Ergonomics and Safety VII* (Bittner, A.C. and Champaney, P.C., eds). Taylor & Francis, London. pp. 169–172.

Tayyari, F. and Ramsey, J.D. (1985) A model for estimating maximum aerobic capacity. In: *The Proceedings of the Human Factors Society – 29th Annual Meeting*, Human Factors and Ergonomics Society (formerly Human Factors Society), Santa Monica, CA, pp. 176–177.

Tayyari, F., Burford, C.L. and Ramsey, J.D. (1989) Guidelines for the use of systematic workload estimation. *International Journal of Industrial Ergonomics* **4**: 61–65.

Weir, J.B. de V. (1949) New methods for calculating metabolic rate with special reference to protein metabolism. *Journal of Physiology*, **109**: 1–9.

Workstation design 7

7.1 INTRODUCTION

A strong relationship exists between the comfort of workers and their productivity. Unfortunately, this fact has not yet been accepted by many industrial organizations, where the management expects the productivity of the organization and the quality of its products to be a function of pay rate only (Johnson, 1985). This is an indication of a lack of understanding of the concepts of ergonomics and the roles of its principles for designing an effective workplace. Management should understand that worker discomfort due to long standing instead of sitting, for example, puts additional energy demands on the employee that by no means contribute to the worker's productivity. In addition to fatigue and the resulting deteriorated worker's performance, an awkward workplace design can result in development of occupational injuries (e.g., cumulative trauma disorders) to the worker.

The goal of ergonomics is not just to reduce effort; it is rather to maximize the worker's productivity at a level of effort which is not harmful to the worker. Hence, as Johnson (1985) agrees, ergonomic solutions do not answer such questions as "whether the organization should have comfort or efficiency" or "how much compromise should be made between comfort and efficiency," but rather attempt to minimize the incompatibilities between the capabilities of workers and the demands of their jobs, with resulting increases in productivity, enhanced safety performance, and reduced overall cost.

7.2 WORKPLACE AND WORK SPACE

Most workers spend a major portion of their time in a small work area, called the work space or "work envelope." A **work space** is a three-dimensional region surrounding the worker, defined by the outmost points touched by various parts of the body and by the controls, tools, or other equipment used by the worker (Damon *et al.*, 1966). The term

workplace is more comprehensive and can be as varied as assembly stations, offices, warehouses, vehicle cabs, or any area where work is performed. The design specifications of the workplace in relationship with workers' physical characteristics and job requirements have significant impact on their productivity, and physical and mental well-being. Both static and dynamic anthropometry of the user population must be considered in workplace design. The required anthropometric data may be found in tables and figures given in Chapter 4 of this book or obtained from other sources (Woodson and Conover, 1964; Damon *et al.*, 1966; Dreyfuss, 1967; Van Cott and Kinkade, 1972; Roebuck *et al.*, 1975; Woodson, 1981; Bailey, 1982; Eastman Kodak Company, 1983).

7.3 WORKPLACE DESIGN PROBLEMS

The workplace should be designed in such a way that employees will be able to perform their jobs effectively. To achieve this crucial goal, the workplace designer should keep two design factors in mind. The first factor is that there is a large variability in size of people in the work-force population. The second factor is to understand the user population; that is, culture, education, training, skills, attitude, physical and mental capability, etc. Therefore, designers should be cautious that the worst design mistake they can make is, probably, to design to their own personal specifications (the syndrome of "If I can use it, it must be designed well"). Such a mistake can be avoided by using the relevant anthropometric data in design of the workplace for the work-force population.

Workstations are typically designed either for seated or standing work. The main factor which determines whether the workstation must be a seated or standing workstation is the nature of the job performed in the station. Sometimes the job requires both sitting and standing postures. In this case the workstation should be so designed to permit alternating between sitting and standing. Ergonomists use the following bases for determining the type of workstations (Ayoub, 1994):

Seated workstations are recommended for the following situations:

- All items needed during the routine task cycle can be easily supplied and handled within the seated workstation.
- The job being performed does not require reaches more than 40 cm (16 in) forward or higher than 15 cm (6 in) above the work-surface.
- The job does not require large forces, such as handling objects heavier than 4.5 kg (10 lb).
- The job involves writing or light assembly for a major part of the shift.
- The job requires precision or fine manipulative movements that need a level of stability.

- The job includes foot control operation, that is performed more easily and safely while sitting and maintaining good postures.

Standing workstations are recommended in the following circumstances:

- The workstation does not have knee clearance (suitable leg-room) for a seated operation.
- The job involves handling objects weighing more than 4.5 kg (10 lb).
- The job requires high, low, or extended reaches frequently.
- The job requires frequent movement from one station to others.
- The job requires the exertion of downward forces, as in packaging and wrapping operations.

Sit–stand workstations are recommended in the following situations:

- The job requires frequent reaches more than 41 cm (16 in) forward or more than 15 cm (6 in) above the work-surface.
- The job consists of multiple tasks, some of which are best performed in the sitting position, and others are best performed while standing.

Prolonged work in the same position, whether sitting or standing, will cause discomfort. Prolonged sitting, without provision for adjustment, can first, affect the natural curvature of the spine, which in turn may disturb the functions of the internal organs of breathing and digestion, and second, weaken the abdominal muscles (Grandjean, 1988). These problems will become profound when awkward sitting postures are assumed and poorly designed seats are used. In prolonged standing, on the other hand, workers are trying to balance their body, which imposes a static load on the muscles involved (especially in the back and legs) and can cause blood pooling in the lower extremities.

Prolonged sitting is probably as bad as prolonged standing. A clear example of prolonged sitting with a variety of postural problems is work at visual display terminals (VDTs). Nevertheless, sitting has certain advantages over standing (Grandjean, 1988; Ayoub, 1994):

- Taking the body weight off the legs.
- Ability to avoid unnatural body postures.
- Lower energy consumption due to less muscular activity for maintaining postures. This helps avoid or delay the onset of fatigue.
- Less demand on the cardiovascular system.
- Providing more stability needed for tasks requiring precision or fine manipulative movements.
- Ability to operate foot controls more easily, precisely, and safely while maintaining good working postures.

With regard to movements, there is a contradiction between methods engineers and ergonomists. For example, methods engineers would consider standing and transporting a part over a short distance as an ineffective movement, while ergonomists believe that movement can be very effective in relieving fatigue caused by static postures (Johnson, 1985). From physiological and orthopedic points of view, a desirable design allows workers to alternate between sitting and standing, when they want. For jobs requiring prolonged sitting, the ergonomist should provide an easily adjustable workstation/chair to allow the operator to make frequent adjustments. If an easily adjustable workstation/chair is not available, the operator will try to compensate by assuming a variety of body postures, such as slouching, to relieve postural stress. If adjustable workstations (e.g., chairs) are provided, supervisors must make sure that workers know how to adjust them and that they are provided with guidelines on recommended seating postures. There are many ergonomic principles and recommendations that should be incorporated in the design of workplaces. Ignoring these principles can lead to designing stressful workplaces.

7.4 GENERAL PRINCIPLES FOR WORKSTATION DESIGN

The general principle for designing individual workstations is to provide an efficient and safe location in which the work can be performed. To establish this general principle of workstation design, the design engineer must answer the following six key questions:

- What must the worker see while on the job? This includes: parts of the workplace that must be visible to the worker; number, types, and locations of controls that must be utilized by the worker and the types of actions required to operate them; interaction with other workers; and necessity to view the job activities performed by other workers.
- What must the worker hear? This includes the oral communication during job performance, auditory signals that must be heard by the worker, and the requirement of hearing mechanical operation of the equipment used.
- What tasks must the worker perform? The designer must determine the required movements and job tasks performed by the worker. It must be determined whether the job requires lifting, carrying, and positioning of materials. The types of tools needed to perform the job must also be considered.
- What is the sequence of job activities? The designer must understand the nature and sequence of the job activities that must be performed by the worker.
- What clearances are required? The designer must determine and make

provisions for clearances that are required in order for the worker to perform the job efficiently and safely. The designer must consider the size of the workers who will perform jobs at the workstation and clearances for their clothing and movements. The designer should also consider any possibility of accidental activation of controls, and injuries by striking against objects in the workplace.

- What storage is required? The designer must make provisions for the storage of raw materials, in-process work-pieces, and finished product, as well as the work-tools and other job aids that must be used and stored at the workstation.

7.5 ERGONOMICS PRINCIPLES FOR WORKPLACE DESIGN

The ergonomic recommendations for determining the dimensions of the workplace (workstations and work spaces) are based on the following three factors, with the first two being the most important:

- anthropometric data;
- the nature of the job;
- behavioral patterns of employees.

An effective workstation for the human operator involves incorporation of certain established design principles from the fields of ergonomics and work study. Conducting systems analyses of the job and function allocations – among operators and between workers and machines – helps determine the types and number of tools and equipment necessary for the operators to perform their functions in the work system. Once the required tools and equipment are determined, they must be so arranged or positioned that operators can effectively perform their functions. In general, the workplace design must satisfy the following important criteria:

- be economical;
- enhance the workers' efficiency;
- allow good working postures;
- minimize fatigue;
- minimize health-and-safety risks, such as stresses on the musculo-skeletal system.

Workplace design based on the dimensions and capabilities of workers can be ergonomically correct. It is, however, not practical to determine universal design specifications for all work space or workstations because their dimensions depend upon the physical characteristics of their users as well as their intended applications. Nevertheless, the following principles of ergonomics and work study (motion and time study) should be used as general guidelines for workplace/work space design. The designer should also apply common sense and empirical principles.

- Work spaces must be designed for the expected user population. Thus, all dimensions should be determined based on relevant anthropometric data.
- Work spaces must have adequate clearance for the user's head, torso, arms, knees, and feet. The largest individuals who will use the work space determine the amount of clearance needed.
- Bent or unnatural postures should be avoided. Bending the trunk or the neck sideways is more harmful than bending forwards.
- Keeping an arm outstretched either forwards or sideways should be avoided. The further the elbows are away from the body, the more fatigue will be developed, as well as loss of precision and level of skill in using the hands and arms. Raised or outstretched arms result in an undesirable static muscle loading of the shoulders.
- Design the work for more sitting than standing. Prolonged work in the same position should be avoided, whether seated or standing. Therefore, some movements should be incorporated into the task whenever possible, and the workplace should permit easy postural changes. This provision minimizes employee discomfort and fatigue.
- Arm movements should be either in opposition to each other or otherwise symmetrical. One has better nervous control in symmetrical movements.
- The location of the working field should be at the best distance from the eyes of the operator.
- All tools and parts needed by the worker should be placed in the order in which they are to be used so that the path of the worker's movement is continuous. Furthermore, the knowledge that a tool or part is always in the same place eliminates the time and annoyance of searching for it.
- Tools should be prepositioned in such a way that they can be conveniently picked up for use. For example, a power screwdriver that is repeatedly used in a job can be suspended just above the task area using a coil spring. In this situation, workers do not even need to raise the head to see the tool; they merely grasp and pull it down into the required position for immediate use.
- All tools and parts should be placed within a comfortable reaching distance. It becomes fatiguing if the worker must repeatedly change positions to over-reach beyond the maximum grasping area, or work beyond the normal working area. All the work activities should be performed within the maximum reach arc. Figure 7.1 depicts the boundaries of the normal and maximum distances. Work activities beyond the maximum reaching distance should not be required. As a

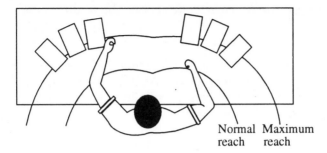

Fig. 7.1 The normal and maximum reaching distances. The normal-reach arc that is the distance from elbow to hand is the boundary of the horizontal working area. The maximum-reach arc that is the distance from shoulder to hand is the grasping area.

general rule, hand grips, operating levers, tools, and materials should be arranged around the workplace in such a way that the most frequent movements are carried out with the elbows bent and near to the body.

- Hand-work can be raised by using supports under the elbows, forearms, or hands.
- Use rubber or padded strips to cover sharp edges or corners of work tables and benches with which the operator's body (e.g., hands or arms) may come in contact.
- Torso twisting action should be prevented. Twisting is especially dangerous when performed in conjunction with lifting.
- The combination of stretching and lifting or placing objects at a distance which will significantly increase the load on the back should be avoided. Hooks or similar devices should be used to pull the object as close to the body as possible, before lifting it.
- Lifting heavy or bulky objects above shoulder height should be prevented. Such objects can generate large torque loads on the lower back.
- Wrist bending in repetitive tasks should be prevented. For example, a container can be tilted so that when objects are repetitively removed from or placed into it, the hand is aligned with the long axis of the forearm. It is especially important to avoid bent wrists when the application of large forces is required.

7.6 USING ANTHROPOMETRIC DATA IN WORKSTATION DESIGN

The designers may deal with several interrelated and/or conflicting criteria and try to combine them in some trade-off fashion. In a complex situation, the designer should follow the routine steps of design procedure. The following are the general steps in a systematic design procedure (Woodson and Conover, 1956):

1. **Preparation**: All necessary information with regard to the jobs to be performed in the workplace. All capabilities and limitations are assessed. In this step the following information should be collected:
   - types of job functions (tasks);
   - the human–machine interfaces;
   - the workplace requirements and constraints;
   - the workplace environmental conditions;
   - the characteristics and requirements of the equipment used in the workplace;
   - the descriptions and capabilities of work populations.
2. **Identification of all feasible design alternatives**: The collected information is assembled to link the design components together to explore all feasible design alternatives that effectively combine components to satisfy the design constraints. The following should be considered in this step:
   - functional characteristics;
   - compatibility of the functional characteristics with design constraints;
   - reliability of the alternative designs under the expected conditions.
3. **Selection of the best design alternative**: All identified alternatives are compared to select the best alternative. The criteria used for comparison of the alternatives and selection of the best alternative should include:
   - economy of production;
   - efficiency of operations;
   - ease of maintenance.
4. **Examination of the final alternative**: The selected final design alternative should be evaluated experimentally to insure that the design objectives have been achieved and the constraints are satisfied. Making a mock-up and evaluation of the overall workplace may be necessary to insure that all design trade-offs are feasible and all problems are resolved. It should be noted that the workplace design problems may not be just of dimensions, they can also be of safety, environmental, psychosocial, or other problems.

When anthropometric data are used, it is important for design engineers to apply their common sense, too. For example, assigning a short operator to a parts assembly task at a very high workstation can lead to the development of cumulative trauma disorders in the operator's upper extremities (i.e., in the shoulder, elbow, and wrist areas) due to the awkward posture assumed during the work. In contrast, assigning a very tall person to a task performed at a low level can lead to that person developing lower-back fatigue and pain due to a bent posture. However, modification of the workplace is not always economically feasible and can be very costly. In such a case, selecting the "right" people for the jobs could be appropriate. This concept is known as "fitting the person to the task", as opposed to the main goal of ergonomics, which is "fitting the task to the person." In summary, the following two general methods can be used for improving workplace effectiveness: modification (redesign) of the workplace ("fitting the task to the person" concept), and selecting the worker who fits the workplace ("fitting the person to the task" concept). The latter is done by assigning a worker who has capabilities appropriate for the job. However, the design engineer should aim at the accommodation of the majority of the work-force population.

There are three basic design criteria for establishing the dimensions of a workplace (Sanders and McCormick, 1993), each of which is appropriate for a particular situation:

- Design for the extreme.
- Design for the average.
- Provide adjustability.

In designing for extremes, reaches are designed for the smallest and clearances are designed for the largest individuals. For example, if the shortest person can reach a push-button control, the tallest person would reach it with no trouble.

Many components of assembly operations are designed for the "average person." This criterion can be applied to workplaces (especially, conveyors) that are fixed and cannot be adjusted. The concept behind designing for the average person is that the resulting design will inconvenience a small number of people a minimum amount. It must be kept in mind that in most cases designing for the average person would inconvenience the majority of the population. For example, when designing a seat height for the average person, it would be too high for the shorter 50% of the workforce and too low for the taller 50%. Facilities used by many different people for relatively short periods of time are appropriate for "designing for the average person."

The third design criterion, providing adjustability, would accommodate a larger proportion of the work-force. The disadvantages of this criterion include cost of design and manufacturing, maintenance costs, time of adjustment, and reliability.

7.7 RECOMMENDATIONS FOR SEATED WORKSTATIONS

It is inappropriate to determine design specifications for a universal seated workstation since its dimensions will vary according to its intended purpose and its user's physical characteristics. Special-purpose seats (pilot seats, school children's chairs, factory workbenches, and their corresponding surrounding work spaces differ significantly in size, shape, and component materials. On the other hand, there are general-purpose seats used in offices, homes, and in passenger vehicles. The design of car-type seats should accommodate as large a portion of the general population as possible.

The recommendations and guidelines that follow will develop specific values for the workplace dimensions and ranges of adjustability. However, no matter how precise the analysis, the workstation should be designed to allow the worker to make frequent adjustments easily. Postural changes are essential to effective and efficient workstations.

7.7.1 RECOMMENDATIONS FOR SITTING POSTURES

It is often seen that, in most sitting positions, the seated person is constantly seeking a more comfortable position. This is done by many position changes of legs, thighs, or buttocks. An ergonomically designed chair is one of the most important parts of a workstation. It can favorably affect posture, circulation, the amount of effort required to maintain a position, and the amount of pressure on the spine. Unless otherwise mandated by the nature of the task, a seated workstation (as illustrated in Fig. 7.2) should generally be so designed that its user can assume a posture characterized as follows:

- The upper arms and the lower legs are vertical.
- The forearms and the thighs are horizontal.
- The feet are flat on the floor.
- The seat backrest supports the inward curvature of the lumbar region of the spine.
- The weight of the upper body is evenly distributed on a large surface area of the buttocks and thighs.

The following are common principles for design of sitting (or seated) workstations (National Safety Council (NSC), 1988; Ayoub, 1994):

- Everything workers need while performing their task must be accessible and easy to handle in the seated position.
- The sitting workstation design should not require the hands to work at more than 15 cm (6 in) above the work-surface.
- Provide mechanical assists or eliminate the requirement of large forces. The worker should not handle objects weighing more than 4.5 kg (10 lb) manually.
- Provide an ergonomically correct chair. The chair should allow users to keep their spine and head upright to prevent back and neck strain.
- Eliminate lifting from the floor.

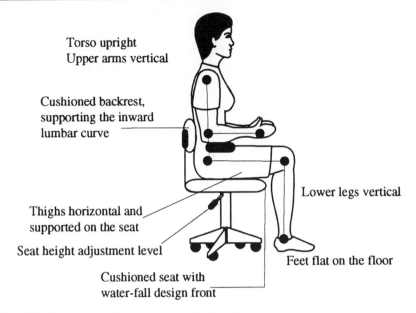

Torso upright
Upper arms vertical

Cushioned backrest,
supporting the inward
lumbar curve

Lower legs vertical

Thighs horizontal and
supported on the seat

Seat height adjustment level

Feet flat on the floor

Cushioned seat with
water-fall design front

Fig. 7.2 Illustration of appropriately designed chair.

7.7.2 RECOMMENDATIONS FOR SEAT DESIGN/SELECTION

The seat should adapt (adjust) to the user, not vice versa. To incorporate this ergonomic fundamental principle, the following criteria are recommended to be considered in designing and/or selecting the work seat:

- Chairs should be stable, and fully and easily adjustable from the seated position. Mobile chairs are not recommended for use on the shop floor where the risk of tipping over is present.
- Where mobility is required (e.g., for office employees), wheels or castors should be fitted to the chair (hard castors for soft floors and soft castors for hard floors to increase stability). Mobility is not recommended for a slippery floor, which makes it difficult to keep the chair in the desired position.
- Where wheels or castors are fitted, the chair should have five legs to decrease the risk of tipping over.
- The chair should have a padded backrest that is at least 18 cm (7 in) high and 33 cm (13 in) wide. The backrest should be adjustable up-and-down (7–15 cm (2.75–6 in) above the seat) and forward-and-backward (35.5–42 cm (14–16.5 in) from the front edge) for proper lumbar (lower-back) support.
- Seat height should be adjustable so that the upper-body weight is distributed over the buttocks, not over the thighs, by maintaining the

thighs horizontal. (A range of adjustability from 38 to 50 cm (15–20 in) above the floor accommodates about 90% (5th to 95th percentiles) of the US civilian population – accounting for shoe allowance.)

- The depth of the seat should be about 38–40 cm (15–16 in) and its width should be 42–46 cm (16.5–18 in), respectively.
- There should be enough clearance between the front edge of the chair and the back of the knees (about 5–15 cm (2–6 in)).
- The front of the seat should be of a waterfall shape, which provides enough clearance for the flesh of the thighs, thus preventing reduction of blood circulation.
- The seat pan should slightly slope backward (1–5°) to prevent the ejection effect of the seat.
- The angle between the backrest and the seat should be about 100°.
- Both seat pan and backrest should be upholstered and covered with a breathable fabric that absorbs perspiration and gives way about 2–2.5 cm (3/4–1 in) when compressed.
- Torso twisting during the task performance should be avoided. If frequent lateral movements are required, the seat should swivel.
- Where the seat height is fixed and excessive, footrests should be provided for users to be able to maintain their thighs horizontally and relieve the pressure under their thighs from the chair front.
- If a footrest is required, it should be slightly angled toward the person. It should support the soles of both feet; a surface of 27–30 cm (10–12 in) in depth by 40 cm (16 in) in width should be adequate. The footrest surface should be of a non-slip material.
- Angle-bar-type footrests, which do not meet the above criteria, should not be used. Bars, brackets, or narrow strips are not adequate footrests since they impose concentrated pressure under the feet.
- Armrests should be provided if the task requires the arms to be held away from the body. The distance between armrests should be about 47 cm (18.5 in) to accommodate most users. If the armrests are too close to each other, it would be difficult for large-size users to get in and out of the seat. On the other hand, the farther the arms are held away from the body, the greater would be the fatigue and the lower the manual dexterity.
- Armrests should be padded and covered with an absorbent non-slip material.
- Handrests should be provided for intricate tasks (such as fine assembly or inspection). When the weight of the arms is supported, the hands are stabilized and hand dexterity is maximized.
- Headrests should be provided where the head must be tilted forward or backward for prolonged periods of time. An example of headrest application is for the use of optical viewing tools such as microscopes.

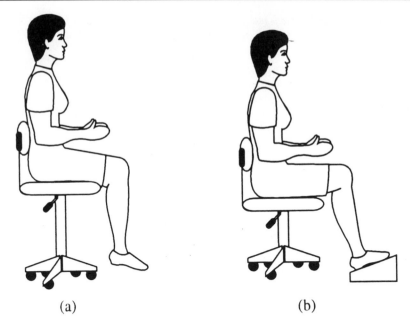

(a)   (b)

Fig. 7.3 Illustration of footrest use. (a) Feet not touching the floor puts pressure on the thighs, causing circulation problem and fatigue. (b) Feet touching footrest relieves pressure on the thighs.

The head is relatively heavy for the weak neck muscles. If the head is not kept straight, the pull of gravity will cause stress and strain in the muscles of the neck.

• In order to achieve a good posture, it may be necessary to adjust the work-surface height. For example, typewriters should be placed on lower work-surfaces at secretarial workstations.

Wherever the seat is too high and its adjustment is not possible or appropriate, a footrest should be provided for users to be able to place their feet comfortably on it in order to prevent under-thigh pressure. Figure 7.3 illustrates the use of a footrest.

7.7.3 RECOMMENDATIONS FOR DESIGNING WORK TABLES/DESKS

The heights of tables/desks (or workbenches and counters) are generally set at elbow-level. An awkward table height reduces efficiency and induces fatigue. The width and depth of a work table/desk depend on the task and its requirements (i.e., tools, parts, computer hardware, reading and writing materials, etc.). In general, the following are the critical dimensions in designing any work-surface (Fig. 7.4):

1. working height;
2. working width;
3. working depth;
4. knee-room height;
5. knee-room depth;
6. knee-room width;
7. kick room.

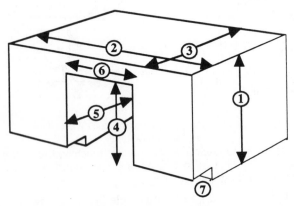

Fig. 7.4 An illustration of critical dimensions in designing work tables/desks.

7.7.4 RECOMMENDATIONS FOR DESIGNING THE LEG-ROOM

Clearances of the leg-room, under the work-surface, are illustrated in Figure 7.5.

where: a = buttock-to-knee length;
  b = sitting abdominal depth = $0.40 \cdot a$ (Diffrient *et al.*, 1974);
  c = buttock-to-popliteal length;
  d = popliteal height;
  A = depth at knee level;
  B = depth at foot level;
  B_1 = $c - b$ = (buttock-to-popliteal length) – (sitting abdominal depth or buttock depth);
  B_2 = amount of the lower leg forward of perpendicular = $d \cdot \operatorname{Sin} \theta°$;
  B_3 = depth required to accommodate foot length (the feet may be placed flat on the floor);

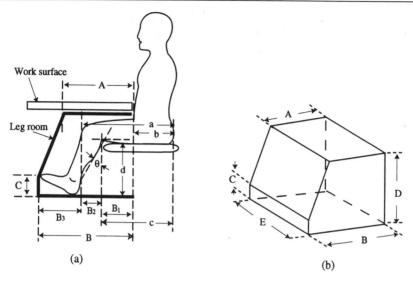

Fig. 7.5 (a) The under-work-surface clearances (b) illustrated as a three-dimensional leg-room envelope.

C = foot length at $\theta°$ = foot length . Sin $\theta°$;
θ = expected maximum angle between the vertical and the lower leg.

With reference to Figure 7.5, the minimum leg-room under the work-surface at the knee level can be found by deducting the sitting abdominal depth (approximated by the buttock depth) from the buttock-to-knee length; or by the following equation:

$$A = 0.60 \cdot a \qquad (7.1)$$

The minimum depth of the leg-room at the foot level can be determined by the following equation:

$$B = B_1 + B_2 + B_3 \qquad (7.2)$$

The minimum height of the leg-room under the work-surface is determined by the sum of the popliteal height plus the thigh clearance plus the shoe allowance. The anthropometric data of the 95th percentile of the 50–50 mixed male–female user population or the 95th percentile of the male population should be used to accommodate at least 95% of the population. Thus, for the 50–50 mixed male–female US work-force, the minimum height of the leg-room should be 67 cm (26.5 in). A minimum toe clearance (indicated by C in Fig. 7.5) of 10 cm (4 in) should also be considered.

The width of the leg-room under the work-surface should be at least 10 cm (4 in) larger than the thigh breadth (approximated by the hip breadth) of the 95th percentile of the expected user population. This provision allows for clothing and freedom of movement. Therefore, for the 50–50 mixed male–female US workforce, the minimum width of the leg-room should be 51.5 cm (20.3 in).

Example 7.1

The application of anthropometric data in workstation design is best explained by the following two example cases. In the first case, the workstation is fully adjustable (i.e., both table and seat are assumed to be adjustable). In the second case, the table is fixed but the seat is adjustable. In both cases design recommendations will be made to accommodate 66% of the population.

Case 1: when both table and seat are adjustable A footrest is not used since the workstation is fully adjustable and the feet rest on the floor. The calculation procedures can be formulated as follows (Tayyari, 1993):

Footrest = 0
Seat height = popliteal height + heel height (about 2.5 cm or 1 in)
Table height (under-table) = seat height + thigh clearance (thickness)
Table-top height = seat height + elbow-rest height − 5 cm or 2 in
            (below elbow level)

For 66% of the population to be accommodated, the 17th to 83rd percentiles must be considered. From the standard normal table (Appendix C), the 83rd percentile is approximately 0.955 standard deviation above the mean (which means that the 17th percentile would be 0.955 standard deviation below the mean). Calculations of dimensions of a fully adjustable workstation (i.e., when both the table height and seat height are adjustable) to fit 66% of the population are made (using data from Table 4.2) as follows:

Footrest = 0
Seat height:
  Minimum (for the shortest person) = 41.9 − 0.955 × (3.4)
                    + 2.5 (heel height)
                    = 41.2 cm
  Maximum (for the tallest person) = 41.9 + 0.955 × (3.4)
                    + 2.5 (heel height)
                    = 47.6 cm
Table height (under-table):
  Minimum (for the shortest person) = 41.2 + [14.0 − 0.955 × (1.8)]
                    = 53.5 cm

Maximum (for the tallest person) $= 47.6 + [14.0 + 0.955 \times (1.8)]$
$$= 63.3 \text{ cm}$$
Table-top height:
Minimum (for the shortest person) $= 41.2 + [23.6 - 0.955$
$$\times (3.0)] - 5 = 56.9 \text{ cm}$$
Maximum (for the tallest person) $= 47.6 + [23.6 + 0.955 \times (3.0)]$
$$- 5 = 69.1 \text{ cm}$$

However, a design to accommodate 66% of the population would not represent the best design strategy; in general, a design to accommodate 90–95% of the population would be more acceptable.

Case 2: when table is fixed, but seat is adjustable The calculation procedures can be formulated as follows (Tayyari, 1993):

Footrest height $=$ the adjustability of the table height when both table and seat are adjustable; that is, footrest replaces the table adjustability
Table height (under-table) $=$ footrest height $+$ heel height $+$ popliteal height $+$ thigh clearance
Table-top height $=$ table height (under-table) $+ 2.5$ cm or 1 in (table thickness)
Seat height $=$ popliteal height $+$ heel height (about 2.5 cm or 1 in) $+$ footrest height $=$ table height (under-table) $-$ thigh clearance

To fit the workstation to 66% of the population when the table height is fixed, but the seat is adjustable, dimensions are calculated as follows:

Table height (under-table) $=$ fixed at 63.3 cm
Table-top height $= 63.3 + 2.5 = 65.8$ cm

Footrest:
Minimum (for the tallest person) $= 63.3 - 2.5 - [41.9 + 0.955(3.4)]$
$$- [14 + 0.955(1.8)] = 0 \text{ cm}$$
Maximum (for the shortest person) $= 63.3 - 2.5$
$$- [41.9 - 0.955(3.4)] -$$
$$[14 - 0.955(1.8)] = 9.9 \text{ cm}$$
Seat height:
Minimum (for the tallest person) $= 41.9 + 0.955(3.4)$
$$+ 2.5 + 0 = 47.6 \text{ cm, or}$$
$$= 63.3 - 15.7 = 47.6 \text{ cm}$$
Maximum (for the shortest person) $= 41.9 - 0.955(3.4)$
$$+ 2.5 + 9.9 = 51.0 \text{ cm, or}$$
$$= 63.3 - 12.3 = 51.0 \text{ cm}$$

The results of calculations for the two workstation options for accommodation of various desired ranges of percentiles are given in Table 7.1.

7.8 ERGONOMIC GUIDELINES FOR STANDING TASKS

It is desirable to design the workplace for sit–stand work, in which workers can perform their assigned job sitting or standing. Due to the nature of the task, sometimes sitting is not possible. For such a situation, the workstation has to be designed for standing work.

7.8.1 RECOMMENDATIONS FOR STANDING POSTURES

- Standing still in one place for long periods of time should be avoided. The activity of the leg muscles acts as a pump and assists the veins in returning blood to the heart. Prolonged standing stops this pumping action and causes swelling of the lower extremities.
- When fully adjustable work tables cannot be provided for standing work or the operating level at a machine cannot be varied, the

Table 7.1 Workstation heights, in centimeters, for various ranges of a mixed population

	Adjustable table and seat				Fixed table/adjustable seat			
Percentiles fitted	17–83	5–95	2.5–97.5	1–99	17–83	5–95	2.5–97.5	1–99
Percentage fitted	66	90	95	98	66	90	95	98
Seat height								
Minimum	41.2	38.8	37.7	36.8	47.6	50.0	51.1	52.3
Maximum	47.6	49.4	51.1	52.0	51.0	55.4	58.1	60.1
Range of adjustability	6.4	10.6	13.4	15.2	3.4	5.4	7.0	7.8
Table height (under-table)								
Minimum	53.5	49.8	48.2	46.6	63.3	66.4	68.6	70.2
Maximum	63.3	66.4	68.6	70.2	63.3	66.4	68.6	70.2
Range of adjustability	9.8	16.6	20.4	23.6	0	0	0	0
Task or table-top height								
Minimum	56.9	52.5	50.4	48.4	65.8	68.9	71.1	72.7
Maximum	69.1	72.9	75.6	77.6	65.8	68.9	71.1	72.7
Range of adjustability	12.2	20.4	25.2	29.2	0	0	0	0
Footrest height								
Minimum	0	0	0	0	0	0	0	0
Maximum	0	0	0	0	9.9	16.6	20.4	23.6
Range of adjustability	0	0	0	0	9.9	16.6	20.4	23.6

working heights should be set to suit the tallest operator, while the smaller operators can be accommodated by giving them something to stand on.

- When the work requires fine or precise manipulations, working heights must be raised to a level at which operators can see clearly while keeping their back in a natural posture.
- When the hand-work calls for great force, or much freedom of movement, it is necessary to lower the working surface.
- A working level which is too high can force the worker to lift the shoulders through taxing the trapezius muscle or to lift the upper arms with the deltoid muscles. Lifting shoulders is strenuous static work that can generate great pains in the shoulder muscles.
- If prolonged standing is inevitable, padded or rubber mats for the operator to stand on should be provided to reduce fatigue.
- Mats provided for the standing operator should have feathered (reduced) edges (Fig. 7.6) to minimize tripping hazard.

7.8.2 RECOMMENDED WORK-SURFACE HEIGHT FOR STANDING WORK

Determining an optimal height of the work-surface is a crucial role the designer can play in enhancing the worker's well-being and productivity. If the working height is too low, the back and neck must be excessively bent forward, which can result in fatigue and pain in the back and neck. On the other hand, if the work height is raised too high, workers have to lift up their shoulders frequently during the job performance, which can result in neck and/or shoulder cramps. In either case, the development of cumulative trauma disorders in the back and neck regions should be expected.

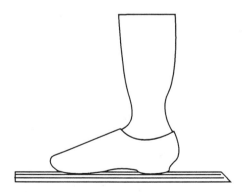

Fig. 7.6 Rubber pad reduces fatigue, and feathered (reduced) edges prevent trip hazard.

In general, the preferred working height for standing manual work is 5–10 cm (or 2–4 in) below elbow level. However, the nature of the work performed at the workstation is a major factor in determining the work height. The recommendations given in Table 7.2 account for the nature of the work.

Example 7.2

Assume that you want to determine what working heights will be convenient for **average** US male workers, performing light work (e.g., assembling small parts) at a standing workstation. According to the anthropometric data given in Chapter 4, the average elbow height of the US male worker population is 109.6 cm with a standard deviation of 4.9 cm. Then, you can calculate the lower and upper limits of the work-surface height (HW_{min} and HW_{max}, respectively) as follows:

HW_{min} = 109.6 − 15 = 94.6 cm (without considering shoe height)
HW_{max} = 109.6 − 10 = 99.6 cm (without considering shoe height)

Thus, including a 3.25-cm provision for work boots, a working height of between 98 and 103 cm should be satisfactory for male workers of average height.

The reader should be warned about setting the work height for an average-size person, without allowance for individual variations. Such a work height is too high for short people, who must frequently lift up their shoulders, which may lead to cramps in the neck and shoulders. It is also too low for tall people, who will have to bend over the work-surface, which will cause backache and neck problems. However, if such a situation is inevitable, shorter workers should use wooden footrests or pallets, or similar supports; and taller workers should use seats. In new workstation design, the height of the work-surface must be established by the tallest user, and the others can use foot supports. Use of adjustable (lift) tables is an ideal solution to work-surface problems.

Example 7.3 (designing workplace for assembly operations)

Static posture can be avoided by a proper task design (including the workplace). Sitting requires about 30% less energy than standing and is particularly appropriate for light assembly tasks that require fine hand–eye coordination (Johnson, 1985). In designing the workplace for a sitting posture, provisions should be made for footrests, possible lateral movement (e.g., glide up the line), and swivel movement (if no lateral force needs to be exerted). If, however, the job can be performed sitting or standing, the workplace should be so designed that the worker can alternate between sitting and standing positions.

Table 7.2 The recommended work-surface height for standing workstations

Type of work	Recommended height*	Work-surface height for 2.5th–97.5th percentile†		
		Males	Females	50–50 Mixed males–females
Precision work (e.g., drawing and sorting fine items)	5–10 cm (2–4 in) above elbow height	110–129.2 cm 43.3–50.9 in	109.4–124.3 cm 43.0–48.9 in	108.2–128.2 cm 42.6–50.5 in
Light work (e.g., assembling small parts)	10–15 cm (or 4–6 in) below elbow height	90–109.2 cm 35.4–43.0 in	89.4–104.3 cm 35.2–41.0 in	88.2–108.1 cm 34.7–42.6 in
Heavy work (e.g., metal cutting)	15–40 cm (6–16 in) below elbow height	75–94.2 cm 29.5–37.1 in	74.4–89.37 cm 29.3–35.1 in	73.2–93.1 cm 28.8–36.7 in

*Recommended by Grandjean (1988).
†Work-surface height = mean elbow height ± 1.96 × (standard deviation) + shoe height ± recommended mid-range height with respect to elbow height.

Figure 7.7 illustrates a workplace design which is appropriate for sit–stand postures in an assembly situation. Standing workplaces provide for increased mobility, greater ability to apply force, and increased reach envelopes. In designing an effective workplace, the height of interest is not just the bench or conveyor – the real height is rather the height of the work-surface (i.e., the heights of the bench, the fixture, the part and the tool used).

In assembly situations, the conveyor heights may be appropriate for sitting or standing, but in many situations the design of conveyors or the placement of parts bins and/or fixtures are so that the worker's legs cannot fit under the conveyor. Figure 7.7a shows provisions for leg-room and movement under a workbench.

The work-surface height must be determined according to the worker anthropometric dimensions and the task requirements. The anthropometric data can be obtained from various ergonomics/anthropometric reference manuals.

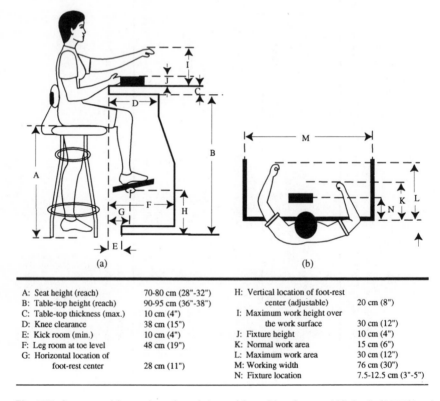

(a)   (b)

A: Seat height (reach)	70-80 cm (28"-32")	H: Vertical location of foot-rest	
B: Table-top height (reach)	90-95 cm (36"-38")	center (adjustable)	20 cm (8")
C: Table-top thickness (max.)	10 cm (4")	I: Maximum work height over	
D: Knee clearance	38 cm (15")	the work surface	30 cm (12")
E: Kick room (min.)	10 cm (4")	J: Fixture height	10 cm (4")
F: Leg room at toe level	48 cm (19")	K: Normal work area	15 cm (6")
G: Horizontal location of		L: Maximum work area	30 cm (12")
foot-rest center	28 cm (11")	M: Working width	76 cm (30")
		N: Fixture location	7.5-12.5 cm (3"-5")

Fig. 7.7 An assembly workstation. Adapted from Van Cott and Kinkade (1972) and Johnson (1985).

7.9 SOME CRITICAL PROBLEMS AND EASY SOLUTIONS

It is crucial for workers to have adequate work space and easy access – both physical and visual – to everything they need. The workstation should therefore provide clearance for the user's head, torso, arms, knees, and feet. Such a workstation permits adequate visibility and space for body movements, and safeguards the user against impacts and discomfort. Generally, the largest expected users determine the amount of clearance needed. If they can fit in the workstation, then the others can too.

Another important aspect of a workstation design is the minimization of direct pressure on the body parts. This is a common problem associated with many existing workstations. It negatively affects the person's comfort, blood circulation, and nerve function. The body parts commonly affected are the palm of the hands, forearms, legs, and feet. Pressure on the palm of the hands is usually due to the tools used. Pressure on the forearms is caused by leaning them against the edge of the workstation (which is usually sharp and/or hard) for support.

When designing the workstation, attempts should be made to eliminate awkward postures, and to eliminate or minimize direct pressure on the body segments. The latter may be accomplished by distributing inevitable contact stress over a larger surface area of the body.

7.9.1 PROVIDING EASY ACCESS

Easy access can be provided by:

- allowing adequate workspaces;
- providing or enlarging the size of openings;
- eliminating barriers and obstructions between the worker and the items needed in the task;
- rearranging the workstation components, such as equipment, tools and shelves;
- removing all objects which interfere with visual tasks or obstruct the sight.

7.9.2 MINIMIZING PRESSURE ON THE FOREARMS

Workstation improvements can be made to reduce the forearm pressure by:

- rounding and padding the edge;
- changing the work-surface height or the seat;
- providing arm rests;
- redesigning the job.

7.9.3 MINIMIZING PRESSURE ON THE LEGS AND FEET

Prolonged standing, especially when leaning against table edges or machines, can interfere with the circulation in the lower extremities. Workstation improvements for reducing pressure on the legs and feet can include the following:

- changing the workstation layout to eliminate leaning against the table edge, machine, or any other obstacles;
- reducing the size of the workbench/table to avoid over-reach and leaning against the table edge;
- rounding and padding the edge;
- changing the work-surface height;
- standing on floor mats;
- wearing cushioned shoe inserts (insoles);
- redesigning the job.

7.9.4 AVOIDING AWKWARD POSTURES

An appropriate work posture reduces the stress on workers' bodies and improves their job performance. An appropriate workstation design and layout allows workers to assume natural back (spine) curvature, to maintain straight wrists, and to keep their elbows at their sides. Awkward postures can be avoided by:

- changing the workstation layout to keep workers' elbows at their sides, and eliminating the requirement of twisting motions, especially during heavy lifting;
- avoiding reaching down into boxes, bins and tubs by providing break-down or removable-side containers, hydraulic tilters, and spring-loaded bottom containers.

7.10 APPLICATIONS AND DISCUSSION

Workstations are as varied as industry itself. One industry or facility may have many seated workers while another may have very few. In general, ergonomists should utilize their training in anthropometry, biomechanics, and work analysis to design an appropriate workstation for an employee or group of employees. Several examples and guidelines are presented in this chapter. However, not all recommendations will be appropriate for all workplaces.

By examining the recommendations and principles outlined in this chapter, the ergonomist should be better able to design a workplace that allows the worker's capabilities and limitations to meet the demands of the job. Many guidelines are common sense, but unless the designer is reminded of them, they are very easy to forget. For example, in the design of a seated workstation, it is very easy to get caught up in the range of adjustability of the chair, the work-surface heights, clearances under the workstation, and forget about providing an adjustable footrest for workers with shorter popliteal heights. At the other extreme, it is easy to get carried away and over-design a workstation. If the workstation is only used by one person, and there is little or no turnover in the job, then the user population is one person. In that case, it is likely uneconomical and unnecessary for the ergonomist to worry about a design to accommodate 90 or 95% of the general working population.

The workstation design principles presented in this chapter will aid the ergonomist in applying the basic scientific principles to workstation design. First the ergonomist must understand the problem. Next, alternatives are suggested and evaluated. Recommendations are made and then the design is evaluated to make certain that the recommendations were appropriate.

REVIEW QUESTIONS

1. What is the difference between work space and workplace?
2. What are the two design factors that the workplace designer should keep in mind?
3. Workstations are typically designed for either seated or standing work. What is the main factor which determines whether the workstation must be designed for sitting or standing?
4. For what situations are seated workstations recommended?
5. For what situations are standing workstations recommended?
6. For what situations are sit–stand workstations recommended?
7. Explain how prolonged sitting, without provision for adjustment, and prolonged standing can affect the person's body.
8. What are the advantages of sitting over standing workstations?
9. Methods engineering analysts believe that a job requirement to stand and transport a part over a short distance is an "ineffective work movement" (motion economy). What is the ergonomist's opinion on this movement?
10. What are the six key questions that must be answered before the design of an individual workstation can begin?
11. Based on what three factors are the ergonomic recommendations for determining the dimensions of the workplace (workstations and work spaces) made?
12. What criteria must be satisfied by the workplace design?
13. What are the ergonomics principles for designing work spaces for the expected user population? List at least six principles.
14. Draw a diagram that shows the normal and maximum-reach distances for a sitting workstation.

15. The designers may deal with several interrelated and/or conflicting criteria and try to combine them in some trade-off fashion. In a complex situation, the designer should take the routine steps of the design procedure. List the four general steps for a systematic design procedure.

16. What are the two general methods for improving the effectiveness of an existing poor ergonomic workplace?

17. What are the three basic design criteria for establishing the dimensions of a workplace?

18. Why is it inappropriate to design for the "average person"?

19. Why is it inappropriate to determine design specifications for a universal seated workstation?

20. List the characteristics of an ergonomically correct sitting posture.

21. List the common principles for design of sitting (or seated) workstations.

22. In order to adapt the seat to the user, several ergonomics criteria should be considered in the design and/or selection of the work seat. The text lists twenty criteria; list at least eight of them.

23. If a footrest is required, what recommendations can you make?

24. Describe when armrests should be provided and how they should be designed.

25. Describe when and why handrests should be provided.

26. Describe when and why headrests should be provided.

27. What are the critical dimensions in designing a work-surface?

28. In order to provide clearances for the leg-room under the work-surface, describe anthropometric data and how they are used to determine the minimum depth at the knee level.

29. Assuming that the abdominal depth is approximately 40% of the buttock-to-knee length and using the anthropometric data given in Chapter 4, determine the minimum depth at the knee level under the work-surface to provide leg-room clearance for 90% of the 50–50 mixed population.

30. To provide clearances for the leg-room under the work-surface, describe what anthropometric data can be used to determine the minimum depth at the foot level.

31. The text recommends seven ergonomics guidelines for standing tasks. List five of the guidelines.

32. Determine the working heights to be convenient for the 90th percentile of the US 50–50 mixed work–force, performing heavy work at the recommended height at a standing workstation.

33. If the standing working height is set based on average body measurements without allowances for individual variation, what ergonomic problems do you expect to occur? If adjustable working

cannot be provided, what ergonomics solutions would you recommend?

34. What are the occupational causes of direct pressure on the palm of the hand and forearms? How can you reduce the pressure on the forearms?

35. What are the methods of providing easy access?

36. Prolonged standing, especially when leaning against table edges or machines, can interfere with the circulation in the lower extremities. What workstation improvements can you suggest to reduce pressure on the legs and feet?

REFERENCES

Ayoub, M.M. (1994) Workplace design guidelines. In: *The Proceedings of International Symposium on Ergonomics*. American Society of Safety Engineers, Des Plaines, Illinois.

Bailey, R.W. (1982) *Human Performance Engineering: A Guide for System Designers*. Prentice-Hall, Englewood Cliffs, New Jersey.

Damon, A., Stoudt, H.W. and McFarland, R.A. (1966) *The Human Body in Equipment Design*. Harvard University Press, Cambridge, MT.

Diffrient, N., Tilley, A.R. and Bardigjy, J.C. (1974) *Humanscale*^{TM} *1/2/3*. Copyright by Henry Dreyfuss Associate. The MIT Press, Massachusetts Institute of Technology, Cambridge, Massachusetts.

Dreyfuss, M. (1967) *The Measure of Man: Human Factors in Design*, 2nd edn. Whitney Library of Design, New York.

Eastman Kodak Company (1983) *Ergonomic Design for People at Work*, vol. I. Lifetime Learning Publications, Belmont, California.

Grandjean, E. (1988) *Fitting the Task to the Man: A Textbook of Occupational Ergonomics*, 4th edn. Taylor & Francis, London.

Johnson, S.L. (1985) Workplace design application to assembly operations. In Alexander, D.C. and Pulat, B.M. eds. *Industrial Ergonomics: A Practitioner's Guide*. Institute of Industrial Engineers, Norcross, Georgia.

NSC (1988) *Ergonomics: A Practical Guide*. National Safety Council, Chicago.

Roebuck, J.A.; Kroemer, K.H.E. and Thomson, W.G. (1975) *Engineering Anthropometry Methods*. John Wiley, New York.

Sanders, M.S. and McCormick, E.J. (1993) *Human Factors in Engineering and Design*, 7th edn. McGraw-Hill, New York.

Tayyari, F. (1993) Design for human factors. In: *Concurrent Engineering: Contemporary Issues and Modern Design Tools* (Parsaei, H.R. and Sullivan, W.G., eds). Chapman & Hall, London, pp. 297–325.

Van Cott, H.P. and Kinkade, R.G. (eds) (1972) *Human Engineering Guide to Equipment Design*. US Government Printing Office, Washington, DC.

Woodson, W. E. (1981); *Human Factors Design Handbook*. McGraw-Hill, New York.

Woodson, W.E. and Conover, D.W. (1964) *Human Engineering Guide for Equipment Designers*, 2nd edn. University of California Press, Berkeley, California.

Cumulative trauma disorders

8

8.1 INTRODUCTION

The fast pace of modern production systems and job specialization have contributed to many occupational health issues. The criticality of these issues has been a fast-growing one, and their costs and complexity have become an obstacle for many businesses. Occupational health problems may be divided into the following four general categories:

- Musculoskeletal disorders, such as:
  - back injury;
  - carpal tunnel syndrome;
  - tendinitis and tenosynovitis;
  - arthritis;
  - vibration white finger (Raynaud's phenomenon).
- Cardiovascular diseases, such as:
  - myocardial infarction;
  - stroke;
  - hypertension.
- Dermatologic disorders, such as:
  - dermatoses;
  - burns;
  - contusions;
  - lacerations.
- Other disorders, such as:
  - hearing loss – induced by noise;
  - psychological impairment;
  - reproductive impairment;
  - neurotoxic disorder;
  - occupational lung disease – lung cancer, and asthma.

The US National Institute for Occupational Safety and Health

(NIOSH) has listed the above health problems and some others as the 10 leading occupational injuries and diseases (NIOSH, 1983a, 1984). NIOSH has ranked the occupational injuries and diseases in order of their national importance (Table 8.1), that is, according to their frequency, severity, and possibility of prevention.

Among the occupational injuries and illnesses listed by NIOSH in Table 8.1, ergonomists and human factors specialists concentrate on the prevention of musculoskeletal injuries, cardiovascular diseases, noise-induced loss of hearing, and psychologic disorders. Single-exposure, traumatic problems (e.g., crush injuries, falls, accidents, fractures) are usually dealt with by safety professionals, and the rest (e.g., toxicological problems) are covered by industrial hygienists.

This chapter deals with work-related musculoskeletal disorders, their risk factors, and ergonomics interventions for minimizing the risk of developing such maladies.

8.2 MUSCULOSKELETAL DISORDERS

Musculoskeletal disorders occur in every kind of occupation and industry. They occur in backs and upper and lower extremities. They occur gradually over a relatively long period of time of exposure to the corresponding contributing factors. They are, therefore, called cumulative trauma disorders (CTDs). Although these disorders have become common problems of the workplace, they still lack precise definitions. The following definitions are suggested to the reader for better understanding of musculoskeletal disorders (Emerson and Taylor, 1946; Putz-Anderson, 1988, Hoaglund, 1990; NC-OSHA, 1991):

- **Strain**: A muscle, ligament, or tendon insertion is strained when it is pulled or pushed to its extreme by forcing the joint beyond its normal

Table 8.1 The 10 leading occupational diseases and injuries in the USA

Occupational lung disease (industrial hygiene-related)
Musculoskeletal injuries (ergonomics-related)
Occupational cancers (industrial hygiene-related)
Amputation, fractures, eye loss, lacerations, and traumatic deaths (safety-related)
Cardiovascular diseases (ergonomics-related)
Disorders of reproduction (industrial hygiene-related)
Neurotoxic disorders (industrial hygiene-related)
Noise-induced loss of hearing (ergonomics-related)
Dermatologic conditions (industrial hygiene-related)
Psychologic disorders (ergonomics-related)

From NIOSH (1983a).

range of motion. It can result from lifting a heavy object or holding against an external force.

- **Sprain**: A ligament is sprained when it is stretched so far that some of its fibers are torn.
- **Tendinitis**: Tendinitis is inflammation of a tendon. The inflammation may be primary, as caused by rheumatoid arthritis; or it may be secondary to a physical injury. It can result from: direct blows to the tendons themselves, strains which over-stretch the tendons, and/or trauma from overuse due to repeated movements over long periods of time. Symptoms of tendinitis include pain, burning sensation, and/or dull ache over the affected area. Swelling and reduced ability to use the affected joint can result.
- **Tenosynovitis**: Tenosynovitis is irritation and inflammation of a tendon sheath. Since the inflammation involves both tendons and their synovial sheaths, the condition may also be called tendosynovitis. Causes and symptoms are similar to those of tendinitis. For example, repetitive flexion of the fingers, forceful flexion and extension of the wrist, and abduction and flexion of the shoulder can cause tendinitis and tenosynovitis at these joints. Other factors associated with tenosynovitis include aging, gender (female), and certain systemic diseases.
- **Ganglionic cyst**: A ganglionic cyst or ganglion is a tendon disorder, which resembles a hernia-like projection near the affected joint. This disorder usually manifests itself as swollen nodules on finger tendons at either the wrist or adjacent to other bony articulations of the finger. The affected tendon sheath swells, creating a bump under the skin. Ganglionic cysts often appear on the dorsal aspect of the wrist. They may be painless, but the affected joint may be moderately painful and weak. Ganglionic cysts have a tendency to rupture and disappear, but can be removed surgically.
- **Bursitis**: Bursae are small connective-tissue sacs lined with synovial membrane that contain some fluid similar to synovial fluid. They are located where pressure is exerted over moving parts; for example, between skin and bone (as at the elbow and over the kneecap), between tendons and bone, or between muscles, or ligaments, and bones, where they allow smooth movement of one part over another (Anthony and Kolthoff, 1975). Bursitis is inflammation of a bursa.
- **Myositis**: Myositis is inflammation of muscles. Myositis may be primary, as in polymyositis; or secondary to physical injury, as in over-stretching muscles.
- **Arthritis**: The term arthritis comes from *arthron*, meaning joint, and *itis*, meaning inflammation. Thus, arthritis is inflammation or an abnormal condition of a joint. Posttraumatic arthritis, osteoarthritis, and rheumatoid arthritis are common types of arthritis.

CTD is a rather comprehensive term which is used to distinguish that group of musculoskeletal disorders which involve injuries to the nerves, muscles, tendons, tendon sheaths, ligaments, bones, and joints in the upper and lower extremities. The main distinction between CTDs and **strain and sprain** injuries is that the latter usually occur in single-exposure situations, and are hence called acute trauma, such as injuries due to falling and slipping (Putz-Anderson, 1988), while the former occur over several days, months, or even years of repetitive exposure.

The term back injury is used to identify that group of musculoskeletal disorders to the vertebrae, intervertebral disks, and other tissues of the back. As previously mentioned, a back injury may be acute (i.e., occurring in a single-exposure incident) or chronic (i.e., developing over some period of time). The chronic type of back troubles is usually due either to assuming awkward postures or to overexertions during manual materials handling. Since these types of back injuries are developed by multiple exposures over some period of time, they are usually classified under CTD.

If workers keep repeating similar movements over a relatively long period of time, their body can wear out its parts, like a mechanical machine. Then, they develop some symptoms that are categorically described as CTDs. These disorders are developed gradually due to repeated stress on the body which exceeds its capacity. CTDs affect neuromusculoskeletal systems, that is, nerves, muscles, tendons, tendon sheaths, ligaments, bones and joints, in the upper extremities (i.e., the hands, wrists, elbows, and shoulders), the lower extremities (i.e., feet, knees, and hips) and the back (i.e., the neck and lower back). CTDs may also be called repetitive motion injuries (RMIs). Therefore, the terms musculoskeletal disorders, neuromusculoskeletal injuries, CTDs, and RMIs are used synonymously to refer to the same group of occupational health impairments.

CTDs affect the parts of the body that are involved in performing tasks. The upper body, specifically the back and arms, is most susceptible to CTDs. Jobs such as assembling parts, data entry using computer keyboards, food packing and soldering all consist of short but highly repetitive cycles that contribute to the development of CTDs. As shown in Figure 8.1, the more frequently occurring types of CTDs include cervical syndrome, chronic low-back pain, de Quervain's disease, trigger finger, Raynaud's syndrome (vibration white finger disease), carpal tunnel syndrome, tendinitis (including tennis elbow), and tenosynovitis.

Unlike cuts and bruises or broken bones, CTDs are not visible and seldom have a dramatic beginning. They occur under the skin, affecting soft tissues such as muscles, tendons, tendon sheaths, nerves, and other tissues. For this reason they are also referred to as musculoskeletal

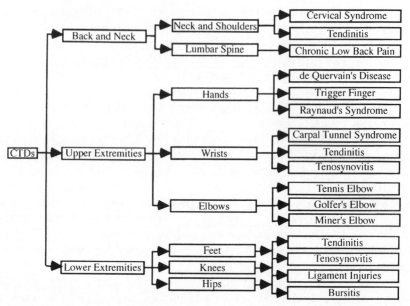

Fig. 8.1 The common types of cumulative trauma disorders (CTDs) and the body parts affected by them.

disorders. They are not noticeable until tendons become inflamed, nerves are pinched or compressed, or blood flow is restricted. They most commonly appear as carpal tunnel syndrome, tendinitis, tenosynovitis, and bursitis.

Awkward posture is one of the common causes of CTDs. Table 8.2 lists several examples of awkward postures that should be avoided to reduce the risk of developing these types of work-related health impairments.

8.3 THE BACK AND BACK PROBLEMS

As described in Chapter 2, the spine forms a double-S shape, when it is balanced (i.e., in natural position), with none of its curvatures exaggerated or flattened. A nerve bundle splits off at each vertebra. There are a total of 31 pairs of nerves that branch out from the vertebrae, threaded between the bony processes and distributed throughout the body.

8.3.1 BACK INJURIES

If an intervertebral disk is ruptured or deteriorated by a disease (e.g., degenerative disk disease), a nerve can be pressed by the bulging disks,

Table 8.2 Examples of awkward postures in various body parts

Body parts	Causing activities
Hands and wrists	Turning the palm of the hand upward (supination)
	Twisting the wrists
	Bending the wrist dorsally (extension) or palmarly (flexion), wrist adduction (bending ulnarly) and abduction (bending radially, towards the thumb), or pinch grip
Shoulders	Raising the upper arms above shoulder level
	Rotating the arms
	Holding the elbows away from the body
Neck and back	Bending the neck downward
	Holding the chin forward
	Bending the back forward to reach an object

or it can be pinched or pressed against the adjacent vertebra and, hence, pain is felt. Back pain is also caused when the muscles or tendons surrounding the spinal column become damaged. However, a back injury can be due to a variety of different problems, including:

- chronic lumbar strain;
- herniated intervertebral disk by sudden or jerky movement, extreme twisting, and/or strong push or pull;
- compression fracture of vertebrae;
- muscle spasm;
- scoliosis (abnormal curvature of the spine – usually in the lateral direction, shown in Fig. 8.2);
- unstable (dislocated) vertebrae;
- kyphosis (backward curvature of the thoracic spine – humpback appearance, shown in Fig. 8.2);
- hyperlordosis (excessive development of a forward curve in the lumbar spine, shown in Fig. 8.2).

8.3.2 CAUSES OF BACK INJURIES

Most low-back problems are due to damage to the lumbar intervertebral disks, especially to the L5/S1 disk, and to degenerative disk disease. For unknown reasons, intervertebral disks may degenerate and lose their strength (Grandjean, 1988). This process is characterized as follows:

- Disks become flattened.
- Disks may even become ruptured and their viscous fluid squeezed out.
- Degenerative processes impair the mechanics of the vertebral (spinal) column.
- Degenerated disks allow tissues and nerves to be strained and pinched, leading to back pain, sciatic problems, and even in severe cases to paralysis of the legs.

The risk of developing a low-back problem increases with aging since the disks become progressively less resilient, and susceptible to degenerative disk disease, allowing the disk to bulge into the spinal canal. This destroys the cushion between the vertebrae, allowing the facet joints to pinch together and on the nerve roots. The pinching causes pain and sensorimotor weakness in the distribution of the nerve roots branching out from the spinal cord.

Figure 8.3 show how forward bending stretches the muscles and ligaments and increases their tension. This forces the intervertebral disk backward, pressing it against the spinal cord and the bony structure.

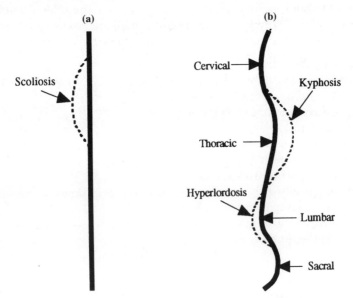

Fig. 8.2 Normal and abnormal curvatures of the human spine. Normal and abnormal curves are, respectively, shown by solid and broken lines. (a) Anterior or posterior view; (b) lateral (left-side) view. Adapted from Crouch (1972).

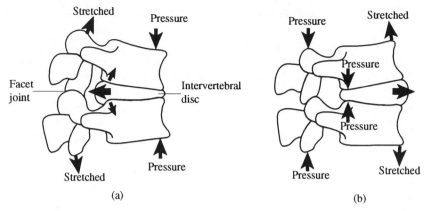

Fig. 8.3 Illustration of pressure onto an intervertebral disk in (a) forward and (b) backward bending.

Thus, the disk may become herniated or forced to bulge into the spinal canal. Figure 8.3 also shows that backward bending relieves the tension in the muscles and ligaments, but causes the facet joints to pinch together, causing pain and/or disturbance in the sensorimotor system, as mentioned above. Understanding this mechanism helps avoid damaging the delicate tissues of the back.

Back injuries are seldom caused by single incidents. They are usually developed by long-term wear and tear. This is why they are often classified as CTDs. The following are common causes of back injuries.

- overexertion in manual material handling, e.g., lifting, lowering, pulling, pushing, and carrying activities;
- poor lifting techniques such as:
  - sudden or jerky movement (to create inertia), instead of smooth lifting;
  - extreme twisting instead of pivoting;
  - lifting objects too far away from the body;
  - lifting with the back flexed;
  - using weak back muscles instead of the leg muscles, which are stronger than back muscles;
- reaching above shoulders, especially when lifting an object;
- awkward workstation design, at which:
  - work is too low;
  - work is too high, requiring reaching;
  - work is far away from the body;
  - the back is unsupported;
- poor postures in sitting and standing. Examples of poor postures are:
  - humped back and rounded shoulders;

  – flattened or sway back (excessive   bending) in the lumbar
    region;
  – holding the chin forward;
  – slumped posture in sitting;
- vertical vibrations, such as those experienced by truck and bus drivers
  or constuction machinery operators;
- accidents, including falling, slipping and tripping incidents;
- poor non-occupational postures in sitting and sleeping. A poor
  sleeping posture may be induced by the type of the mattress used. For
  example:
    – a soft, sagging mattress puts the back in an unbalanced position;
    – a hard mattress does not support the back's natural curvatures;
- large abdomen with weak muscles;
- weak muscles (especially back muscles) due to lack of exercise and
  stretching.

8.3.3 BACK-INJURY PREVENTION

Efforts have continuously been made to redesign workplaces, tools and
equipment, and/or work methods to fit the task to the workers, and
hence reduce work-related injuries, including back injuries. However,
job redesign cannot alleviate all back injuries. This is evidenced by the
incidents in which back pains have been induced by a simple bending
over to pick up, for example, a phone or sheet of paper from a desk, or a
light-weight tool from the floor. These tasks, obviously, are neither
heavy nor complicated to require modification or simplification. The
most probable causes in such cases are the victim's lack of flexibility and
loss of physical fitness.

   A variety of measures have been explored in attempts to reduce the
risk of back injury. Pre-employment physical examinations are often
recommended to avoid placing workers in a job that would exceed their
capabilities (Taylor, 1987). However, the most effective program would
likely be one in which ergonomists, medical personnel, and the workers
themselves assess the demands of the jobs, redesigning them if
necessary, and then monitoring new employee progress on the job to
insure that proper methods and tools are being utilized. The following
recommendations can be made for preventing back injuries:
- Recommendations for working at a seated workstation:
  – sitting as close to the work as possible;
  – lowering the seat height enough (but not too low) to place the feet
    flat on the floor. If this is inappropriate, the use of a footrest is
    recommended;
  – supporting the low back and avoiding sitting slumped;
  – avoiding forward and downward leaning over the task;
  – avoiding prolonged sitting, even in good postures;

– maintain good sitting postures during long driving (or equipment operation).
● Recommendations for working at a standing workstation:
  – placing work at a comfortable height;
  – changing positions frequently;
  – placing one foot up, for example, on a foot-rail and alternating the feet frequently in prolonged standing;
  – avoiding standing on hard (e.g., concrete) floors. Cushioned mats should be used for standing;
  – avoiding bending forward at the waist during prolonged standing. The requirements for working at low heights should be eliminated;
  – avoiding standing with swayed back and relaxed stomach muscles, and knees locked.
● Static postures should be frequently changed or interrupted. Such postures are assumed in jobs requiring forward bending (e.g., nurses in making beds and caring for patients) or working above shoulders (e.g., painting, mechanics working over head). Changing to opposite positions are recommended; for example, if the job requires forward bending, to reduce stress, the person should also periodically perform backward bending.
● Exercises for restoring flexibility and physical fitness are highly recommended.

Preventive techniques should be considered in developing a program for controlling occupational injuries and health problems. The reader is also referred to Chapter 9 for specific recommendations for manual materials handling.

8.4 THE NECK AND NECK PROBLEMS

Another part of the spine which is as important as the lumbar spine, but has not received as much attention, is the cervical spine (neck region). The cervical spine can be characterized as follows:

● It consists of the top seven vertebrae of the spinal column (C1–C7; Chapter 2).
● It is very mobile.
● It exhibits a lordosis (forward curvature) when standing upright.
● It is a delicate part of the spine and prone to degenerative processes and arthrosis.

8.4.1 NECK INJURIES – CERVICAL SYNDROME

Problems with the neck structure can cause pain that may become quite severe. A great number of adults are afflicted with neck troubles due to injuries of the cervical vertebrae and their intervertebral disks (Grandjean, 1988). Cervical syndrome (or cervical sprain) is a general term referring to several conditions dealing with the head, neck, and shoulders. It is caused by an injury to the cervical vertebrae and cervical intervertebral disks, and irritation of the cervical nerve roots, usually those branching out from the C5–C6 vertebrae. Common symptoms of cervical syndrome are:

- painful cramps in the shoulder muscles;
- pain and reduced mobility in the cervical spine;
- sometimes painful radiation in the arms (hence, also called cervico-brachial syndrome).

Cervicobrachial syndrome can be considered an occupational illness since it is often observed in occupations such as key-punchers, assembly-plant workers, typists, cashiers, telephone operators, and VDT operators (Maeda *et al.*, 1982).

8.4.2 CAUSES OF NECK INJURIES

Discomfort in the neck often increases with the degree of forward bending of the head, which is observed in many occupations. A common cause of a large proportion of cervical syndrome injuries is traumatic whiplash injury to the cervical spine caused by a sudden jerking motion of the head, either forward or backward (Durham, 1960). However, awkward postures, poor sleeping positions, osteoarthritis of the cervical spine, and athletic injuries are also among common causes of strains or sprains involving the muscles, ligaments, and joints in the neck.

Irritation, injury, or compression of the nerve roots branching out of the cervical vertebrae can manifest symptoms in the arms and/or hands that include pain, numbness, tingling and weakness.

8.5 CTDs IN THE UPPER EXTREMITIES

There are a large number of CTDs that may develop in the upper extremities. This section reviews those that are cited as occupational-related musculoskeletal disorders. This section draws on many published works (especially Emerson and Taylor, 1946; Putz-Anderson, 1988; Hoaglund, 1990; Tayyari and Sohrabi, 1990; NC-OSHA, 1991).

8.5.1 CARPAL TUNNEL SYNDROME

The structure of the hand was briefly discussed in Chapter 2. In this section the structure of the wrist is examined. Due to its widespread development, a special emphasis is placed on carpal tunnel syndrome (CTS) in this section.

8.5.1.1 Carpal tunnel structure and syndrome

The carpel tunnel is a small tunnel-like structure in the wrist which is enclosed by eight carpal bones and the transverse carpal ligaments. Nine tendons, which function as finger flexors, and the median nerve pass through the tunnel, from the forearm to the hand. Figure 8.4 illustrates the carpel tunnel (Tayyari and Sohrabi, 1990).

The median nerve derives its fibers from the lowest three, sometimes four, cervical and the first thoracic segments (C6–C7 and T1) of the spinal cord, and is formed by the union of the medial and lateral cords of the brachial plexus (Chusid, 1979). It innervates the pronators of the forearm, the long finger flexors, and the abductor and opponent muscles of the thumb. It is the sensory nerve to the palmar aspect of the hand and the thumb, the index and middle fingers, and the radial half of the ring finger.

Complete interruption of the median nerve results in an inability to pronate the forearm or bend the hand in a radial direction, in paralysis of flexion of the index finger and terminal phalanx of the thumb, in weakness of flexion of the remaining fingers, in weakness of abduction and opposition of the thumb, and in sensory impairment over the radial

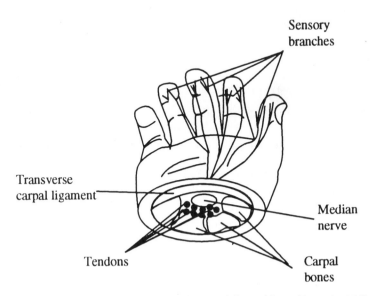

Fig. 8.4 Illustration of carpal tunnel syndrome. Adapted from Tayyari and Sohrabi (1990).

two-thirds of the palmar aspect of the hand and over the dorsal aspect of the distal phalanges of the index and middle fingers.

The median nerve may be injured in the axilla by dislocation of the shoulder and in any part of its course by severing, pinching, or other traumatic wounds. As described by Adams and Victors (1981), incomplete lesions of the nerve between the axilla (at the shoulder) and the wrist may result in causalgia, a painful post-traumatic condition, which begins several weeks following an injury to a nerve. However, the wrist is the most common site of injury to the median nerve.

When the wrist tendons become irritated and swollen, the resulting compression on the median nerve impairs the sensory and motor function of the hand. Nerve compression can, however, be induced by wrist fractures, rheumatoid arthritis, tumors, and diabetes mellitus (NIOSH, 1989). The increased pressure in the bony tunnel can trap and/or pinch the median nerve, causing occasional numbness, pain and tingling in the thumb, index, and middle fingers. This condition is called CTS.

Compression of the nerve at the wrist (CTS) can result from occupational exposure to repeated trauma; from tendon irritation, especially when high pinch forces are used with the wrist in a deviated position; from infiltration of the transverse carpal ligament with amyloid (a hard protein deposit resulting from tissue degeneration, as occurs in multiple myeloma); or from thickening of the connective tissue in cases of rheumatoid arthritis, acromegaly (enlargement of the bones in the hands caused by prolonged over-activity of the pituitary gland), and hypothyroidism. However, the exact causes of CTS are often not apparent.

CTS is becoming more common among individuals as an occupational CTD. Any combination of the following **symptoms** can indicate the presence of CTS (Tayyari and Sohrabi, 1990):

- numbness, tingling and/or painful burning in the thumb, index and middle fingers in one or both hands. These symptoms occur most frequently at night and may be so severe that they awaken the patient. They may also be present in the morning upon waking;
- maintaining the arms in certain static positions may cause the symptoms to occur;
- difficulty in moving the fingers;
- loss of grip strength, particularly when trying to squeeze bottles or open jars;
- loss of sensation in fingers;
- subjective feeling of swollen fingers with little or no swelling apparent;
- atrophy of the thenar muscles at the base of the thumb in advanced cases.

As Durham (1960) suggested, CTS must be differentiated from several other conditions that may affect the median nerve. Examples of these conditions are progressive atrophy, syringomyelia, multiple sclerosis, protrusion of a cervical intervertebral disk, periarthritis of the cervical spine or shoulder joint, and scalenus anticus syndrome.

8.5.1.2 Methods of diagnosis

Putz-Anderson (1988) has compiled a volume of useful information including medical diagnosis and treatment of CTDs. The following are general clinical methods for diagnosing CTS with a brief explanation for each (Tayyari and Sohrabi, 1990; Pottorff *et al.*, 1992):

- **Tinel's test**: This method is performed by tapping the median nerve at the wrist. Tapping a compressed median nerve induces tingling in one or more fingers in the hand afflicted with CTS.
- **Phalen's test**: In this test patients are asked to put the backs of their two hands together and acutely flex the wrists for 1 min. Development of tingling in the fingers is a positive sign of CTS.
- **Electromyogram (EMG) test**: EMG test is administered to measure the conduction velocity of motor nerves. In this test, an electrical stimulus is given in the forearm; the transmission time across the wrist is measured to determine the sensory function of the median nerve. Measurements of the conduction velocity or the conduction time between two points on a nerve and the duration of the action potential of the muscles are of great value in diagnosis of neuropathies. Certain nerve disorders reduce the conduction velocity of the motor nerve fibers.
- **Vibrometry**: This test is very helpful for early detection of CTS, using a vibrometer that measures vibration sense in the range from 8 to 500 Hz (cycles per second).
- **Neurometry**: In this method, which is similar to the EMG test, a neurometer is used to assess the person's sensory impairment. The condition of CTS is diagnosed by detecting abnormalities of conduction in motor fibers of the median nerve within the carpal tunnel.

8.5.1.3 Other conditions similar to CTS

CTS must be differentiated from several other conditions that affect the median nerve (Durham, 1960). Examples of those conditions are listed below:

- progressive atrophy;
- syringomyelia syndrome (a condition caused by formation of a cavity or cyst in the spinal cord, usually starting in the cervical region, that causes spasms and sensory disturbances);

- multiple sclerosis;
- protrusion of cervical intervertebral disk (pinching the median nerve at the cervical region);
- periarthritis of the cervical spine or shoulder joint;
- scalenus anticus syndrome (also known as Naffziger's syndrome), which is a common cause of compression of brachial plexus and subclavian artery.

Because of the similarities in symptoms, the diagnosis of CTS must be performed by medical specialists in the areas of occupational medicine and musculoskeletal disorders.

8.5.2 VIBRATION WHITE FINGER

The worst disease associated with segmental vibration is vibration white finger. It is also called Raynaud's syndrome and involves the narrowing of blood vessels in the hand; this can lead to progressive collapse of the blood vessels and recurrent episodes of blanching of the fingers.

Vibration white finger is caused by exposure to hand–arm vibration from hand-held vibrating tools, especially under cold climatic conditions. Examples of such tools are chipping hammers, grinders, buffers, and pneumatic tools. The vibration generated by the tool is transmitted through the worker's hand where it damages the blood vessels in the fingers. At certain frequencies and with prolonged duration, blood vessels and soft tissues are literally shaken apart. The symptoms depend on its progressive level, as summarized below:

- The first symptoms of vibration white finger are tingling and/or numbness in the fingers during the vibration exposure which may continue after the exposure has been discontinued.
- The second symptom is blanching (whitening) of one fingertip because of a temporary constriction of blood flow. The restoration of normal blood circulation in the white fingertip is a painful process. This pain occurs intermittently and will be aggravated and increased with exposure to vibration and cold temperature. It should be noted that exposure to cold temperatures results in blood vessel constriction.
- As another symptom, other fingers also blanch during attacks in the affected finger.
- The intensity of the pain and frequency of the attacks increase in time.

According to the AFS Segmental Vibration Task Force (1983), the two common triggering factors for a vibration white finger attack are exposure to cold and emotional distress. The reason for a response to cold is the sensitivity to cold, especially by a person who has had frostbite, in which damage to capillaries in the fingers has already taken place. The attack induced by emotional distress is due to restricted blood flow, from tense constriction of blood vessels, to already damaged capillaries. When a vibration white finger attack is triggered, the severity of the attack is determined by the classification of this progressive disease, as shown in Table 8.3 (Taylor and Pelmear, 1975; NIOSH, 1983b).

Another aspect of the classification is that a patient with vibration white finger with stage 03 or 04 classification will have attacks during the summer and/or winter. A person with less severe vibration white finger will usually only have attacks during cold weather. Attacks can last from a few seconds to 15 min or longer. Up to and including stage 01, there is no interference with work or social activities. By stage 04 the person's occupation may have to be changed and hobbies restricted. Continued exposure to vibration causes the disease to be progressive.

The following are among other factors that may contribute to the development of vibration white finger (AFS Segmental Vibration Task Force, 1983):

- intensity and duration of vibration exposure;
- temperature of workplace;

Table 8.3 Stages of vibration white finger

Stage	Condition of fingers	Work and social interference
00	No tingling, numbness, or blanching of the fingers	No complaints
0T	Intermittent tingling	No interference with activities
0N	Intermittent numbness	No interference with activities
TN	Intermittent tingling and numbness	No interference with activities
01	Blanching of one fingertip with or without tingling and/or numbness	No interference with activities
02	Blanching of one or more fingers beyond the tips, usually during winter	Possible interference with non-work activities; no interference at work
03	Extensive blanching of fingers, during summer and winter	Definite interference at work, at home, and with social activities; restriction of hobbies
04	Extensive blanching of most fingers, during summer and winter	Occupation usually changed because of severity of signs and symptoms

- smoking habits;
- caffeine intake;
- alcohol consumption;
- history of frostbite;
- hypertension;
- stress;
- vitamin deficiencies;
- over-the-counter, prescription and illegal drugs;
- diabetes;
- family history of cold sensitivity;
- fluid retention;
- other diseases.

Recommendations for reducing vibration exposure will be provided in Chapter 15.

8.5.3 RAYNAUD'S PHENOMENON

Although it is discussed here as a separate disorder from vibration white finger, Raynaud's phenomenon is actually a broad disorder, with vibration white finger being one of its subcategories. The symptoms for Raynaud's phenomenon and Raynaud's disease are the same – episodes of lowered surface temperature of the extremities and blanching of the skin, accompanied by localized redness of the skin after several minutes to several hours. The ear lobes, the tip of the nose, and the cheeks may also show color changes.

If these are primary symptoms and not secondary to any other diseases, the disorder is called Raynaud's disease. If an underlying disorder is detected after 2 years of follow-up, the process is considered to be Reynaud's phenomenon. This cut-off point is a rather arbitrary period practiced by physicians. For example, a patient can develop rheumatoid illness over a 14-year period that can cause the phenomenon. The bases for distinction between Raynaud's disease and phenomenon are summarized in Table 8.4 (Hoffman, 1980).

8.5.4 EPICONDYLITIS

Epicondylitis is a form of tendinitis which involves an inflammation of the tendons of the elbow. There are two common types of epicondylitis:

- **Lateral epicondylitis**: Because it is a common complaint among tennis players, lateral epicondylitis is commonly called **tennis elbow**. It involves the tendons on the outer side of the elbow.
- **Medial epicondylitis**: Medial epicondylitis is often reported by golf players and, therefore, is commonly called **golfer's elbow**. It involves the tendons on the inner side of the elbow.

Table 8.4 Distinction between Raynaud's disease and phenomenon

Findings	Raynaud's disease	Raynaud's phenomenon
Symptoms due to other underlying diseases (after at least a 2-year follow-up)	No	Yes
Laboratory evaluation (e.g., blood count, rheumatoid factor, antinuclear antibodies, serum viscosity)	Normal	Abnormal
Angiography (assessment of blood and lymph vessels)	Normally no structural changes in vessels	Structural changes often present
Trophic changes (necrosis, atrophy calcinosis)	Uncommon	Common

Adapted from Hoffman (1980).

Epicondylitis can occur by repeated or sustained rotation of the forearm in combination with bending (especially, dorsiflexion–extension) of the wrist, as when using a screwdriver (NC-OSHA, 1991). A worker whose job requires repetitive, forceful wrist extension is at risk of developing epicondylitis.

The symptoms of epicondylitis include pain, burning sensation, or ache at the elbow. It also causes swelling and weakness in the elbow. The pain due to tennis elbow radiates into the dorsal aspect of the forearm. There is also local tenderness over the lateral humeral epicondyle or distal to it in the common extensor origin. The symptoms may appear at night at rest, although they usually appear following activities, especially dorsiflexion of the wrist or grasping (Hoaglund, 1990). The radiation of pain due to golfer's elbow mostly travels in the palmar aspect of the forearm.

8.5.5 BURSITIS

Bursitis is inflammation of a bursa. It results from repeated irritation or friction due to repetitive abduction and rotation of joints, or long-continued pressure over the bursa. The symptoms of bursitis are pain, swelling, tenderness, and limitation of motion. The most common examples of this disorder are those involving the following:

- **Subacromial bursitis** involves the subacromial or subdeltoid bursa, located between the rotator cuff and the coracoarcromial ligament in the shoulder. It hinders the free movement of tendons and limits the joint mobility in the shoulder. This type of bursitis is informally called hod carrier's shoulder. (A hod is a bin carried over the shoulder for transporting loads, such as mortar or bricks.)
- **Olecranon bursitis** involves the olecranon bursa, that lies over the point of the elbow – the large point on the proximal end of the ulna that projects behind the elbow joint and forms the point of the elbow, which is informally called the crazy bone. Olecranon bursitis is also called miner's elbow, and sometimes student's elbow.
- **Prepatellar bursitis** affects the bursa that lies over the patella at the knee. This form of bursitis may be informally called housemaid's knee and can affect carpet layers.

8.5.6 TRIGGER FINGER

Trigger finger (gamekeeper's thumb or trigger thumb) results from a thickening of tendon and tendon sheaths in fingers or thumbs, or formation of a nodule on the tendon (swelling). Its symptoms include the following: snapping action as the affected finger or thumb flexes or extends; and the affected finger or thumb may lock in the flexed or extended position, as an obstruction occurs when the tendon can no longer move through the sheath because of the thickening or formation of a nodule.

8.5.7 DE QUERVAIN'S DISEASE

De Quervain's disease is a thumb disorder that affects the long abductor and short extensor tendons of the thumb, on the wrist side of the base of the thumb. These tendons are connected to muscles on the dorsal aspect of the forearm, whose contraction pulls the thumb back and away from the hand (Putz-Anderson, 1988). The symptoms of de Quervain's disease include pain during any motion involving the affected thumb, swelling on the side of the wrist, near the base of the thumb, and/or a cracking noise in the joint when the thumb is moved.

This disorder, which was named after the French physician who first described it, is usually caused by overuse of the thumb, and repetitive grasping. Repetitive friction between thumb tendons and their common sheath causes thickening of the fibrous sheath and constriction of the tendons. A major risk factor is repeated ulnar deviation (bending the wrist toward the little finger), in combination with forceful exertions of the thumb. Such a hand–wrist position may be observed in using scissors, using straight-handled tools against a vertical surface, and using straight-handled needle-nose pliers.

8.5.8 RADIAL NERVE PARALYSIS

Pressure against the trunk of the radial nerve in the axilla can cause **radial nerve syndrome**. Radial nerve syndrome can result from the use of crutches. This syndrome is also known as **Saturday-night palsy** since it is commonly developed in individuals sitting in a bar, while drinking, with an arm resting over the backrest of a chair for a long period of time. It can be caused by striking the outer aspect of the upper arm, where the radial nerve is in an unprotected position. Radial nerve paralysis can be caused by a tourniquet (a rubber band wrapped around the upper arm for detecting the artery) applied to the arm too tightly or remaining on for a prolonged period of time. It can also appear after a few weeks following a fracture in the humerus.

8.5.9 ULNAR NERVE PARALYSIS

The ulnar nerve can become traumatized at the elbow, where the nerve lies in an unprotected position. Resting on the elbow may cause a pressure paralysis of the muscles, which are innervated by the ulnar nerve, that lasts several weeks. Dislocation of the elbow or fracture of the bones near the elbow can cause this paralysis.

8.5.10 ROTATOR CUFF TENDINITIS

Rotator cuff tendinitis is the most common tendinitis (bursitis) of the shoulder, which is the inflammation of one or more of the four rotator cuff tendons in the shoulder joint. The rotator cuff is made up of four tendons fused over the joint to provide stability and mobility for the shoulder. These tendons rotate the arm inward and outward and assist the shoulder in abduction (i.e., moving the arm away from the side).

8.6 FUNDAMENTAL RISK FACTORS OF CTDs

An exact cause-and-effect relationship cannot be established for the development of CTDs. The most probable risk factors associated with each type of CTDs were mentioned previously. There are certain risk factors which are persistently associated with or contribute to the development of CTDs. These risk factors may be classified into three categories: occupational, non-occupational, and other factors (Tayyari and Sohrabi, 1990; Tayyari and Emanuel, 1993).

8.6.1 OCCUPATIONAL RISK FACTORS

The following are known to be occupational factors that contribute to the development of CTDs:

- **Highly repetitive movments**: Performing some manual tasks involves highly repetitive use of the limbs. Repeating the same movements over and over will eventually fatigue the involved muscles. Highly repetitive tasks have cycle lengths of 30 s or less, while a cycle length of 1.5 min is considered optimal for fast-paced tasks (Eastman Kodak Company, 1986). Prolonged use of tense and fatigued muscles increases the risk of CTDs. Frequent movements become much more risky if they are combined with poor postures and excessive force. However, high frequencies of repetition, even with small forces, can cause or contribute to the development of CTDs.
- **Awkward postures**: No matter whether sitting or standing, a poor body position may place the person at risk of developing CTDs. Fixed postures, even good ones, are also harmful if held unchanged for so long that muscle tension builds up and circulation is reduced. Excessive bending of the wrists in any direction, that is, palmarly, dorsally, ulnarly, or radially can lead to CTDs. Extreme positions increase pressure on muscles, tendons, and nerves. Pinch grips in holding an object by pressing the fingers against the thumb, or holding an object at a point far from its center of gravity can also lead to CTDs.

  Performing tasks requiring the use of the hands and arms in deviated positions, in which the tendons will have to operate while bent over bones or tough ligaments, should be avoided. When the hand is used with the wrist in a deviated posture, the tendons are bunched together and are forced to operate while stretched over the carpal bones when bending the wrist dorsally (Fig. 8.5a) or stretched

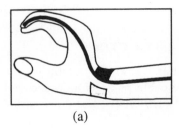 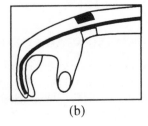

(a)   (b)

Fig. 8.5 When the wrist is bent (a) dorsally or (b) palmarly, the tendons will have to operate while bent over the carpal bones or the transverse carpal ligaments, respectively. The resultant friction may cause the tendons or their sheaths to swell. Adapted from Tayyari and Sohrabi (1990).

over the transverse carpal ligaments when bending the wrist palmarly (Fig. 8.5b).

- **Excessive forces**: Performing certain manual tasks requires the worker to exert excessive force. Activities requiring excessive force strain muscles and tendons, and hence increase the risk of developing CTDs. Force is required in gripping a tool; rotating a control; lifting, pushing or pulling an object; and cutting a work-piece. Applying contact force by pressing against a hard surface, such as a lever, work-tool, or work table increases the risk of CTDs.
- **Inadequate tool handles**: Using tools that dig into the palm of the hand and irritate the tendons can lead to CTDs.
- **Pinch-hand position**: Carrying thin items (e.g., books) held in a pinch position between the thumb and fingers creates a strain in the hand and can lead to CTDs.
- **Low-frequency vibration**: Using vibrating tools (e.g., air- or motor-powered drills, chippers, sanders, saws, and drivers), especially under cold conditions, can cause CTDs.
- **Gloves**: Using work gloves that do not fit properly may result in undue force or stress on the hand or wrist.
- **Low temperature**: Working in the cold can lead to CTD injuries as blood flow may be reduced or restricted in the extremities.

Combinations of these occupational risk factors are expected to intensify their individual ill-effects. For example, Silverstein *et al.* (1987) reported that CTS has a strong correlation with high-force/high-repetitive jobs and/or to a lesser extent with low-force/high-repetitive jobs and high-force/low-repetitive jobs. A lessening of either the force or the repetition appeared to reduce the occurrence of CTS. Repetitive motions are particularly a problem when a part of the body is used in an awkward posture or position.

CTDs have not only been associated with such physically demanding jobs as meat packing, poultry processing, and automobile assembly, but also with less physically demanding, and seemingly harmless, tasks such as merchandise scanning and computer data entry. Armstrong (1981) reported that for a given force exertion and posture, 15–25% more stress will be imposed upon the tendons and median nerve of a small female hand than of a large male hand.

8.6.2 NON-OCCUPATIONAL RISK FACTORS

The following non-occupational factors have been observed to be associated with many CTD cases:

- **Certain medications or therapies**: Estrogen–progestin medication, or high-dose progestin medication, danazol therapy, or other steroidal medication for the treatment of endometriosis can lead to CTDs. Sikka *et al.* (1983) have stated that medication may lead to fluid retention which may lead to greater nerve compression.

- **Age**: The risk of developing CTS is higher among individuals over 40 years of age.
- **Gender**: Risk is higher in women than men.
- **Wrist size and wrist structural variations**: Individuals with small wrist size may be more susceptible to CTS.

8.7 PREVENTION OF CTDs

As shown in Figure 8.6, prevention measures that minimize the risk of CTDs may be grouped into the following three categories: engineering solutions; administrative controls; and personal protective equipment.

8.7.1 ENGINEERING SOLUTIONS

Engineering solutions are based upon ergonomic principles which are used to analyze repetitive motion tasks, identify the stressful ones, and reduce such stressors. For example, workstations, work-tools and work methods can be modified to eliminate repetitive movements, excessive forces and/or awkward postures. There are various engineering solutions, each of which may be suitable for certain situations.

8.7.1.1 Job Redesign

The jobs performed by workers in a problem area should be investigated to pinpoint job elements which may be responsible for problems. Some repetitive tasks may not be necessary at all, and should be eliminated. Some tasks can be made easier to perform. If the job requires a limited number of repetitive tasks, then the job should be enlarged so that the worker can perform a variety of jobs tasks, different in nature, during the course of the work-shift. Highly repetitive manual tasks should be automated. The following ergonomics guidelines should be considered in job design and job modification:

- All repetitive tasks performed in awkward postures or that require excessive exertion should be eliminated.
- Jobs should be enlarged to consist of more tasks, requiring different movements. This provision helps avoid performing the same few job activities all shift long, every day. A job consisting of various types of tasks requires the workers to perform different body motions and postures.
- Design fast-paced tasks with a cycle length of longer than 30 s, preferably 1.5 min.

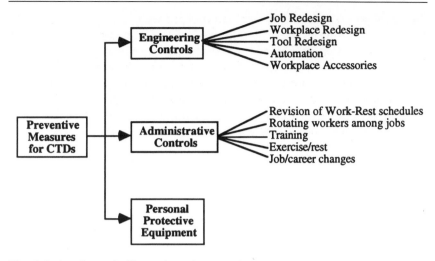

Fig. 8.6 A schematic illustration of preventive measures for cumulative trauma disorders (CTDs).

- Avoid keeping the same posture, even a good posture, for too long. Allow and encourage postural changes.
- Minimize pinch grips. Use the whole hand as much as possible when grasping is required. Hold objects in cupped hands instead of using pinch grips.
- Hold an object near its center of gravity, so that its weight is balanced.
- Provide flanged support on tool handles that require the user to exert forces in the direction of the tool handle.
- Avoid activities requiring excessive forces. Try to minimize lifting, pushing, pulling, and grasping.
- Avoid contact forces, for example, pressing against a hard surface, such as a work-tool or work table.
- Avoid tasks requiring the same movements repeated over and over for a long period of time. Frequent rest pauses help tensed and fatigued muscles to recover their normal effectiveness.
- Minimize exposure to vibration and cold temperatures.
- Frequent movements become much riskier if they are combined with poor postures and excessive force.
- Highly repetitive and strenuous tasks should be automated if economically and technologically feasible.

8.7.1.2 Workplace redesign

Many musculoskeletal disorders are caused by awkward postures due to poor workplace (workstation/work space) designs. All workstations in the problem area should be evaluated based on ergonomics principles. The key solution is to design or redesign for neutral postures, as described in Chapter 7. Many situations merely need simple changes; others may require significant changes to the workplace.

The workstation should allow workers to perform their job tasks in a comfortable body position. The body joints should be kept as close to the neutral position as possible during the job performance. This will minimize static loads on joints and the risk of developing CTDs caused by extreme or frequent movement. Maintaining a neutral posture will also maximize the body's efficiency and strength in performing the job. The following are some ergonomics guidelines for workstation design or redesign:

- Correct working posture. Although excellent tools have been designed, great tool design is not enough. The design of the workstation, where the tool is used, is as important as tool design. Inappropriately designed workstations force workers to assume awkward postures affecting the back, neck, shoulders, and wrists (Fig. 8.7). The seat and task (table) heights are two very crucial dimensions in designing a workstation (see Chapter 7 for workstation design guidelines).
- Elimination of extended reaches. The most frequently used items in task cycles should be kept within the worker's **normal reach**. Occasional safe over-reaches, however, may be allowed to relieve static loads and fatigue (such over-reaches should not involve high forces, twisted or awkward postures, or movement into unsafe areas of the workplace).
- It is important to provide employees with work-surfaces that do not cause friction or rubbing on any body part while permitting ease of movement.
- When different workers use the same workstation, the workers

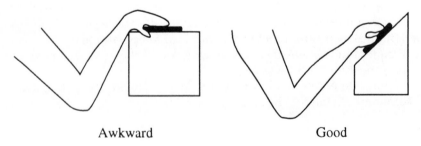

Awkward   Good

Fig. 8.7 Proper workstation design can prevent awkward wrist postures. Adapted from Armstrong (1983).

should be able to change or adjust the workstation setting during the work-shift to fit various postures to avoid an awkward or fixed posture throughout the shift.

- Avoid a poor body position that puts the person at risk of developing CTDs.
- Avoid excessive wrist deviation from its neutral position. This includes bending the hand palmarly, dorsally, ulnarly, and radially.
- Frequently used work items should be within the normal reach.
- Whether standing or sitting, the body should be above the work so that the hands, wrists, and forearms would be able to move straight out from the body.
- The elbows should be kept as close to the body as possible.
- Tilt part bins, boxes, etc., so that the wrist is not bent, the elbow is not raised, and the shoulder is not abducted when reaching such items.
- Adjust the work-surface (table-top or the work-piece itself) to slant toward the body to bring the work into an easier reach.
- Provide armrests to support the forearms, which can enhance the manipulative ability of the hand.
- Store parts in containers designed and situated within the worker's reach to minimize hand flexion and extension.

8.7.1.3 Tool and equipment redesign

A poor tool design is another crucial factor that imposes awkward postures and/or unnecessary exertion upon the tool users. When multiple workers use the same workstation and equipment, the equipment should be adjustable. Comprehensive recommendations for tool design and selection are given in Chapter 10.

8.7.1.4 Workplace accessories

There are situations in which engineering solutions and administrative controls are either insufficient or not feasible. A great number of ergonomics products are marketed that seem to be useful for minimizing the risk of developing musculoskeletal disorders. Two general classes of such products are described below:

- **Computer accessories**: There are a large number of computer accessories commercially available for use at computer workstations. The types of accessories available for computer users are described in Chapter 17. Examples are: mouse pads/nests, wristrests, footrests, armrests, document holders, and keyboard holders/trays.
- **Floor covering**: Where prolonged standing is required, concrete and hard floors should be covered by a mat. This improves comfort,

relieves pressure in the back, knees and feet, and improves circulation in the soles of the feet. Such provisions help reduce fatigue and the potential risk of developing CTDs in the lower extremities. Floor covering is also recommended for wet and greasy industrial areas or where chemicals are used. Some mats have drainage holes for liquid and material chips to pass through. Care should be taken in choosing and designing mats that they should have feathered (reduced) edges to minimize tripping hazards.

8.7.2 ADMINISTRATIVE CONTROLS

When engineering solutions are not immediately available or fail to be economically feasible, administrative controls should be tried. These controls are used to limit exposure. For example, work–rest cycles can be modified to allow muscles to recover from work-induced fatigue and tension. Examples of administrative controls are:

8.7.2.1 Revision of work–rest schedules

Modify work–rest regimens to provide for more frequent rest breaks. As a general rule, more work breaks of shorter duration are preferred to fewer breaks of longer duration. As discussed in Chapter 6, remember that a rest break means performing a different activity that allows for recovery of muscles utilized in the activity of concern.

8.7.2.2 Rotating workers among jobs

Rotate employees within the work group exposed to stressful and repetitive tasks. Rotation should be among tasks not requiring use of the same muscles or motions.

Workers should share their jobs to avoid repeating the same motions over and over. This may be implemented in daily or weekly schedules so that workers perform various tasks instead of performing the same repetitive tasks every day. However, rotation of jobs within each day is preferred over rotating them beween days. For example, if there are four different tasks to be performed by four different employees, each employee should take a turn in performing each task for 2 h, as shown in Table 8.5.

8.7.2.3 Training

Prevention is the best method of coping with CTDs. Workers should be trained to protect themselves by learning how to reduce the risk of developing CTDs. The training should make them aware of what they

Table 8.5 An example of rotating four workers among four industrial tasks

	Time			
	8–10	10–12	1–3	3–5
Worker 1	Machining	Stocking	Hammering	Welding
2	Welding	Machining	Stocking	Hammering
3	Hammering	Welding	Machining	Stocking
4	Stocking	Hammering	Welding	Machining

can do to achieve this goal. Workers should understand the importance of proper positioning of the body and using the right tools in their jobs. They should learn to give their bodies time to recover from fatigue and heal. If they can, they should take short breaks more often as opposed to long breaks less frequently. They should also be instructed to use different muscle groups throughout the work shift by sharing tasks with their co-workers. When workers are at risk, they should communicate the matter with their supervisor. Together, they may be able to find a better way to do the job and improve the work-life. If workers already have symptoms such as tingling, numbness, or pain, they should immediately report the matter to the supervisor and seek a medical evaluation. It should always be remembered by everyone within the organization that a CTD will not disappear by itself, unless the causing problem is identified and resolved. Workers should be instructed to practice the following:

Good working postures

- aligning the ears, shoulders, and hips;
- aligning the elbows, wrists, and hands to maintain the wrists in a neutral position, with the hands slightly below the elbows;
- keeping the elbows as close to the body as possible;
- keeping the feet flat on the floor or a footrest with the thighs horizontal when sitting;
- supporting one foot during standing on a footrest and alternating the feet periodically to reduce strain in the low back region, and to allow for frequent, easy postural changes.

Good work habits

- using whole hand grasps, but avoiding pinch grips;
- holding an object at about its center of gravity to balance its weight;

- tilting the work-surface or containers may help maintain the wrists in a straight posture and bring the task within an easier reach.

Good tool use

- using tools with the right grip size for the hand;
- using tools with the right handle lengths for the hand;
- using tools which are shaped to minimize the wrist bending. Depending on the task orientation, pistol-grip tools may be the best to use;
- using tools with textured or cushioned handles for a better grip;
- maintaining work tools in good repair for easier use. Sharp blades and bits require less effort than dull or broken ones.

Use of gloves

- wearing gloves that fit properly and allow flexibility so that gripping will not become more difficult;
- when gloves are too bulky or constricting, using finger tapes (or cots) can improve grasping by increasing friction to reduce required effort;
- wearing anti-vibratory gloves with padding can protect the palm of the hand from vibration injuries.

Providing ergonomically designed products and workstations is not enough to safeguard workers against the risk of developing CTDs if they do not use them properly. Therefore, workers should be trained and supervised to use their workstations, equipment, and work tools properly. They should also be trained to learn about body mechanics and harmful postures and job activities.

The following are examples of occupational activities that have been cited to have connections with CTDs (Tayyari and Emanuel, 1993) and workers should understand them:

- using the hand in an awkward position, in which the tendons will have to operate while bent over bones or tough transverse ligaments;
- using tools that dig into the palm of the hand and irritate the tendons (e.g., short-handled screwdrivers);
- holding thin items (e.g., books, glass) by a pinch grip between the thumb and fingers causes strain in the hand and increased tension in the wrist (Fig. 8.8a). In contrast, such items should be held and carried in a cupped hand (Fig. 8.8b);
- repetitive use of small hand tools;
- performing job tasks involving highly repetitive manual acts;
- performing job tasks necessitating excessive wrist deviations or other stressful wrist postures;

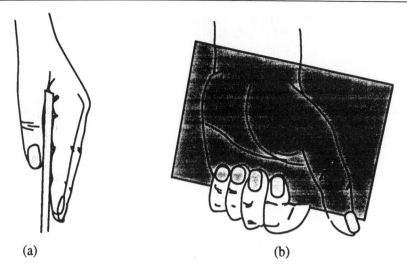

(a)   (b)

Fig. 8.8 Carrying a thin item in a cupped hand (b) is less straining than holding it by (a) a pinch grip between the thumb and fingers, which causes strain on the hand. Adapted from Tayyari and Emanuel (1993).

- performing job tasks involving hand and arm movements requiring force;
- using vibrating tools, especially under cold conditions;
- using gloves that do not fit properly.

8.7.2.4 Exercise/rest provision

Resting and relaxation can reduce tension and mental fatigue and improve the morale of workers. Implementing exercise programs in the workplace can improve worker health and productivity. This goal is achieved through the following improvements:

- Improved flexibility: Exercises can improve body flexibility and should be used for warming the body up prior to performing work.
- Improved strength: Implementing an effective exercise program can help employees strengthen their muscles and stay fit on the job.
- Relaxation: When muscles are fatigued or become sore, they also become tense. Such a tense condition may cause injuries to the body. When workers relax and rest at work, the fatigue and pain are reduced or eliminated. The body can repair itself if it is given a chance. Workers should periodically allow their body and mind to relax by sitting down, closing the eyes and taking several deep breaths, without worrying or even thinking about work (Pottorff et al., 1992). This procedure is especially recommended for VDT users and should

be repeated until the person feels relaxed. Relaxation programs are available that use recorded audio or video tapes, or electronic devices for helping employees learn to relax.

8.7.2.5 Job/career change

When all engineering solutions and administrative controls are exhausted, and CTD cases are still a problem, workers may elect to change their job or career. However, before self-selecting out of a job, the worker should have exhausted engineering and administrative controls with ergonomics, engineering, and medical personnel. It is usually not in the best interest of either the worker or the company to have employees self-select out of jobs rather than resolving the issues associated with the job.

8.7.3 PERSONAL PROTECTIVE EQUIPMENT

Personal protective equipment (PPE) is normally recommended as a last resort when engineering and administrative controls have not been successful. However, in the case of CTDs, there are very few personal protective devices that have been developed. PPE should not be confused with medical devices. PPE is a preventive measure, while medical devices are used in posttraumatic situations. Examples of PPE for hand–arm protection include gloves, mitts, finger cots, thumb and fingerguard thimbles, finger guards, hand pads, sleeves (usually used with gloves), and protective hand creams. The following characteristics of PPE should be considered for hand protection:

- Gloves may be used to reduce vibrations entering the hands and arms, and to protect the hands from extreme temperatures.
- Gloves should fit properly and be flexible to avoid making gripping more difficult.
- Finger cots may be used to increase friction to reduce gripping force, and to protect the fingertips from cuts.
- Fingerless, padded gloves may be used to protect the palm against impact and vibration while maintaining the dexterity in manipulation using the fingertips. These reduce shock to the hand during use of vibrating tools, such as pneumatic tools, or pounding and impacting against hard surfaces.
- Long gloves that support the wrists would not only protect the hands, but also reduce wrist deviation during work.
- Back-support belts may help reduce back injuries. Although to date no scientific evidence has been presented to show that back belts are either beneficial or harmful, they may, however, encourage better

posture and, at least, keep the user alert of the risk of bending the back during lifting. Nevertheless, back-support belts should not be considered a protective device.

8.8 TREATMENT OF CTDs

When the symptoms of CTDs appear, it becomes necessary for the afflicted worker to seek medical help immediately. Medical treatment can be effective in remedying the illness, or in preventing further deterioration of such a problem or any other problems. The type of medical treatment depends on the stage of the problem. It can range from joint immobilization to surgery. The following are methods of treating CTDs:

- **Medical devices**: Medical devices may be used to minimize body strains due to external stresses. For example, splints and braces are used to immobilize the affected joint to restrict the patient from performing repetitive motions or excessively bending the joint. This provides an opportunity for the affected area to heal.
- **Medication**: Medication may be prescribed in conjunction with the use of splints to relieve the symptoms of the CTD. The relief may be temporary or permanent. The medication usually includes pain relievers and anti-inflammatory drugs.
- **Surgery**: Surgery should be the last resort, when splints and use of medications fail to alleviate the pain and/or symptoms of CTDs or the muscles of the affected region begin to atrophy. As with any surgical procedure, the worker should seek a second opinion before electing surgery.
- **Post-surgery therapy**: When a surgery is performed, the affected body part will usually lose its pre-surgery strength. Post-surgery physical therapy and work hardening are necessary in order for the patient to return to normal activity and work. This process may take up to a year (Duncan et al., 1987). However, some patients whose jobs require low muscular activities may need only a few weeks to recover and return to work (Pottorff et al., 1992).

8.9 APPLICATIONS AND DISCUSSION

With the maladies of the modern age, CTDs have become widespread across industry and worldwide. The common causes of CTDs have been shown to be repetitive movements, excessive forces, and awkward postures. However, there are other various factors (e.g., low temperature, unfit gloves, vibration) that either compound the effects of these causes or significantly affect the development of CTDs. Ergonomics principles through workplace design, work-tool and work-method designs can be applied to reduce or prevent these occupational disorders.

As the ergonomist conducts job analyses, it is important to look at both the repetition and posture in addition to the forces required to perform the job. Highly repetitive jobs may lead to CTDs. Workers should be given adequate rest and should be assigned tasks that will utilize a variety of muscle groups. Remember that rest can take several forms, ranging from a break in a designated rest area to simply performing a less demanding task (hopefully one that does not require high levels of use of the muscles used in the fatiguing or repetitive task).

Work should be designed so that fatiguing and strenuous work postures and movements are eliminated. Ergonomic workstation and work-method designs should allow operators to change their working posture and movements. If operators have no opportunity for such posture and movement changes, they should be given appropriate rest breaks to recover from the work-induced fatigue and strain. Suggestions related to workstation design and calculations of rest breaks have been covered in previous chapters.

After a preliminary diagnosis by medical personnel, those with positive signs of CTDs should immediately be referred to medical experts for further evaluation and necessary early treatment before the problem reaches an advanced level, where treatment is costly to both the worker and the organization.

REVIEW QUESTIONS

1. Name the four categories of occupational health problems, and some examples of each.
2. Among the 10 leading occupational diseases and injuries recognized by NIOSH (listed below) which ones are of ergonomic concerns?
   (a) occupational lung disease;
   (b) musculoskeletal injuries;
   (c) occupational cancers;
   (d) amputation, fractures, eye loss, lacerations, and traumatic deaths;
   (e) cardiovascular diseases;
   (f) disorders of reproduction;
   (g) neurotoxic disorders;
   (h) noise-induced loss of hearing;
   (i) dermatologic conditions;
   (j) psychologic disorders.
3. In what kind of occupations and industries do musculoskeletal disorders occur?

4. In what parts of the body do musculoskeletal disorders occur?
5. How rapidly do musculoskeletal disorders occur when the person is exposed to CTD-contributing factors?
6. Define the following terms: strain, sprain, tendinitis, tenosynovitis, ganglionic cyst, bursitis, myositis, and arthritis.
7. What is the difference between cumulative trauma disorders (CTDs) and strain and sprain?
8. What type of tissues and what parts of the body can be affected by CTDs?
9. What are the most common forms of CTDs?
10. Give two examples of activities that cause awkward postures in the hands and/or wrists.
11. Give two examples of activities that cause awkward postures in the shoulders.
12. Give two examples of activities that cause awkward postures in the neck and back.
13. Describe and sketch the following abnormal curvatures in the spinal column:
    (a) scoliosis;
    (b) kyphosis;
    (c) hyperlordosis.
14. Back pain may be due to a ruptured or deteriorated intervertebral disk, or due to a damage in the muscles or tendons surrounding the spinal column. A back injury can be due to a variety of different changes, deformities, and other problems in the back structure. Give five examples of such problems. If necessary, draw diagrams when explaining your answer.
15. Give five common causes of back injuries.
16. What recommendations can be made for preventing back injuries in those who work at seated workstations?
17. What recommendations can be made for preventing back injuries in those who work at standing workstations?
18. What are the characteristics of cervical spine?
19. What is cervical syndrome? What is it caused by? What are its symptoms?
20. Being recognized as an occupational illness, in what occupations is cervicobrachial syndrome often observed?
21. What are the common causes of most cervical syndrome and strains or sprains involving the muscles, ligaments, and joints in the neck?
22. Describe the structure of the carpal tunnel.
23. What is the function of the median nerve, and which parts of the body are innervated by it?
24. What is the result of complete interruption of the median nerve?
25. What happens when the wrist tendons become irritated and swollen?

26. List the symptoms of carpal tunnel syndrome.
27. Name and briefly describe the general clinical methods of diagnosing carpal tunnel syndrome.
28. What is syringomyelia syndrome?
29. What is scalenus anticus syndrome or Naffziger's syndrome?
30. List the stages of vibration white finger, and the corresponding conditions of fingers and the work or social interference.
31. Name at least eight factors other than exposure to hand–arm vibration and cold that may contribute to the development of vibration white finger.
32. What is the relationship between Raynaud's phenomenon and vibration white finger?
33. Describe the similarities and differences between Raynaud's phenomenon and Raynaud's disease.
34. What is epicondylitis? What are the two common types of epicondylitis called? What are their common causes and symptoms?
35. What is Saturday-night palsy? How can it develop? Why is it called by this name?
36. What is bursitis? What are its causes and symptoms?
37. What is hod carrier's shoulder?
38. What are miner's elbow and student's elbow?
39. What is housemaid's knee?
40. What is trigger finger? What are its symptoms?
41. What is de Quervain's disease? What are its symptoms? How does it develop?
42. What is ulnar nerve paralysis? How can it develop?
43. What is rotator cuff tendinitis? Briefly describe the function of the rotator cuff tendons in the shoulder joint.
44. Tasks of what cycle length are considered highly repetitive? What is an optimal cycle length for fast-paced tasks?
45. List the occupational risk factors of CTDs.
46. List four non-occupational risk factors associated with CTD cases.
47. List the three groups of prevention measures for minimizing the risks of CTDs and their corresponding strategies.
48. Under what circumstances are administrative controls used to minimize the risks of CTDs? How can they minimize such risks?
49. Describe how implementing exercise and relaxation programs in the workplace can improve worker health and productivity.
50. List the methods of treating CTDs.

REFERENCES

Adams, R.D. and Victors, M. (1981) *Principles of Neurology*, 2nd edn. McGraw-Hill, New York.

AFS Segmental Vibration Task Force (1983) Vibration white finger. *Modern Casting*, 73: 38.

Anthony, C.P. and Kolthoff, N.J. (1975) *Textbook of Anatomy and Physiology*, 9th edn. C.V. Mosby, St Louis.

Armstrong, T.J. (1981) Carpal tunnel syndrome and the female worker. In: *The Proceedings of the American Conference of Governmental Industrial Hygienists*. Portland, Oregon, p. 28, ACGIH, Cincinnati, Ohio.

Armstrong, T.J. (1983) *Ergonomics Guides: An Ergonomics Guide to Carpal Tunnel Syndrome*. American Industrial Hygiene Association, Akron, Ohio.

Chusid, J.G. (1979) *Correlative Neuroanatomy and Functional Neurology*, 17th edn. Lange Medical Publications, Los Altos, California.

Crouch, J.E. (1972) *Functional Human Anatomy*, 2nd edn. Lea & Febiger, Philadelphia, Pennsylvania.

Duncan, K.M., Lewis, R.C., Foreman, D.A. and Nordyke, M.D. (1987) Treatment of carpal tunnel syndrome by members of the American Society for Surgery of the Hand: results of a questionnaire. *American Journal for Hand Surgery* **12**: 384–391.

Durham, R.H. (1960) *Encyclopedia of Medical Syndromes*. Paul B. Hoeber. Medical Division of Harper & Brothers, New York.

Eastman Kodak Company (1986) *Ergonomic Design for People at Work*, vol. 2. Van Nostrand Reinhold, New York.

Emerson, C.P. Jr and Taylor, J.E. (1946) *Essentials of Medicine*, 15th edn, J.B. Lippincott, Philadelphia.

Grandjean, E. (1988) *Fitting the Task to the Man: A Textbook of Occupational Ergonomics*. Taylor & Francis, London.

Hoaglund, F.T. (1990) Musculoskeletal injuries. In: *Occupational Medicine*, (LaDou, J., ed.). Appleton & Lange, Norwalk, Connecticut, pp. 58–71.

Hoffman, G.S. (1980) Raynaud's disease and phenomenon. *American Family Physician*, **21**: 91–97.

Maeda, K., Horiguchi, S. and Hosokawa, M. (1982) History of the studies on occupational cervicobrachial disorder in Japan and remaining problems. *Journal of Human Ergology*, **11**: 17–29.

NC-OSHA (1991) *Cumulative Trauma Disorders*. Industry guide no. 16. Division of Occupational Safety and Health, North Carolina Department of Labor, Raleigh, North Carolina.

NIOSH (1983a) Leading work-related diseases and injuries – United States, pp. 24–26. In: *Morbidity and Mortality Weekly Report*, 32, National Institute for Occupational Safety and Health (NIOSH), U.S. Department of Health and Human Services (DHHS), US Government Printing Office, Washington, DC.

NIOSH (1983b) Vibration syndrome. *Current Intelligence Bulletin 38*, National Institute for Occupational Safety and Health (NIOSH), US Department of Health and Human Services (DHHS), Publication No. 83–110, US Government Printing Office, Washington, DC.

NIOSH (1984) *Engineering Control of Occupational Safety and Health Hazards: Recommendations for Improving Engineering Practice, Education, and Research*. National Institute for Occupational Safety and Health (NIOSH), US Department of Health and Human Services (DHHS), publication no. 84–102, US Government Printing Office, Washington, DC.

NIOSH (1989) *Carpal Tunnel Syndrome: Selected References*. National Institute for Occupational Safety and Health (NIOSH), Centers for Diseases Control,

Public Health Service, US Department of Health and Human Services (DHHS), Cincinnati, Ohio.

Pottorff, T., Rys, M. and Konz, S. (1992) Toward a knowledge based system of cumulative trauma disorders for office work. In: *Advances in Industrial Ergonomics and Safety IV* (Kumar, S., ed.). Taylor & Francis, London, pp. 731–738.

Putz-Anderson, V. (ed.) (1988) *Cumulative Trauma Disorders: A Manual for Musculoskeletal Diseases of the Upper Limbs*. Taylor & Francis, London.

Sikka, A., Kemmann, E., Vrablik, R.M. and Grossman, L. (1983) Carpal tunnel syndrome associated danazol. *American Journal of Obstetrics and Gynecology* **147**: 102–103.

Silverstein, B.A., Fine, J.L. and Armstrong, T.J. (1987) Occupational factors and carpal tunnel syndrome. *American Journal of Industrial Medicine* **11**: 343–358.

Taylor, B.B. (1987) Low back injury prevention training requires traditional, new methods. *Occupational Health and Safety* **56**: 44, 48, 50.

Taylor, W. and Pelmear, P.L. (1975) *Vibration White Finger in Industry*. Academic Press, London.

Tayyari, F. and Emanuel, J.T. (1993) Carpal tunnel syndrome: an ergonomics approach to its prevention. *International Journal of Industrial Ergonomics* **11**: 173–179.

Tayyari, F. and Sohrabi, A-K. (1990) Carpal tunnel syndrome update. In: *Advances in Industrial Ergonomics and Safety II* (Das, B., ed.). Taylor & Francis, London, pp. 201–206.

Manual materials handling

9

9.1 INTRODUCTION

Injuries associated with manual materials handling (MMH) have grown substantially and are currently estimated to exceed several billion dollars annually in the USA. In addition to the compensation costs are the tremendous costs associated with the suffering of the impaired workers. MMH injuries can result from lifting, lowering, pushing, pulling, or carrying objects, as well as interactions with the environment (e.g., slipping and falling on a wet floor) while performing activities.

The subject of MMH concentrates on the identification and control of injury-causing conditions associated with MMH, and minimizing the health hazards by employing administrative controls (e.g., proper personnel selection, training in good material handling techniques, and worker rotations) and engineering controls (e.g., job redesign, mechanical assists). The application of biomechanics (specifically, statics and dynamics) in MMH is tremendous.

MMH places strains on both the cardiovascular system and the musculoskeletal system. The strain on the cardiovascular system is revealed by increased oxygen consumption and heart rate to deliver more oxygen and chemical energy to the involved muscles and removing the waste products (carbon dioxide and water) from them. Such a problem is not usually very critical and can be resolved by resting. However, a musculoskeletal strain, which can be injurious, is of a greater concern.

Some of the most traumatic and costly MMH injuries impact on the back. More specifically, the lower back (the fifth lumbar vertebra, the disk, and the first sacral vertebra) has been the area of concern in most studies examining low back pain associated with MMH. Back injuries were discussed in Chapter 8.

9.2 MANUAL LIFTING TASK EVALUATION

MMH evaluations are conducted in a variety of ways. Biomechanical, physiological, and psychophysical approaches have been used for many years to evaluate the MMH stresses imposed on workers. However, since the National Institute for Occupational Safety and Health (NIOSH) published its *Work Practices Guide for Manual Lifting* (NIOSH, 1981) and its revision (Waters *et al.*, 1994), those two documents have been widely used to assess MMH activities.

9.2.1 THE NIOSH ORIGINAL LIFTING MODEL

In 1981, NIOSH published the *Work Practices Guide for Manual Lifting* (*WPG*). The *WPG* defines a manual lifting task as the act of manually grasping and raising an object of a definable size without mechanical aids. The *WPG* was based on the thought that "an overexertion injury is the result of job demands that exceed a worker's capacity," (Putz-Anderson and Waters, 1991). This thought, which has also been used as the basis of the revised lifting equation, can be expressed by a strain index, as shown in the following equation:

$$\text{Strain index (SI)} = \frac{\text{job demands}}{\text{worker capacity}} \tag{9.1}$$

Therefore, any lifting situation in which this strain index exceeds 1.0 would present a potential for overexertion injury.

The *WPG* was developed with the input of many experts. Considerations for the *WPG* involved epidemiological, biomechanical, physiological, and psychophysical studies of the capabilities and limitations of people while performing MMH activities. In establishing the scope and limit of the recommendations of the *WPG*, the following assumptions were presented (NIOSH, 1981):

- smooth lifting;
- two-handed, symmetric lifting in the sagittal plane (directly in front of the body, no twisting during lift);
- moderate width (i.e., maximum 75 cm or 30 in);
- unrestricted lifting posture;
- good couplings (handles, shoes, floor surface);
- favorable ambient environments.

The 1981 *WPG* presented a mathematical equation for determination of an action limit (AL) for manual lifting tasks. In its guide, NIOSH divided lifting tasks into three classes (Fig. 9.1):

- acceptable (below the AL);
- unacceptable for some individuals (between the AL and maximum

permissible limit, or MPL), with administrative controls recommended;

● unacceptable for most individuals (above the MPL), with engineering controls recommended to redesign the work to eliminate or reduce the MMH hazard.

9.2.1.1 Action limit

Because of the large variability in capacity of individuals, loads falling between the AL and MPL may be lifted if administrative controls (e.g., personnel selection and training) are applied since:

● Musculoskeletal injury and severity rates increase moderately when workers perform a lifting task up to the AL (epidemiologic criterion).
● A 350-kg (3430-N or 770-lb) biomechanical compression force on the L5/S1 intervertebral disk imposed by the conditions described by the AL can be tolerated by most young, healthy workers (biomechanical criterion).
● Metabolic rates would exceed 3.5 kcal · min^{-1} for most individuals performing a lifting task above the AL (physiologic criterion).
● Lifting loads up to the AL are acceptable to over 99% of male and over

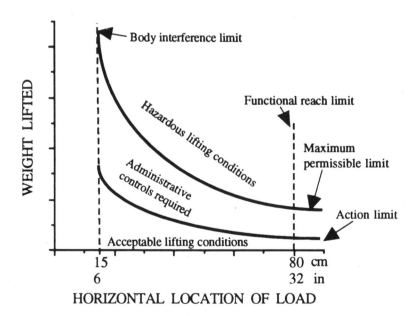

Fig. 9.1 Illustration of the three classes of lifting tasks based on object weight and horizontal distance of the object.

75% of female populations with a nominal risk of back injury (psychologic criterion).

9.2.1.2 Maximum permissible limit

Lifting loads beyond the MPL should not be permitted, but engineering controls must be applied to lower the load within the acceptable range. This limit has been set based on four criteria (epidemiologic, biomechanic, physiologic, and psychologic) as summarized below:

- Musculoskeletal injury and severity rates increase significantly when a lifting task is performed above the MPL.
- Biomechanical compression forces on the L5/S1 intervertebral disk above 650 kg (6370 N or 1430 lb) are not tolerable by most people.
- Metabolic rates would exceed 5.0 kcal · min^{-1} for most individuals performing a lifting task above the MPL.
- Only about 25% of male and fewer than 1% of female workers would find a lifting task above the MPL acceptable.

9.2.1.3 The original NIOSH lifting equation

The original NIOSH lifting equations for AL and MPL are as follows:

$$AL(kg) = 40 \times \left(\frac{15}{H}\right) \times (1 - 0.004|V - 75|) \times \left(0.7 + \frac{7.5}{D}\right)$$

$$\times \left(1 - \frac{F}{F_{max}}\right) \text{(metric units)} \qquad (9.2)$$

$$AL(lb) = 90 \times \left(\frac{6}{H}\right) \times (1 - 0.01|V - 30|) \times \left(0.7 + \frac{3}{D}\right)$$

$$\times \left(1 - \frac{F}{F_{max}}\right) \text{(US customary units)}$$

$$MPL = 3 \times AL \qquad (9.3)$$

where: H = horizontal location of lift centerline;
V = vertical location of the hands at origin of lift;
D = vertical travel distance from origin to destination of lift;
F = frequency of lifting, average number of lifts per minute;
F_{max} = maximum frequency of lifting which can be sustained (from Table 9.1);
AL = action limit;
MPL = maximum permissible limit = 3 × AL.

Table 9.1 Maximum lifts per minute (F_{max})

Period	Average vertical location (V) in cm (in)	
	$V > 75$ (30) Standing	$V \leqslant 75$ (30) Stooping
1 h	18	15
8 h (>1 h)	15	12

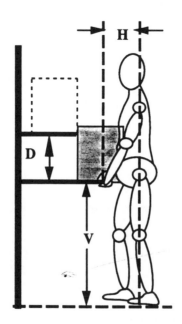

Fig. 9.2 Illustration of the independent variables in the National Institute for Occupational Safety and Health (NIOSH) lifting model.

9.2.1.4 Lifting task variables

The primary lifting task variables that affect the AL and MPL, as presented in equation 9.2 and shown in Figure 9.2, are as follows:

- Horizontal location (H) of the hands at origin of lift, measured from the midpoint between ankles (in centimeters or inches). H must be between 15 and 80 cm (6–32 in). The minimum 15 cm (6 in) is due to body interference.
- Vertical location (V) of the hands at origin of lift, measured from floor level (in centimeters or inches). V must be between 0 and 175 cm (0 and 70 in), which is the range of vertical reach for most individuals.

- Vertical travel distance (D) from origin to destination of lift (in centimeters or inches). D must be between 25 cm (10 in) and $(200 - V)$ cm $[(80 - V)$ in]. If the travel distance is less than 25 cm (10 in), then $D = 25$ cm (10 in) must be used.
- Frequency of lifting (F), average number of lifts per minute. F must be between 0.2 (one lift every 5 min) and F_{max} (shown in Table 9.1). If the frequency of lift is less than once per 5 min, then set $F = 0$.
- Maximum frequency of lifting (F_{max}) which is determined based on the duration or period of the task during the work-shift. Lifting is assumed to be occasional (less than 1 h) or continuous (more than 1 h, up to 8 h). Table 9.1 provides the F_{max} values.

In calculating the four modifying factors (horizontal factor, vertical factor, distance factor, and frequency factor), it should be noted that each factor has to be less than or equal to 1. If a factor exceeds 1 (or falls its lower bound), an error has been made. A common error is to use travel distances of less than 25 cm (10 in). If the actual travel distance is 10 cm, the distance factor could be improperly calculated as $D = 0.7 + 7.5/10$, or $D = 1.45$. Examination of the constraints of the variables shows that if $D < 25$, use $D = 25$ in the distance factor calculation. A factor having a value of 1.00 indicates that the factor has no effect on the lift being analyzed.

Graphical illustrations of the four factors for calculation of the AL using the NIOSH original model (NIOSH, 1981) are given in Figure 9.3. To calculate the action limit, the load constant (40 kg) is multiplied by each of the four adjustment factors obtained from the graphs. The advantage of using the graphs is that no factor will be out of the range of limits determined from the variable constraints. The primary disadvantage of using the graphs is inaccuracy from misreading the graph, or misinterpreting the value of the factor.

9.2.1.5 Evaluation of lifting task variables

As shown in equation 9.2, the NIOSH model for the AL consists of four multiplicative factors. These factors are explained for the equation given for the use of metric units as follows;

- **H factor = (15/H).** This factor indicates the amount of adjustment necessary as a function of the horizontal location (H) and ranges from 0.1875 (for $H = 80$ cm) to 1 (for $H = 15$ cm). If $H = 15$ cm, this factor is equal to 1 and no adjustment for the horizontal location is necessary. When $H = 60$ cm, this factor is $(15/60) = 0.25$, which means that the AL is reduced from 40 kg to $40(0.25) = 10$ kg.
- **V factor = $(1 - 0.004|V - 75|)$.** The factor for vertical location (V) deals with the absolute deviation of V from 75 cm (which is approximately

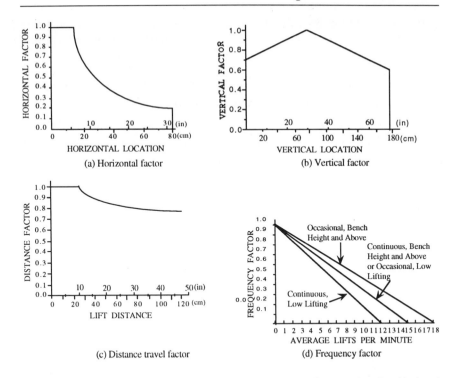

Fig. 9.3 Graphical illustration of the four adjustment factors for the National Institute for Occupational Safety and Health (NIOSH, 1981) action limit.

equal to the knuckle height) and ranges from 0.6 (for $V = 175$ cm) to 1 (for $V = 75$ cm). If $V = 75$ cm, then the V factor $= (1 - 0.004|V - 75|)$ $= (1 - 0.004|75 - 75|) = 1$, and, hence, no adjustment for the vertical location is necessary. If $V = 25$ cm, the V factor $= (1 - 0.004|25 - 75|)$ $= 1 - 0.004(50) = 0.8$. Also, for $V = 125$ cm, the V factor $= (1 - 0.004|125 - 75|) = 1 - 0.004(50) = 0.8$. Thus, when $V = 25$ or 125 cm, the absolute deviation of V from 75 cm is 50 cm and the V factor will reduce the AL from 40 kg to $40(0.8) = 32$ kg.

- **D factor $= (0.7 + 7.5/D)$.** The vertical travel distance (D) ranges from 0.7375 (for $D = 200$ cm or lifting from the floor to 200 cm above the floor) to 1 (for $D = 25$ cm, the minimum allowed value). If $D = 25$ cm, D factor $= (0.7 + 7.5/D) = 0.7 + 7.5/25 = 1$ and, hence, no adjustment is necessary. If $D = 50$ cm, D factor $= (0.7 + 7.5/50) = 0.85$, which reduces the AL from 40 kg to $40(0.85) = 34$ kg.
- **F factor $= (1 - F/F_{max})$.** The frequency factor is slightly more

complicated than the other three factors. If the observed frequency is 9 lifts per minute (i.e., $F = 9$) and the lifting point is below 75 cm (i.e., $V \leqslant 75$ cm), then F factor $= (1 - F/F_{max}) = 1 - 9/12 = 0.25$, which reduces the AL from 40 kg to $40(0.25) = 10$ kg. The F factor ranges from 0 (for $F = F_{max}$) to 1 (for $F = 0$, which is set for frequencies $\leqslant 0.2$ lifts per minute or less than 1 lift every 5 min).

To illustrate the combined effects of these factors, let's assume that a continuous (all-day-long) lifting activity is performed above the knuckle height ($V = 90$ cm) with an average $H = 30$ cm and an average rate of 6 lifts per minute. The load is placed on shelves of an average height of 120 cm above the floor. This information is summarized as follows:

$$H \quad = 30 \text{ cm};$$
$$V \quad = 90 \text{ cm};$$
$$D \quad = 120 - 90 = 30 \text{ cm};$$
$$F \quad = 6 \text{ lifts} \cdot \text{min}^{-1} \text{ and}$$
$$F_{max} = 15 \text{ lifts} \cdot \text{min}^{-1} \text{ (for 8 h in standing position).}$$

Then;

$$\begin{aligned}
\text{AL} \quad &= 40 \times (15/30) \times (1 - 0.004|90 - 75|) \times (0.7 + 7.5/30) \\
&\quad \times (1 - 6/15) \\
&= 40 \, (0.5) \, (0.94) \, (0.95) \, (0.6) = 10.7 \text{ kg} \\
\text{MPL} &= 3 \times 10.7 = 32.1 \text{ kg}
\end{aligned}$$

As can be seen, the H factor, followed by the F factor, has had the greatest effect on reducing the AL value.

The job demand (typically the weight of the object being handled) would then be compared to the calculated AL and MPL. If the task in question involved lifting objects weighing 5 kg (or anything less than 10.7 kg) then the task should represent a minor risk to the workers. However, if previous experience (e.g., accident/injury data) indicates a problem with the job, then further analysis should be conducted. Perhaps the problem occurs during a part of the activity not considered in the AL or MPL calculation. Remember, the AL and MPL calculations are tools and do not represent absolutes.

If the actual weight of the object being handled were between 10.7 and 32.1 kg, the task presents some risk and administrative or engineering controls are recommended. If the object weight exceeds 32.1 kg, engineering controls are needed.

9.2.1.6 Administrative controls

An administrative control is appropriate only for lifting jobs falling between the AL and MPL. As in any ergonomic analysis, administrative controls are recommended when engineering controls are not feasible.

Examples of administrative controls for MMH activities include: two (or more) persons lifting, limiting exposure through job rotation, training, and employee selection and placement. The ergonomist should continue to look for engineering solutions to reduce or alleviate the potential work problem.

9.2.1.7 Engineering controls

The ultimate goal of the ergonomist in job redesign is to eliminate job hazards through engineering design. However, when lifting jobs fall above the MPL, lifting is not permitted and engineering controls must be applied to bring the job within the acceptable zone. Examples of engineering controls include: automation (letting a machine do the heavy work), hoists and other mechanical assists, materials handling aids such as conveyors, hand trucks, and lift trucks, job and workstation redesign (e.g., removing barriers) and reducing the size and weight of the object being handled (an opposite strategy, that can be effective, is to increase the weight beyond what a person could lift and thereby forcing the use of lifting aids).

9.2.1.8 Limitations of the NIOSH lifting model

In using the NIOSH *WPG* in real-world situations, the following considerations should be taken into account:

- Other MMH activities (e.g., holding, carrying, pushing, and pulling) are assumed to be minimal.
- When lifting activities are not performed, the individual is assumed to be at rest.
- The work force is physically fit and accustomed to physical labor.
- Safety factors commonly used by engineers to account for the unexpected conditions are not included.

9.2.2 THE NIOSH REVISED LIFTING MODEL

The original NIOSH equation was criticized by practitioners because it was difficult to use and many of its underlying assumptions were inflexible (DeClercq and Lund, 1993). A revised lifting equation (Waters *et al.*, 1994) has been developed to account for the quality of the couplings (handles) and twisting action during lifting. However, the revised NIOSH manual lifting equation has not yet been accepted and its future implementation is in question. The revised equation has three principal components:

- the recommended weight limit (RWL);
- the load constant (LC);
- the multipliers.

9.2.2.1 Recommended weight limit

The revised lifting equation multiplies the LC by the six multipliers to determine a RWL as shown in equation 9.4.

$$\text{RWL} = \text{LC} * \text{HM} * \text{VM} * \text{DM} * \text{FM} * \text{AM} * \text{CM} \qquad (9.4)$$

where the asterisk symbol indicates multiplication.

9.2.2.2 Load constant

The LC is the maximum load that can be lifted safely, given that the conditions of lift, as defined by the multipliers, are all optimal. The load constant has been reduced from 40 kg (about 90 lb) in the 1981 equation to 23 kg (about 51 lb) in the revised equation.

9.2.2.3 Multipliers

The revised NIOSH lifting equation consists of six multipliers (Waters *et al.*, 1994). Depending on the lifting conditions, these multipliers decrease the LC to determine the amount of weight that can be lifted safely, or the RWL. The multipliers of the new equation are presented in Table 9.2. The task-descriptive variables used in the multipliers are as follows:

- H: The horizontal location of the hands (load center) from the midpoint between the ankles, measured at the origin and the destination of the lift (in centimeters or inches). In those cases where H cannot be measured, H can be estimated as:

    $H = 20 + L/2$ for $V \geqslant 25$ cm (or $H = 8 + L/2$ for $V \geqslant 10$ in)
    $H = 25 + L/2$ for $V < 25$ cm (or $H = 10 + L/2$ for $V < 10$ in)

Table 9.2 The multipliers for the NIOSH 1991 revised lifting model

Multiplier	British units	Metric units
Horizontal multiplier (HM)	$(10/H)$	$(25/H)$
Vertical multiplier (VM)	$(1-0.0075|V-30|)$	$(1-0.003|V-75|)$
Distance multiplier (DM)	$(0.82 + 1.8/D)$	$(0.82 + 4.5/D)$
Frequency multiplier (FM)	From Table 9.3	From Table 9.3
Asymmetry multiplier (AM)	$(1-0.0032A)$	$(1-0.0032A)$
Coupling multiplier (CM)	From Table 9.4	From Table 9.4

where L is the length (in the sagittal plane) of the container being handled and V is the location of the hands (relative to the standing surface) at the origin of the lift. If H is measured to be less than 25 cm (10 in), then use a minimum value of $H = 25$ cm (10 in). The maximum value that H should assume is 63 cm (25 in).

- V: The vertical location of the hands from the standing surface (floor), measured at the origin of the lift (in centimeters or inches). V can range from $V = 0$ (lifting an object that is resting on the floor) to $V = 175$ cm (70 in), which represents the upper safe limit for lifting.
- D: The vertical travel distance between the location of the hands at the origin and the destination of the lift (in centimeters or inches). D must fall in the range $25 \leqslant D \leqslant 175$ cm ($10 \leqslant D \leqslant 70$ in). If the measured value of D is less than 25 cm (10 in), then the minimum value of $D = 25$ cm ($D = 10$ in) must be used in the RWL calculation.
- A: Angle of asymmetry, the angular displacement of the load from the sagittal plane, measured from the origin and/or the destination of the lift (degrees). Ergonomic job design should eliminate twisting while lifting. However, if twisting cannot be eliminated, the asymmetric multiplier is used to account for load reduction due to the angle of twist while lifting. The range of twisting (asymmetry) ranges from $A = 0$, when the load is lifted directly in front of the body in the sagittal plane, to $A = 135°$, which represents the twisting limit for the body. The angle of twist (A) is measured at the origin of the lift. If significant control is required at the termination of the lift, then the angle of twist (A) should be determined by the terminating posture (if A is greater than it was at the origin of the lift).
- F: The average frequency rate of lifting, measured in lifts per minute. The duration of the lifting task is defined as:
  - short duration refers to continuous lifting up to 1 h, followed by a recovery period of at least 120% of work time;
  - moderate duration refers to continuous lifting up to 2 h, followed by a recovery period of at least 30% of work time;
  - long duration refers to continuous lifting up to 8 h, without additional fatigue allowances other than normally given during the work-shift (e.g., mid-morning, lunch, and mid-afternoon breaks).

  The short- and moderate-duration tasks require an adequate recovery time (as indicated above) before a similar lifting activity occurs. If adequate recovery time is not provided, the task duration becomes the sum of the consecutive lifting bouts that were not followed by an adequate recovery time. For example, if an employee were required to lift for 45 min, recovered for 15 min and then continued the lifting activity for another 45 min, before being assigned to a non-demanding task, the lifting duration would be of moderate duration (1–2 h) and should not be considered as two short-duration bouts.

For frequencies of less than once every 5 min ($F = 0.2$ lifts · min^{-1}), a minimum of $F = 0.2$ lifts · min^{-1} should be used in the RWL calculation.

The frequency multiplier (FM) values are given in Table 9.3, the coupling multiplier (CM) values in Table 9.4, and the hand-container coupling quality classifications are found in Table 9.5.

Example 9.1

An employee is required to unload 30-cm cardboard cubes from a pallet and place them on a conveyor that is 80 cm high. The cartons are unloaded at a rate of 6 cartons · min^{-1}. Each pallet has three layers of cartons, with each layer containing 16 cartons. The operation requires the worker to twist approximately 90° while turning to unload the carton on to the conveyor. A maximum of four pallets are unloaded during the first hour of each shift (the activity takes place only once per shift). Find the recommended weight limit for this activity.

Table 9.3 The frequency multiplier (FM) values

Frequency lifts · min^{-1}	Work duration (continuous)					
	≤8 h		≤2 h		≤1 h	
	$V < 75$ cm $V < 30$ in	$V \geqslant 75$ cm $V \geqslant 30$ in	$V < 75$ cm $V < 30$ in	$V \geqslant 75$ cm $V \geqslant 30$ in	$V < 75$ cm $V < 30$ in	$V \geqslant 75$ cm $V \geqslant 30$ in
0.2	0.85	0.85	0.95	0.95	1.00	1.00
0.5	0.81	0.81	0.92	0.92	0.97	0.97
1	0.75	0.75	0.88	0.88	0.94	0.94
2	0.65	0.65	0.84	0.84	0.91	0.91
3	0.55	0.55	0.79	0.79	0.88	0.88
4	0.45	0.45	0.72	0.72	0.84	0.84
5	0.35	0.35	0.60	0.60	0.80	0.80
6	0.27	0.27	0.50	0.50	0.75	0.75
7	0.22	0.22	0.42	0.42	0.70	0.70
8	0.18	0.18	0.35	0.35	0.60	0.60
9	0.00	0.15	0.30	0.30	0.52	0.52
10	0.00	0.13	0.26	0.26	0.45	0.45
11	0.00	0.00	0.00	0.23	0.41	0.41
12	0.00	0.00	0.00	0.21	0.37	0.37
13	0.00	0.00	0.00	0.00	0.00	0.34
14	0.00	0.00	0.00	0.00	0.00	0.31
15	0.00	0.00	0.00	0.00	0.00	0.28
>15	0.00	0.00	0.00	0.00	0.00	0.00

Table 9.4 Coupling multiplier values for NIOSH 1991 equation

Quality of coupling	Coupling multiplier (CM)	
	$V < 75$ cm (30 in)	$V \geq 75$ cm (30 in)
Good	1.00	1.00
Fair	0.95	1.00
Poor	0.90	0.90

Table 9.5 Hand-to-object coupling quality classification

Good	Fair	Poor
For containers of optimal design, such as some boxes, crates, etc., a *good* hand-to-object coupling would be defined as handles or hand-hold cut-outs of optimal design (see notes 1–3 below)	For containers of optimal design, a *fair* hand-to-object coupling is defined as handles or hand-hold cut-outs of less than optimal design (see notes 1–4 below).	The term *poor* hand-to-object coupling is used for containers of less than optimal design with no handles or hand-hold cut-outs or for loose parts or irregular objects that are bulky or hard to handle (see note 5 below)
For loose parts or irregular objects, which are not usually containerized, such as castings, stock, supply materials, etc., a *good* hand-to-object coupling is defined as a comfortable grip in which the hand can be easily wrapped around the object (see note 6 below)	For containers of optimal design with no handles or hand-hold cut-outs or for loose parts or irregular objects, a *fair* hand-to-object coupling is defined as a grip in which the hand can be flexed about 90° (see note 4 below)	Lifting non-rigid bags (i.e., bags that sag in the middle)

Notes
1. An optimal handle design has 1.9–3.8 cm (0.75–1.5 in) diameter, ≥ 11.5 cm (4.5 in) length, 5 cm (2 in) clearance, cylindrical shape, and a smooth non-slip surface.
2. An optimal hand-hold cut-out has ≥ 3.8 cm (1.5 in) height, 11.5 cm (4.5 in) length, semioval shape, ≥ 5 cm (2 in) clearance, smooth non-slip surface, and ≥ 1.1 cm (0.43 in) container thickness.
3. An optimal container design has ≤ 40 cm (16 in) frontal length, ≤ 30 cm (12 in) height, and a smooth non-slip surface.
4. A worker should be capable of clamping the fingers at nearly 90° under the container, such as required when lifting a cardboard box from the floor.
5. A less than optimal container has a frontal length > 40 cm (16 in), height > 30 cm (12 in), rough or slipping surface, sharp edges, asymmetric center of mass, unstable containers, or requires gloves.
6. A worker should be able to wrap the hand comfortably around the object without causing excessive wrist deviations or awkward postures, and the grip should not require excessive force.

Solution The example is solved under the following assumptions:

- The operator can get close to each carton while picking it up.
- The cardboard boxes do not have handles; they are gripped from the bottom.
- Personnel are well-trained in this activity.
- The top of the pallet (bottom of the carton on the first layer) is 15 cm above the floor.

In order to analyze the activity, the job must be broken down into components of similar demand. In this case, the job will be broken down into unloading each of the three different layers of cartons on the pallet. This is done in order to apply the multipliers to a consistent job demand. Therefore, the analysis will involve three separate activities which relate to three different lifting tasks. The first step is to determine the values for the RWL multipliers for each of the three layers.

- *Horizontal multiplier*: The Lifting Guide suggests that the horizontal multiplier can be estimated based on the location of the origin of lift. For those cases in which $V \geqslant 25$ cm, the horizontal distance (H) can be estimated as $H = 20 + L/2$, where L is the carton length in the sagittal plane. In this example $H = 20 + 30/2$, or $H = 35$ cm. For $V < 25$ cm, $H = 25 + L/2$. In this case, for the bottom layer of cartons, $H = 25 + 30/2$, or $H = 40$ cm.

  The horizontal multipliers can now be determined for the three layers using the formula $HM = 25/H$.

  $HM_1 = 25/40 = 0.625$ (for layer 1, the bottom layer)
  $HM_{2,3} = 25/35 = 0.714$ (for layers 2 and 3, the middle and top layers)

- *Vertical multiplier*: The vertical multiplier is based on the location of the gripping surface on each carton. Since these are cardboard cartons, with no handles, we assume that the cartons are gripped from the bottom. Therefore, for the three layers the vertical locations are:

  $$V_1 = 15 \ cm, \ V_2 = 45 \ cm, \ and \ V_3 = 75 \ cm$$

  The vertical multipliers are calculated from: $VM = (1 - (0.003|V{-}75|))$.

  $$VM_1 = 0.82, \ VM_2 = 0.91, \ and \ VM_3 = 1.00$$

- *Distance multiplier*: The vertical travel distance is determined as the distance from the origin of the lift to the destination of the lift. In this example the designation of the lift is constant (the conveyor height of 80 cm). Therefore, the vertical travel distance (D) is the difference between V and 80 cm.

  $D_1 = 65$ cm, $D_2 = 35$ cm, and $D_3 = 5$ cm
  (however, if $D \leqslant 25$ cm, then use $D = 25$ cm)

The distance multipliers are determined by DM = $(0.82 + 4.5/D)$

$$DM_1 = 0.89, DM_2 = 0.95, \text{ and } DM_3 = 1.00$$

- *Frequency multiplier*: At 6 lifts: min^{-1} (and three layers of cartons on the pallet), a frequency rate of 2 lifts $\cdot$ min^{-1} per layer will be used for the analyses. Since separate calculations are made for each layer of cartons, the total job frequency of 6 lifts $\cdot$ min^{-1} must be broken down equally to a frequency relative to each layer being unloaded (in this case, 2 cartons $\cdot$ min^{-1} per layer). The frequency multiplier is obtained from Table 9.3 using $\leqslant 1$ h and the appropriate value of V (layers 1 and 2, $V < 75$; for layer 3, $V = 75$).

$$FM_{1,2} = 0.91, \text{ and } FM_3 = 0.91$$

- *Asymmetry multiplier*: For all three layers, a 90° twist is required as the worker twists to place the carton on to the conveyor. The asymmetric multiplier for all three layers is AM = $1 - 0.0032$ A, or

$$AM \quad = 1 - 0.0032 \ (90); \text{ that is,}$$
$$AM_{1,2,3} = 0.712$$

- *Coupling multiplier*: The coupling multiplier is obtained from Table 9.4. For cardboard cartons, the quality of the coupling is considered fair. Therefore, the coupling multiplier is obtained from the $V < 75$ cm column for layers 1 and 2, and $V \geqslant 75$ cm for layer 3 (the top layer).

$$CM_{1,2} = 0.95 \text{ and } CM_3 = 1.00$$

Table 9.6 summarizes the multipliers for the three lifting tasks (three layers) that comprise the job. The product of the six multipliers is listed

Table 9.6 The multiplier values for the three lifting tasks

Multiplier	Layer 1	Layer 2	Layer 3
Horizontal	0.625	0.714	0.714
Vertical	0.82	0.91	1.00
Distance	0.89	0.95	1.00
Frequency	0.91	0.91	0.91
Asymmetry	0.712	0.712	0.712
Coupling	0.95	0.95	1.00
Product of multipliers	0.28	0.38	0.46
RWL	6.44 kg	8.74 kg	10.58 kg

RWL = Recommended weight limit.

at the bottom of the table. To calculate the RWL for each task of the job, the product of the multipliers is multiplied by the LC of 23 kg.

Discussion The RWL is now compared to the task demand (weight of the cartons being lifted). If the calculated RWL exceeds the weight of the containers, the job (or task) should represent minimal risk to the worker. However, if there have been injuries on the job, the analysis should not be ignored just because the RWL is greater than the demand of the job (in this case, the values used in the RWL calculation might be in error, or the injury could be the result of some component of the task that has been ignored). If the task demand (container weight) exceeds the RWL, then some type of ergonomic intervention is needed. The preferable intervention is engineering controls to eliminate the hazard. The multipliers should be examined to determine the areas for greatest impact. In this example, the horizontal multiplier and the asymmetry multiplier are the two smallest values, so they have the potential for greatest impact. If the twisting could be eliminated and/or the package redesigned to make the center of the load closer to the body during the lift, the RWL values can be increased. If these changes are not feasible, then alternatives such as mechanical assists are recommended.

9.2.2.4 Lifting index

In conjunction with the revised lifting equation, Waters *et al.* (1994) suggested a lifting index (LI) for estimation of the hazard of overexertion injury for manual lifting tasks.

$$LI = \frac{\text{load weight}}{\text{recommended weight limit}} = \frac{L}{RWL} \qquad (9.5)$$

where load weight (L) is the weight of the object lifted (in kilograms or pounds, depending on the unit used for RWL). The lifting tasks with LI values greater than 1.0 present a risk of overexertion injury.

9.3 ORIGINAL VERSUS NEW LIFTING GUIDE

The original and new equations are compared in Table 9.7.

Entering the data from Example 9.1 into the 1981 NIOSH equation would yield the following action limit values:

For the bottom row: AL = 8.13 kg
For the middle row: AL = 11.98 kg
For the top row: AL = 15.31 kg

In all cases, the AL exceeded the RWL. However, the original problem had a 90° angle of twist. Since one of the assumptions of the 1981 guide

Table 9.7 Comparison of NIOSH lifting equation multipliers

| Load constant | Multiplier (value) | |
and multipliers	1981 Equation	1991 Equation
Load constant (LC)	40 kg	23 kg
Horizontal (HM)	$15/H$	$25/H$
Vertical (VM)	$1-0.004\|V-75\|$	$1-0.003\|V-75\|$
Distance (DM)	$0.7 + 7.5/D$	$0.82 + 4.5/D$
Frequency (FM)	$1 - F/F_{max}$	From Table 9.3
Asymmetry (AM)	NA	$1-0.0032A$
Coupling (CM)	NA	From Table 9.4

NA = Not applicable.

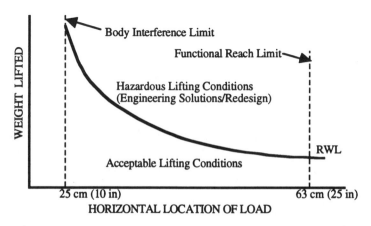

Fig. 9.4 The recommended weight limit at various horizontal distances based on the National Institute for Occupational Safety and Health (NIOSH) revised lifting model. RWL = Recommended weight limit.

was "symmetric sagittal plane lifting", the AL calculations are not appropriate. With the asymmetry removed from the problem, the following RWL values are obtained:

For the bottom row: RWL = 9.04 kg
For the middle row: RWL = 12.28 kg
For the top row: RWL = 14.86 kg

The elimination of the asymmetry resulted in RWL values that were very close to the AL values calculated from the *WPG* (NIOSH, 1981).

Figure 9.4 shows the revised NIOSH equation curve (RWL) which determines the cut-off points or zone of acceptable lifting conditions and hazardous lifting conditions.

9.4 OTHER MMH ACTIVITIES: PUSHING AND PULLING

Although lifting and lowering are the most common of the MMH activities and account for the majority of occupational injuries, other MMH activities should not be ignored. Pushing and pulling activities are also frequently encountered in industry and should be addressed by the ergonomist. Comprehensive data on pushing and pulling, as well as other MMH activities can be found in Mital *et al.* (1993), Snook and Ciriello (1991), and Ayoub *et al.* (1987).

Before standard data are used to assess pushing/pulling activities, caution is advised. Obviously, the shoe–floor interface and the person's posture play a key role in the levels of push/pull force that can be exerted on a load. Therefore, one of the first ergonomic controls can be to insure a non-contaminated floor, good footwear (sufficient to insure a static coefficient of at least 0.50 between the floor surface and the worker's shoe sole), and space to allow for postural changes to find the best posture for pushing/pulling the load.

For general guidelines, the data from Snook and Ciriello (1991) have been summarized in Table 9.8. Note that only two distances are presented: a short distance (up to 2.1 m or 7 ft) and a longer distance (15.2 m or 50 ft). Three frequencies are used in Table 9.8: frequent handling of once per minute; moderate handling of once every 5 min; and infrequent handling of once per shift (8 h). High-frequency pushing/pulling over long distances should not be performed manually. The variability of the data in Table 9.8 is presented in terms of the coefficients of variation in Table 9.9. The coefficient of variation (CV) is the ratio of the standard deviation to the mean. Therefore, to find the standard deviation, multiply the mean value by the CV.

To find push/pull forces acceptable to various percentages of the

Table 9.8 Maximum acceptable pushing/pulling forces*

Task	Gender	Distance	Initial forces (kg)			Sustained forces (kg)		
			$1 \cdot min^{-1}$	$1 \cdot 5\ min^{-1}$	$1 \cdot shift^{-1}$	$1 \cdot min^{-1}$	$1 \cdot 5\ min^{-1}$	$1 \cdot shift^{-1}$
Push	Males	2.1 m	43	45	54	28	34	40
		15.2 m	35	38	45	19	23	28
	Females	2.1 m	25	30	32	18	21	26
		15.2 m	20	24	25	13	15	19
Pull	Males	2.1 m	36	39	47	26	32	37
		15.2 m	31	33	40	19	23	27
	Females	2.1 m	25	30	32	16	20	25
		15.2 m	19	23	25	12	15	18

*The mean values are based on shoulder height push/pull. Also, note that 2.1 m = 7 ft and 15.2 m = 50 ft.

Table 9.9 Coefficients of variation for push/pull data

		Coefficients of variation	
Task	Gender	Initial	Sustained
Push	Males	0.30	0.33
	Females	0.24	0.38
Pull	Males	0.24	0.28
	Females	0.22	0.32

population, use the same basic procedure as was described in Chapter 4 for anthropometric population percentiles (see Table 4.1). The maximum acceptable (push/pull) force is calculated as follows:

- The standard deviation must be estimated by using the mean value of the maximum acceptable push or pull force ($\bar{X}$) from Table 9.8.
- The appropriate CV is selected from Table 9.9.
- The standard deviation (s) is estimated by multiplying the mean by the appropriate coefficient of variation: $s = CV \cdot \bar{X}$.
- The maximum acceptable (push/pull) force is determined for the weakest person in the desired range of the population. That is, the maximum force acceptable to $100\alpha\%$ of a population is the same as the $100(1 - \alpha)$ percentile. Therefore, the force acceptable to $100\alpha\%$ of the population is calculated using the following equation:

$$X_{(1-\alpha)} = \bar{X} - Z_\alpha \cdot s$$

where:

$X_{(1-\alpha)}$ = the maximum (pushing/pulling) force acceptable to $100\alpha\%$ of the population;

$\bar{X}$ = the mean value of maximum acceptable pushing/pulling force obtained from Table 9.8 for the population;

Z_α = the standard normal value corresponding to $100\alpha\%$ of the population (note that $Z_{(1-\alpha)} = -Z_\alpha$);

s = the standard deviation of the maximum (pushing/pulling) force for the population.

Example 9.2

Find the sustained pushing force acceptable to 90% of the female population for a short (2.1 m or less) pushing task performed once every 5 min.

- Mean push force for females ($\bar{X}$) (2.1 m at once every 5 m, from Table 9.8) = 21 kg (or 206 N).
- CV (from Table 9.9) = 0.38.
- Standard deviation (s) = (0.38) · (21 kg) = 8 kg (or 78 N).
- Then the pushing force acceptable to 90% (or the 10th percentile) of the female population, $X_{0.1}$, is calculated as follows:

$$X_{0.1} = \bar{X} - Z_{0.9} \cdot s = (21 \text{ kg}) - 1.282 \cdot (8 \text{ kg})$$
$$= 10.74 \text{ kg (or 105 N)}$$

Note that, by definition, the mean would be acceptable to 50% of the population. For a force to be acceptable to more than 50% of the population, the magnitude of the force must decrease. A force higher than the mean would be acceptable to less than half of the population.

9.5 BACK-INJURY-PREVENTIVE TECHNIQUES

There is no magic pill or formula to eliminate back injuries. With careful application of engineering and administrative controls, exposure and injuries should both be reduced. In order to have a successful back-injury-reduction program, the workers must become involved in the program. The old safety posters reminding workers to "keep your back straight and lift with your legs" are not an adequate solution to the problem of increased back injuries. In their place are awareness and training principles that instruct employees in the following techniques of back-injury prevention:

Develop proper MMH techniques

- Avoid lifting as much as possible.
- Use carts, lift trucks or other mechanical assists.
- Push instead of pull–pushing allows the spine to remain in a neutral posture.
- Get help from fellow employees.
- Know the destination or where the load is to be placed in advance.
- Know the shortest and safest routes to get to the destination of loads carried.
- Get a good grip of the object being lifted.
- Keep the object being handled as close to the body as possible.
- Tuck the stomach in.
- Lift smoothly and avoid jerk motions.
- Avoid twisting while lifting.
- Lift with the leg muscles, which are much stronger than the back muscles.
- Avoid overextension.

- Reduce the load and make more trips of lighter loads.
- Keep the back straight while lifting.

Develop proper MMH strategies

- Get help if the load is too heavy.
- Don't take more than you can carry safely.
- Keep hands clear of pinch points to prevent crushing hands between objects.
- Cross-tie when stacking.
- Don't stack too high.
- Don't carry a load so high that you can't see over it. Never block your vision.
- Pushing is safer than pulling.
- Always know your limitations.
- Protect your hands with proper gloves.
- Get a good grip of the load.
- Never stand under a suspended load. Stand clear of overhead loads.
- Be aware of and keep off all power lines – contact with power lines can be fatal.
- Wear proper foot protection.

Develop proper workstation design

- Keep the spine in a neutral posture during sitting.
- Get support for the lower back by using the seat backrest or a pillow in sitting.
- Adjust the seat height so the thighs are horizontal and feet flat on the floor. If the seat is not adjustable, a footrest should be used.
- Get up and stretch for a few moments every 15–20 min of work.
- Tuck the stomach in and fill the lungs during physical work and exertions.
- Choose a workstation that allows you to alternate between sitting and standing.
- If the work is only performed in a standing posture, a footrest or box should be used to raise one foot over the other. The feet should be alternated when the person feels like it.
- Stand on a mat or carpet rather than on a bare concrete floor.
- Avoid bending over the task. If the employee is bending to read a label, magnifiers can be used to enlarge the label, if necessary, to avoid bending.
- Keep the task in front of the body, not on the sides.
- Warm up before starting the work.
- Exercise to keep the muscles toned.

9.6 APPLICATIONS AND DISCUSSION

Injuries associated with MMH represent a significant cost to industry and society in terms of lost production, medical costs, and human suffering. This chapter focused on the two most common methods for the assessment of MMH problems: the 1981 NIOSH *WPG*, and the 1991 revised guide. Both are tools for the ergonomist to use in designing less stressful work environments. The 1991 revision changed some of the multipliers and addressed the issues of asymmetry and couplings. However, to date the 1991 revision has not received universal endorsement to the exclusion of the 1981 *WPG*. Therefore, the ergonomist would be wise to utilize both tools in the analysis of jobs.

In addition to the NIOSH guides, other tools are also available to the ergonomist. The NIOSH guides attempted to incorporate biomechanical, physiological, and psychophysical research results into their recommendations. However, there are many cases where the separate approaches are more appropriate. The obvious starting point is with the assumptions used in the development of the guidelines. For example, if a task required a worker to hold an object in place with one hand while in an unusual posture (perhaps mechanics lying on their backs holding a component in place while retaining bolts are inserted and fastened), the NIOSH guides would not be appropriate. In this case, either psychophysical data from the literature or a biomechanical model would be appropriate. Changes in job design could be evaluated through models to arrive at a more ergonomically sound design.

REVIEW QUESTIONS

1. What are the causes of manual materials handling (MMH) injuries?
2. What is the focus of MMH analysis?
3. How is the strain on the cardiovascular system revealed?
4. What type of MMH strain is of primary concern to ergonomics?
5. Write the job strain index (SI) equation. How would this index indicate a potential for overexertion injury?
6. What assumptions had been made in establishing the scope and limit of the recommendations of the National Institute for Occupational Safety and Health (NIOSH) Work Practices Guide (*WPG*)?
7. How did NIOSH divide lifting tasks for determination of an action limit (AL) for manual lifting tasks in its original (1981) model? Indicate these divisions on a diagram.
8. Describe the epidemiologic, biomechanic, physiologic, and psychologic criteria used by NIOSH in determination of the AL in its 1981 lifting guide.

9. Describe the epidemiologic, biomechanic, physiologic, and psycho-logic criteria used by NIOSH in determination of the maximum permissible limit (MPL) in its 1981 lifting guide.
10. Write the original NIOSH lifting equations for the AL and MPL.
11. Describe and give the limits of the primary lifting task variables that affect the AL and MPL which are used in the NIOSH's original (1981) lifting equation.
12. What factor of the 1981 NIOSH lifting model does Figure 9.5 represent?
13. Based on the NIOSH original lifting guide, when is it appropriate to apply administrative controls? Give examples of such controls for MMH activities.
14. Based on the NIOSH original lifting guide, under what conditions can lifting above the MPL be performed? What is required to be done? Give examples of such controls for MMH activities.
15. A worker must lift 50-lb boxes of canned food from an 8-in-high pallet to a nearby truck bed, which is 41 in from the floor. During loading, 6 boxes per minute will be lifted. The horizontal distance of the handles is 12 in from the worker's ankles. Answer the following questions:

(a) Is the box within the AL? Why?
(b) Which factor has the greatest effect on reducing the AL?
(c) What do NIOSH lifting guidelines recommend for a safe lifting in this case?
(d) How much can the person lift when working all day?

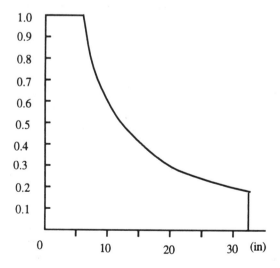

Fig. 9.5 For question 12.

16. A worker must lower 50-lb boxes of canned food from a 61-in-high shelf to a nearby conveyor, which is 41 in from the floor. During loading, 5 boxes per minute will be lowered. The horizontal distance of the handles is 12 in from the worker's ankles. Answer the following questions:

    (a) Is the box within the AL? Why?
    (b) Which factor has the greatest effect on reducing the AL?
    (c) What do NIOSH lifting guidelines recommend for safe lowering in this case?
    (d) How much can the person lower when working all day?

17. In using the NIOSH *WPG* in real-world situations, what considerations should be taken into account?

18. List the three principal components of the NIOSH revised (1991) lifting equation.

19. Write the lifting index (LI) equation used in conjunction with the NIOSH revised lifting equation.

20. Certain items arrive at a packaging station by a conveyor that is 55 cm high. Each item is comfortable to grip and the hand can easily be wrapped around it. Each item weighs 10 kg, and is 30 cm wide and 20 cm deep. The station is staffed by a worker, lifting the items from the conveyor and placing them into cardboard boxes that are located on a 75-cm-high counter. When a box is full, the worker slides it over another conveyor on his side at a rate of one box per minute. The task does not seem to require significant control of the box at the destination. However, the worker twists his body about 45° to pick up the item to be packaged. The worker performs this job over an 8-h shift, while taking rest breaks. Calculate the LI for this activity.

21. Find the initial maximum pulling force acceptable to 90% of the male population for a pulling task performed once per minute, over an approximately 20-m distance.

22. Find the sustained pushing force acceptable to 95% of the female population for a short (2.1 m or less) pushing task performed once per shift.

23. List 10 recommendations for developing proper MMH techniques.

24. List eight recommendations for developing proper MMH strategies.

25. List eight recommendations for developing proper workstation design for safer MMH.

REFERENCES

Ayoub, M.M., Smith, J.L., Selan, J. *et al.* (1987) *Manual Materials Handling in Unusual Postures.* Technical Report. Texas Tech University, Lubbock, TX.

DeClercq, N.G. and Lund, J. (1993) NIOSH lifting formula changes scope to calculate maximum weight limits. *Occupational Health and Safety* **61**: 45–48, 51–53, 61.

Mital, A., Nicholson, A.S., and Ayoub, M.M. (1993) *A Guide to Manual Materials Handling.* Taylor & Francis, London.

NIOSH (1981) *Work Practices Guide for Manual Lifting.* Technical report no. 81–122. National Institute for Occupational Safety and Health (NIOSH), US Department of Health and Human Services, Cincinnati, OH.

Putz-Anderson, V. and Waters, T. (1991) *Revisions in NIOSH Guide to Manual Lifting.* Paper presented at a national conference entitled "A National Strategy for Occupational Musculoskeletal Injury Prevention – Implementation Issues and Research Needs," held in University of Michigan, Ann Arbor, Michigan.

Snook, S.H. and Ciriello, V.M. (1991) The design of manual handling tasks: revised tables of maximum acceptable weights and forces. *Ergonomics* **34**: 1197–1213.

Waters, T., Putz-Anderson, V. and Garg, A. (1994) *Applications Manual for the Revised NIOSH Lifting Equation.* DHHS (NIOSH) publication no. 94–110. Centers for Disease Control and Prevention, NIOSH, Cincinnati, OH.

Work-tool design   10

10.1 INTRODUCTION

Work-tools are such a common part of work-life that workers do not think they may pose hazards. Although work-tools are manufactured with safety in mind, tragic accidents often occur. Tool hazards may be due to improper tool design or lack of worker training in recognizing the hazards associated with the various types of tools and the required safety precautions to avoid the hazards. The latter includes misuse and improper maintenance of tools, which are the greatest source of hazards (Occupational Safety and Health Administration (OSHA, 1988)).

Work-tools are extensions of human hands. They enhance the capability of the users in performing their tasks so efficiently that the tasks would otherwise be difficult, if not impossible. However, the ill-effects of an improperly designed work-tool can be devastating, yet so subtle as to remain blameless. Poorly designed tools are one of the common factors contributing to the development of cumulative trauma disorders (CTDs). They can also impair the productivity of workers. Improper use of a work-tool, regardless of its design, can also lead to CTD problems. Tools can also cause bodily injuries in single-exposure incidents. Ergonomically well-designed and properly used tools can reduce or prevent such problems.

Most work-tools are purchased from external vendors rather than designed and manufactured internally by the employers. They are rarely designed to fit perfectly the specialized needs of a particular manufacturing situation while considering the anthropometric attributes of the users. The principles of ergonomics and biomechanics should be considered in the selection, evaluation, and use of work-tools. General ergonomics principles of work-tool design, and guidelines for the design, evaluation, and proper use of work-tools are presented in this chapter. A great deal of work related to the design, ergonomics, and safety of work-tools is available in the literature (Tichauer, 1978; Fraser,

1980; Eastman Kodak Company, 1983; Freivalds, 1987; Grandjean, 1988).

10.2 GRIP STRENGTH AND ENDURANCE

Grip strength and endurance are two terms related to the design and use of hand tools which have been defined by Bazar (1978). **Grip strength** is the maximum momentary squeeze force exerted on a hand dynamometer. **Endurance** is the length of time a person can exert a specified force. Consideration of grip strength and endurance is very important in the design and use of hand-tools requiring gripping force, and in manual materials handling (MMH). A person with lower grip strength and/or shorter endurance is likely to be more susceptible to development of CTDs than those with higher grip strength and longer endurance.

Grip strength and endurance can be measured using a hand dynamometer. A hand dynamometer is a device which measures the force an individual exerts through the hand grip. The fingers in a normal closed fist form an elliptical shape (Fig. 10.1a). Expectedly, in such a natural posture, the joints are properly aligned and consequently the largest muscle mass is engaged to achieve the hand's maximum grip strength. This phenomenon can easily be examined by holding a straight-handle tool (e.g., a hammer) in a naturally clenched hand, where the tool handle is inclined if the wrist is kept straight. The grip strength in this situation is found to be greater than when the person tries to hold the tool handle perpendicular to the hand–arm axis without changing the wrist posture, as illustrated in Figure 10.1b.

An investigation by Swanson *et al.* (1970) revealed that most males and females can exert their maximum grip force when the spacing of the

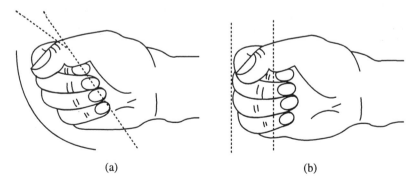

(a)   (b)

Fig. 10.1 Illustration of (a) a normal closed fist and (b) an abnormal closed fist for a vertically held handle.

dynamometer is 3.8–6.3 cm (1.5–2.5 in). However, the following factors are known to affect both grip strength and endurance, and should be considered in hand-tool design:

- **Size of grip span and diameter**: The tool handle should fit the hand of the user. A handle of a too large or too small grip diameter cannot be held comfortably and reduces the grip strength.
- **Type of grasp/grip**: Power grip strength is about four times greater than the maximum pinch strength. Pinch grip requires significantly higher muscle force than power grip.
- **Age**: Maximal muscle strength is reached at the age of about 20 for men, and a few years earlier for women (Rodahl, 1989). Grip strength may reach its maximum value in the middle to late 20s and then declines as age advances. According to Rodahl (1989), the strength of a 65-year-old person is about 75–80% of that attained at the age of 20–30 years.
- **Gender**: The grip strength of men is greater than that of women (Falkel *et al.*, 1985). On average the grip strength for women (24.5–35.0 kg for US females) is about 60% of men's (41.9–59.8 kg for US males; Fraser, 1983).
- **Handedness**: The grip strength and endurance of the dominant hand are higher than those of the non-dominant hand. The average percentage difference between dominant and non-dominant hands for men is in the range of 3.2–11.5% (Schmidt and Toews, 1970).
- **Vibration**: Grip strength requirements for grasping and controlling vibrating tools are higher than those required by non-vibrating tools.
- **Wrist posture**: Grip strength is reduced with the deviation of the wrist from its neutral posture.
- **Gloves**: Gloves increase grip strength requirements. They also increase the size of the hand. For example, woolen or leather gloves add 5 mm to hand thickness and 8 mm to hand breadth at the thumb (Fraser, 1983).
- **Surgery**: Grip strength is very likely to be reduced after wrist surgery to relieve carpal tunnel syndrome.

10.3 TYPES OF WORK-TOOLS

Work-tools may be divided into two categories:

- **Hand tools**: Hand tools are non-powered and include axes, hammers, screwdrivers, hand saws, pliers, etc.
- **Power tools**: There are several different types of power tools, based on the power source they use. They include electric, pneumatic, liquid fuel, hydraulic, and power-actuated tools.

The greatest hazards posed by work-tools result from their misuse and improper maintenance. Safety precautions for preventing the potential hazards associated with work-tools will be presented later in this chapter.

10.4 NEW DEVELOPMENT IN TOOL HANDLE DESIGN

A slightly bent handle better fits the natural contour of the hand and minimizes the need for a tight grip to maintain the tool. Based on this principle, in the early 1970s, John Bennett introduced a handle for push brooms with a 19° bend at its stem (Emanuel *et al.*, 1980). He selected this angle as the basis for his new design to keep the user's wrist straight. It is also approximately equal to the angle formed by the lifeline under the ball of the thumb and the index finger (Fig. 10.2). Bennett has patented the 19 ± 5° angles on virtually all tool handles and named them BioCurve tools. When a BioCurve tool is gripped, the third (middle) finger is locked into the center of the curve while the other fingers naturally fall into the curve of the handle (Fig. 10.3).

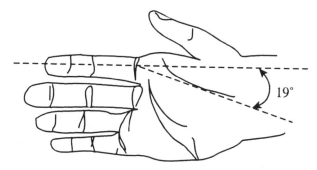

Fig. 10.2 A 19° angle formed by the lifeline under the ball of the hand and the index finger.

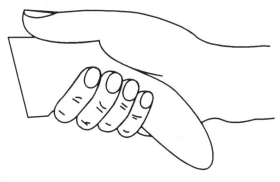

Fig. 10.3 A bent handle allows the wrist to be maintained straight.

The key idea in the design of tools and workstations is to maintain the body parts in their neutral (natural) postures. For example, when a traditional, straight-handled file is used to file horizontally, the wrist must be bent toward the ulna. This ulnar deviation causes tension and stress in the wrist as well as in the forearm muscles, which are the force source of the fingers. According to Tichauer (1975), bending the wrist either in the ulnar direction (outward) or in the radial direction (inward) reduces the rotational ability of the wrist by 50%, and holding the wrist in such awkward positions may frequently cause inflammation of tendons and/or tendon sheaths. Thus, bending the handle of the tool slightly may eliminate the requirement of bent wrist.

The maximum grasping force can be reduced by up to 75% by changing over from holding an object with the whole hand to grasping it with the fingertips (Grandjean, 1988).

10.5 ERGONOMICS GUIDELINES FOR HAND-TOOL DESIGN

Hand tools are designed in many different shapes and characteristics to enhance the effectiveness of the human limbs in performing certain tasks. It is impossible to develop universal specifications for all hand-tools to fit the shape of the hand ergonomically. In general, the tool handle should be designed in such a way that the hand can be kept in line with the forearm as much as possible (to minimize wrist deviation), and the job can be performed as easily and effectively as possible. Three sets of recommendations, based on first, the anatomy of the hand; second, the biomechanics of the hand; and third, ergonomics principles, are presented in the following subsections.

10.5.1 RECOMMENDATIONS FOR TOOL DESIGN BASED ON HAND ANATOMY

Many tasks require awkward anatomical postures that can lead to CTDs (UAW, 1982; Thompson, 1990). Therefore, work-tools must be so designed that the following situations are avoided:

- **Unidirectional wrist deviation**: Grip strength is generally reduced by wrist deviation in any direction. This can be experienced by flexion, extension, or ulnar or radial deviation of the wrist. Such deviations can damage the nerves or tendon sheaths, resulting in carpal tunnel syndrome or tenosynovitis. In addition, the reduced grip strength increases fatigue, and also the risk of losing control of the tool and dropping it. This can cause injury and property damage. Hence, wrist deviation should be avoided by ergonomic tool and job designs.
- **Combined ulnar deviation and supination of the wrist**: Repetitive ulnar deviation and supination of the wrist, as performed in turning a

knob clockwise by the right hand, increases the risk of impact shocks and irritation of the tendon sheaths.

- **Combined ulnar deviation and palmar flexion of the wrist**: Combining ulnar deviation and palmar flexion forces the tendons to bend, bunch together, and rub against the transverse ligaments of the carpal tunnel. Repetition of such a motion can cause tendinitis. Clothes wringing is a classical example of ulnar deviation and palmar flexion of the wrist that can damage the nerves and tendons. Examples of occupational tasks requiring such a wrist motion are tightening screws, looping wires with pliers, operating motorcycle-type hand controls, and operating hand-held chippers and grinders.
- **Dorsiflexion (extension) of the wrist**: Examples of occupational tasks demanding the dorsiflexion of the wrist are scrubbing with towels and rags, and painting or polishing using a hand brush. To avoid dorsiflexion and the risk of developing tenosynovitis, a rag or cloth holder can be used, and the brush handle can be modified (e.g., pistol-grip handles may be used for this purpose) to maintain neutral (straight) wrist posture.
- **Dorsiflexion (extension) and pronation of the wrist**: Performing a combination of dorsiflexion (extension) and pronation of the wrist over an extended time increases the risks of tenosynovitis, tendinitis, and/or carpal tunnel syndrome. In addition, as explained by Thompson (1990), performing rapid and repetitive motions increases the pressure between the head of the radius and the capitulum (the knob or head-shape part at the end of a bone) of the humerus in the elbow. The pressure in the elbow joint causes friction and heat. Forces exerted during lifting, pushing, or pulling worsen the condition.
- **Elbow flexion and inward rotation of the hand**: Elbow flexion when loaded by lifting a heavy object or pulling will pull the head of the radius into the capitulum of the humerus. The situation becomes worse if elbow flexion is combined with inward rotation of the forearm. This combination applies additional pressure and twisting forces to the head of the radius due to the biceps' secondary function as an outward rotator of the forearm (Thompson, 1990). Therefore, inward rotation of the forearm during lifting or pulling should be avoided.
- **Combined radial deviation of the wrist, forearm pronation, and dorsiflexion of the wrist**: Radial deviation also can cause bunching and compression of the tendons and nerves in the wrist. When radial deviation is combined with the forearm pronation and the wrist dorsiflexion, it increases the pressure on the head of the radius in the way explained for elbow flexion (see above). Extensive radial deviation can cause epicondylitis (tennis elbow). Such a motion should be eliminated during workplace design or redesign.

- **Use of the hand as a tool**: Frequent tapping with the heel of the hand, even lightly, can injure the nerves, arteries and tendons of the hand and wrist. In addition to the hand and wrist, the impact shocks can be transmitted to the elbow and shoulder and damage these joints. Examples of occupational tasks in which the hand is used as a hammer for tapping can be seen in many assembly operations, including hitting electrical parts to tighten them after insertion, and hitting doors or wheel covers to align them after assembly. Small or rubber hammers should be provided for these purposes.
- **Repetitive finger action and pinch grips**: Repetitive use of the index finger to operate triggers on many hand tools can cause fatigue in the finger, and a condition called trigger finger. Trigger finger causes jerky motion of the afflicted finger during extension. Sometimes the finger cannot be extended without help. This condition most often occurs through the use of hand tools that are so large that the operation of the trigger requires the flexion of the distal phalanx while the middle phalanx must be kept straight. However, it can also occur with repetitive use of small handles.
- **Repetitive hand–arm vibration**: Operation of hand-held power tools (e.g., power saws, drills, riveting hammers, sanders, grinders, pneumatic wrenches) can cause CTDs. Exposure to hand–arm vibration, especially in cold climates, can lead to vibration white finger. The tool vibration causes constriction of the blood vessels, resulting in reduced blood flow to the hand and fingers. The fingers and hand start to blanch, tingle and feel numb. (See Chapter 8 for details.)

10.5.2 BIOMECHANICAL PRINCIPLES FOR TOOL DESIGN

Tichauer (1978) recommends the following fundamental principles of biomechanics and ergonomics for prevention and solutions of problems associated with use of hand tools. They are based on technical, anatomical, kinesiological, anthropometric, physiological, and hygienic considerations. Many of these principles are also useful in the analysis of machine controls.

- The force required for operating a tool should be sufficient to provide proper sensory feedback to the musculoskeletal system, and particularly to the tactile surfaces of the hand. Therefore, the required force must be optimized. For example, in performing a certain task on a given work-piece, if the ratio of force output to force input is too large, the work-piece can be damaged and/or the worker injured. On the other hand, if this ratio is too small, the task must be repeated a large number of times, which makes the task fatiguing with a high risk of developing CTDs and low productivity.

- The tool should provide a precise and optimal amount of stress concentrated at a specific location on the work-piece. The tool should be so shaped that it will be automatically guided into the intended position where it will do its job best without either injuring the worker or damaging the work-piece. An example of such a design is the Phillips screwdriver as opposed to an ordinary, flat-bladed one. Tool blades (e.g. axes and scrapers) should be adequately sharpened so that the tasks are performed with the minimum number of repetitions, but not so sharp that the blades require frequent sharpening or become fragile.

- The tool should provide a contact surface area between its handle and the user's hand large enough to avoid concentration of a high compressive stress. Otherwise, pressure and impact acting on the hand can squeeze the vessels and/or nerves between the handle and the bones in the hand and wrist, preventing proper blood supply and damaging the nerves, which can cause numbness and tingling of the fingers.

10.5.3 ERGONOMICS PRINCIPLES FOR HAND-TOOL DESIGN AND EVALUATION

The following ergonomic guidelines for tool design, evaluation, and selection have been prepared based on published recommendations (Nemeth, 1985; Eastman Kodak Company, 1986; Grandjean, 1988; St John et al., 1993; Sanders and McCormick, 1993; Tayyari and Emanuel, 1993) and those developed by the authors:

- Avoid rigid, form-fitting handles with grooves for each finger. Such handles do not improve the grip strength and function unless they are sized to a particular user's hand. These types of handles, as one-size-fits-all, are usually designed to fit the average hand size (i.e., the 50th percentile). On one side, users with a small grip size (e.g., the 5th percentile) have to spread the hand too far apart to use the tool efficiently. On the other side, a person with a large grip size (e.g., the 95th percentile) will get uncomfortable ridges in the fingers.

- Avoid hand tools which require awkward movements, or cannot be operated effectively with neutral wrist posture and low force (Fig. 10.4a). Tool handles should be designed so that the user can maintain the hands in line with the forearms as much as possible (Fig. 10.4b).

- Avoid tool handles with sharp corners, edges, or pinch points. Tool handles should be either round or oval. All pinch points should be eliminated or effectively guarded.

- The tool-handle surface should be compressible, non-conductive, and

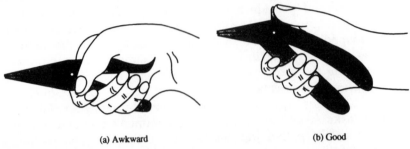

(a) Awkward　　　　(b) Good

Fig. 10.4 Comparison of (a) an inappropriate tool handle that requires an awkward wrist posture and creates high forces in the unprotected palm of the hand and (b) an improved tool handle that allows the wrist to be in a more neutral posture and distributes forces over a larger, more muscular area of the hand. Adapted from Tayyari and Emanuel (1993).

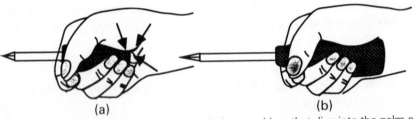

(a)　　　　(b)

Fig. 10.5 Comparison of (a) a short-handled screwdriver that digs into the palm of the hand with (b) a long-handled screwdriver. Adapted from Tayyari and Emanuel (1993).

smooth. However, handles should have enough coefficient of friction to minimize hand-gripping forces required for tool control.

- Avoid hand tools that impose concentrated pressure over the soft tissues of the hand which can impair circulation and the nerve function in the hand. Figure 10.5a shows a short-handle screwdriver digging into the palm of the hand. Long-handle screwdrivers are more comfortable to use (Fig. 10.5b).
- Tool handles should fit the hand. This means that handles should be of the proper thickness, shape, and length so that the stress-bearing area of the hand is as large as practical. The optimal grip with fingers, palm, and the hand is achieved by a span size (diameter) between 6.25 and 9.00 cm (2.5 and 3.5 in).
- Choose or design tools that can be used by either hand. Left-handed workers should not be forgotten in the design of the tool. Many power tools (e.g., chain saws, sanders and drills) are equipped with two handles, one of which is the primary handle (usually with a trigger and used by the dominant hand) and the other is a secondary handle, used as the stabilizing handle. The handles should be designed so that the stabilizing handle can be adjustable to either side of the tool to

accommodate both right-handed and left-handed individuals. Such a design will also permit the user to alternate the triggering hand from time to time to avoid excessive fatigue and to reduce the risk of chronic hand injuries.

- Choose or design tools that can be used effectively by both men and women.
- Choose or design special-purpose tools for specific tasks. However, care must be taken to avoid too many specialized tools that get lost or are not available to the worker, who will most likely try to improvise with an inadequate tool.
- Consider the angles of the grip, forearm, and tool to minimize wrist deviations.
- Consider provisions for tool safety.
- Avoid tools presenting excessive vibration.
- Avoid tool handles (e.g., putty knives, paint scrapers and chisels) that place concentrated pressures on the pressure-sensitive areas over-lying the blood vessels and nerves in the base of the palm of the hand (Fig. 10.6a). The use of this type of handle obstructs blood flow, leading to fatigue, numbness, tingling, and pain in the hand. The long-term effect is the development of chronic hand injuries (e.g., carpal tunnel syndrome). Such handles should be modified to transmit the force through a tougher area of the hand – the area between the thumb and index finger (Fig. 10.6b).

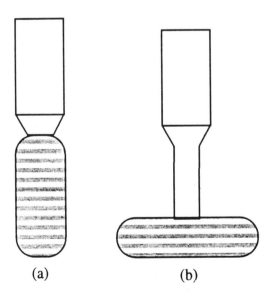

(a)   (b)

Fig. **10.6** Illustration of (a) a common scraper handle and (b) an improved handle. Adapted from St John *et al.* (1993).

- Choose ergonomically designed tools to maintain a more natural position of the wrist and to insure better distribution of grip forces during task performance. Examples are bent-handled tools, such as bent-handled pliers (designed by Western Electric), or Bennett's Bend handles (Emanuel *et al.*, 1980). The wrist should be maintained straight during the tool use. Tools with pistol-grip handles should be used in horizontal tasks, and tools with straight handles in vertical operations.
- Design and locate fixtures and jigs to permit tasks to be performed using the normal range of arm movements.
- Provide vibration dampening for vibrating hand tools.
- Use long-handled screwdrivers (Fig. 10.5b) because short handles (Fig. 10.5a) dig into the palm of the hand, which may cause nerve and blood vessel damage and/or irritation of the tendons.
- Avoid hand tools with fluted handle surfaces. They concentrate stress over a small surface in the hand.
- Fixtures should be used to mount work-pieces at angles which reduce the requirement of wrist deviations.
- Avoid static muscular loading to minimize fatigue. The arms and shoulders should be kept in a normal position, especially when the tool is relatively heavy or the tool is used for prolonged periods of time. Use counter-balancers to minimize static loads.
- Provide workers with tools and machine controls which can be operated most effectively, with neutral body part postures and low forces (Figs. 10.4b, 10.5b, 10.6b and 10.7).
- Use power grips for power, and use precision grips for precision.
- Substitute power tools for hand tools that require high force levels. However, make certain that the power tool is properly designed. It should have adequate grip size (6–9 cm circumference), not develop excessive vibration, be light enough to handle easily (or be counter-balanced), should not require excessive trigger action and should be easy to use. The power tool must be properly maintained to remain effective. Care must be exercised when using power tools in cold and hot environments.

Fig. 10.7 Special-purpose tools help reduce excessive force and prevent wrist deviation.

10.6 SAFETY GUIDELINES FOR TOOL USE

Since work-tools can be hazardous when improperly used, employees should be trained in the proper use of all tools, especially power tools. They should also learn to recognize the potential hazards associated with the different types of tools and the safety precautions for avoiding those hazards. Employees and employers must work together to establish safe working procedures, otherwise safety goals will not be achieved. Whenever employees, for example, encounter a hazardous situation, they should report it to the proper individual immediately.

10.6.1 BASIC SAFETY RULES FOR TOOL USE

All safety hazards associated with the use of work-tools can be controlled by the following five basic safety rules (OSHA, 1988):

- Keep all tools in good condition with regular maintenance.
- Use the right tool for the job.
- Examine each tool for damage before use.
- Operate tools according to the manufacturer's instructions.
- Provide and use the right protective equipment.

10.6.2 GENERAL SAFETY PRECAUTIONS FOR TOOL USE

The following are some general precautions that should be observed by work-tool users. For more detailed safety guidelines, the reader is referred to the publication OSHA 3080 (OSHA, 1988).

- Do not use a chisel as a screwdriver. Otherwise, the tip of the chisel may break and fly, hitting the user or another person.
- If a wooden handle on a tool such as a hammer or an ax is loose, splintered, or cracked, the head of the tool may fly off and strike the user or another person.
- A wrench must not be used if its jaws are sprung, because it can slip.
- Impact tools such as chisels, wedges, or drift pins are unsafe if they have mushroomed heads. The heads may shatter on impact, sending sharp fragments flying.
- Knives and scissors must be sharp. Dull tools can be more hazardous than sharp ones.
- Saw blades, knives, or other tools should be directed away from the aisle areas and other employees working in close proximity.
- Appropriate personal protective equipment (e.g., goggles, gloves) should be used to safeguard the tool users against potential hazards associated with using power tools and hand tools.

- Floors must be kept as clean and dry as possible to avoid accidental slips with or around dangerous work-tools.
- Sparks produced by iron or steel tools can be a dangerous ignition source around flammable substances. Where this hazard exists, spark-resistant tools made from brass, plastic, aluminium, or wood are safe to use.
- Never carry a tool by the cord or hose.
- Never yank or jerk a power-tool cord or hose to disconnect it from the receptacle.
- Keep cords and hoses away from heat, oil, and sharp edges.
- Disconnect tools from their power source when not in use, before servicing, and when changing accessories such as blades, bits, and cutters.
- All observers or visitors should be kept at a safe distance away from the work area.
- Secure the work-piece with clamps or a vise, freeing both hands for operating the tool.
- Prevent accidental starting. The tool user should not hold a finger on the switch button while carrying a plugged-in, portable power tool.
- Tools should be kept in good repair. They should be kept sharp and clean for the best performance. Follow instructions given in the user's manual for lubrication and changing accessories.
- Keep good footing and maintain good balance.
- Proper apparel should be worn. Loose clothing, ties, or jewelry can become caught in moving parts of the work-tool.
- All portable power tools that are damaged should be removed from use and tagged **do not use**.
- Guards, as necessary, must be provided to protect the operator and others from point of operation, in-running nip points, rotating parts, and flying chips and sparks. Safety guards must never be removed when a tool is being used.

10.7 APPLICATIONS AND DISCUSSION

Although numerous types of hand tools have been designed for specific applications, their variety is still limited and many are poorly designed. Due to the fact that each type of tool is frequently used by a large number of operators, for a variety of uses, work stress and CTDs have become occupationally epidemic problems. These problems are disabling many workers, impairing organizational productivity, and disturbing employee–management relations.

The ergonomist has several options with regard to hand-tool design and usage. First, the ergonomist should be involved in the evaluation and selection of hand tools, and in the training of employees in the proper use of those tools. There are usually multiple vendors of hand tools. If one vendor is not interested in providing an ergonomically designed tool, then other vendors should be visited. Recently ergonomics has found its way into advertising, so the ergonomist is cautioned to conduct an evaluation to make sure that an "ergonomic tool" actually lives up to its billing. The ergonomist should also look at modifications to existing tools to allow them to be used more effectively, safely, and efficiently. The solution may be as simple as putting a compressible, non-conducting, non-slip sleeve over a tool handle to reduce the required grip strength and vibration to the hand.

The stress induced by hand-tool usage may be vibratory, circulatory, or mechanical. The stress often propagates to various points within the body quite remote from the actual location of application of the force (Tichauer, 1978). For example, the use of vibrating hand tools held in the right hand can cause numbness and tingling sensation in the fingers of the left hand or pain in the neck muscles by the resonance of the transmitted vibration from the tool. Thus, whenever a specific region of the body becomes the location of frequent manifestations of symptoms of trauma, which are usually associated with the use of poorly designed hand tools, the situation should be carefully analyzed for the potential causes, and the problem should not be ignored due to the fact that that region of the body has not been directly exposed to the stress.

The ergonomics principles and recommendations, as well as safety precautions, provided in this chapter are applicable to alleviating problems caused by work-tools, regardless of their specific types or areas of applications.

REVIEW QUESTIONS

1. What are tool hazards due to?
2. Define grip strength and endurance.
3. With what spacing of a hand dynamometer can most males and females exert their maximum grip force?
4. Name six factors that are known to affect both grip strength and endurance, and which should be considered in hand-tool design. Briefly explain each factor.
5. What are the two categories of work-tools?
6. Based on what principles did John Bennett introduce tool handles with a 19° bend?
7. Why did John Bennett choose the 19 ± 5° angles for his bent-handle designs? Explain how a Bennett bent (BioCurve) handle should be held to serve its purpose effectively.
8. What is the purpose of designing hand tools in many different shapes?

9. Briefly describe how tool handles should be designed.
10. Describe the adverse effects of unidirectional wrist deviation during the use of work-tools.
11. Describe the ill-effect of the repetitive use of combined ulnar deviation and supination of the wrist during a work-tools usage. Give a situational example in which such a wrist motion is performed.
12. Describe the ill-effect of combined ulnar deviation and palmar flexion of the wrist in using work-tools. Give the classical example and an occupational example of such a wrist motion.
13. Give two occupational task examples in which dorsiflexion (extension) of the wrist is used. How can this awkward posture of the wrist be avoided?
14. What type of cumulative trauma disorders (CTDs) may be developed by performing a combination of dorsiflexion (extension) and pronation of the wrist over an extended time?
15. Describe the adverse effects of performing combined elbow flexion and inward rotation of the hand.
16. Explain what radial deviation can do to the tendons and nerves in the wrist, and what happens when this motion is combined with the forearm pronation. What type of CTDs may be developed by extensive radial deviation?
17. What are the ill-effects of frequent tapping with the heel of the hand (using the hand as a hammer)? Give examples of occupational tasks in which the hand is used as a hammer for tapping. Offer a solution for avoiding such situations.
18. Explain the type of CTD associated with the repetitive use of the index finger to operate triggers on hand tools. What are the symptoms of the CTD? In what situations would this condition most often occur, and how?
19. Explain the biological changes due to exposure to the hand–arm vibration which can finally lead to vibration white finger.
20. List the three fundamental principles of biomechanics and ergonomics recommended by Tichauer to reduce hand-tool problems.
21. What can happen if, in performing a certain task on a given work-piece, the ratio of force output to force input is too large? What can happen if this ratio is too small?
22. How sharp should tool blades be in order to be considered adequate?
23. What could happen to the hand and its tissues if the tool does not provide a contact surface area between its handle and the user's hand large enough to avoid concentration of a high compressive stress?
24. Explain the main problem associated with the "one-size-fits-all" type of tool handles.

25. How can you avoid sharp corners, edges, or pinch points on tool handles?
26. What type of material is preferred for tool handles?
27. What is meant by "tool handles should fit the hand"?
28. What is the recommended range of span sizes for hand tools?
29. Many power tools (e.g., chain saws, sanders and drills) are equipped with two handles, one of which is the primary handle (usually with a trigger and used by the dominant hand) and the other a secondary handle, used as the stabilizing handle. Why should the stabilizing handle be adjustable to either side of the tool?
30. What problems can result from providing too many specialized tools, or not providing them at all?
31. When should tools with pistol-grip handles be used? How about tools with straight handles?
32. List the Occupational Safety and Health Administration's (OSHA's) five basic rules for controlling all safety hazards associated with the use of work tools.
33. Why should a chisel not be used as a screwdriver?

REFERENCES

Bazar, A.R. (1978) Grip strength of cerebral palsied. *Human Factors* **20**: 741–744.

Eastman Kodak Company (1983) *Ergonomic Design for People at Work*, vol. 1. Lifetime Learning Publications, Belmont, CA.

Eastman Kodak Company (1986) *Ergonomic Design for People at Work*, vol. 2. Van Nostrand Reinhold, New York.

Emanuel, J.T., Mills, S.J. and Bennett, J.F. (1980) In search of a better handle. In: *Proceedings of the Symposium of Human Factors and Industrial Design in Consumer Products* (Polydar, H.R., ed.), Tufts University, Medford, MA, May 28–30, 1980, pp. 34–40.

Falkel, J.E., Sawka, M.N., Levine, L. and Pandolf, K.B. (1985) Upper to lower body muscular strength and endurance ratios for women and men. *Ergonomics* **28**: 1661–1670.

Fraser, T. (1980) *Ergonomics Principles in the Design of Hand Tools*. Occupational Safety and Health Series no. 44. International Labour Office, Geneva.

Fraser, T.M. (1983) Hand tools, ergonomic design of. In: *Encyclopaedia of Occupational Health and Safety, Volume 1*, 3rd edn. (L. Parmeggiani, ed.), International Labour Organisation, Geneva.

Freivalds, A. (1987) The ergonomics of tools. In: *International Reviews of Ergonomics: Current Trends in Human Factors Research and Practice* (D.J. Osborne, ed.), vol. 1. Taylor & Francis, London, pp. 43–75.

Grandjean, E. (1988) *Fitting the Task to the Man: A Textbook of Occupational Ergonomics*, 4th edn. Taylor & Francis, London.

Nemeth, S.E. (1985) Handtool design. In: *Industrial Ergonomics: A Practitioner's Guide*. (Alexander, D.C. and Pulat, B.M., eds). Industrial Engineering and Management Press, Institute of Industrial Engineers, Norcross, GA.

OSHA (1988) *Hand and Power Tools*. OSHA 3080 (revised). Occupational Safety and Health Administration (OSHA), US Department of Labor, Room N3101, 200 Constitution Ave, NW, Washington, DC 20210.

Rodahl, K. (1989) *The Physiology of Work*. Taylor & Francis, London.

St John, D., Tayyari, F. and Emanuel, J.T. (1993) Implementation of an ergonomics progam: a case report. *International Journal of Industrial Ergonomics* **11**: 249–256.

Sanders, M.S. and McCormick, E.J. (1993) *Human Factors in Engineering and Design*. McGraw-Hill, New York.

Schmidt, R.T. and Toews, J.V. (1970) Grip strength as measured by the Jmar dynamometer. *Archives of Physical Medicine and Rehabilitation* **51**: 321–327.

Swanson, A.B., Matev, I.B. and de Groot, G. (1970) The strength of the hand. *Bulletin of Prosthetics Research* 145–153.

Tayyari, F. and Emanuel, J.T. (1993) Carpal tunnel syndrome: an ergonomics approach to its prevention. *International Journal of Industrial Ergonomics* **11**: 173–179.

Thompson, D. (1990) Ergonomics & the prevention of occupational injuries. In: *Occupational Medicine* (LaDou, J., ed.). Appleton & Lange, Norwalk, Connecticut, pp. 38–57.

Tichauer, E.R. (1975) *Occupational Biomechanics*. Rehabilitation monograph no. 51. New York University, Center for Safety, New York.

Tichauer, E.R. (1978) *The Biomechanical Basis for Ergonomics*. John Wiley, New York.

UAW (1982) *Strains and Sprain: A Worker's Guide to Job Design*. International Union, United Auto Workers (UAW), Detroit, Michigan.

Human–machine systems

11

11.1 INTRODUCTION

Ignoring the human operator's needs, behavior, and preferences results in human errors, which are usually due to poor design. The complexity of a human–machine interface creates confusion, errors, accidents, and loss of resources.

In a human–machine system the human operator interacts with the machine. This interaction is a closed-loop system in which humans play the key role since they make the decisions. In this loop, the human operator first, perceives the status of the machine through **displays**; second, interprets and mentally processes the perceived information; third, makes a decision and, then, fourth, conveys the decision to the machine through manually operated **controls**. Figure 11.1 shows the pathways and direction of information in a human–machine system.

Poor ergonomic designs lead to waste of time and money, and often tragedies. For example, errors in operating a complex machine result in loss of production, and maintenance time wasted in just getting to the parts that need to be replaced. Therefore, for a human–machine system to be efficiently operated, the designers must consider ergonomics principles and concepts during all stages of their system development, not as an add-on "quick fix" to a failing system.

Displays and controls are the principal communication devices interfacing the operator with machines. The role of ergonomics in this interface is to facilitate the effective use of these devices by the human operator and to minimize errors in their use.

There are two modes of receiving information from the environment (Sanders and McCormick, 1993) – **direct** and **indirect** sensing. Direct sensing applies to situations in which the information is sensed directly from the environment through the actual sensory mechanisms (e.g.,

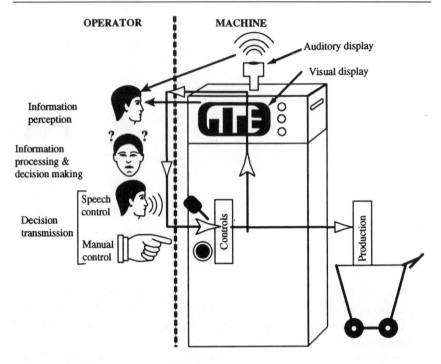

Fig. 11.1 Illustration of the human–machine interface.

direct observation of a container being filled, or hearing noise from a malfunctioning machine, or sensing temperature by touching an object).

Often it is impossible to use the sensory mechanisms to obtain the desired information. For example, it is not possible to use tactile sense to measure or monitor the temperature of a chemical due to the safety hazard. Even in the case of harmless materials, quantification of temperature using the human tactile sensor is impractical. Hence, in such cases the information must be obtained by indirectly sensing (e.g., reading a gauge, or hearing audible signals). For example, thermal sensing probes and transducing devices are used to monitor the exact temperature of various substances. Ergonomists are mainly concerned with this mode of receiving information from the environment.

11.2 THE WORKPLACE CHARACTERISTICS

Most workplaces have the following five characteristics in common:

- The workplace consists of various mechanical and electrical equipment. The industrial workplace usually includes such equipment as lathes, power saws, power presses, grinding machines, and drills. The non-industrial (service) workplace also includes mechanical and electrical equipment. Examples of equipment used in the non-industrial environment include typewriters, computers, and copy machines.

- Manual tools are often required to perform the job. Examples of such tools include screwdrivers, hammers, dollies, pliers, and keyboards.
- Jobs are described by standard sets of procedures that must be followed to produce a desired output.
- The workers interact with the equipment to perform many necessary tasks that are involved in producing the desired output.
- The workplace is serviced by various other facilities. Examples of such facilities are waste removal, water, and power, which are necessary to carry out the work activities in the workplace.

11.3 THE WORKER'S FUNCTIONS

Workers generally perform the following three types of functions (each of which utilizes the sensory processes, information processing, decision making, and control):

- Performing various physical tasks that are required to complete the job. Examples of such tasks are lifting, carrying and lowering materials, positioning the work-pieces for work operations, using tools to perform the job function, and packaging products.
- Controlling the processes that are performed by machines. In this conjunction, the worker sets the machines to work. The worker monitors and controls the work processes so that the job is completed properly. Finally, the worker stops the machine at the completion of the process or whenever something in the system goes wrong. To control the process, it is necessary for the worker to monitor the work being performed.
- Sharing the work environment with the machine in the workplace. Consequently, the worker is exposed to the potential hazards due to the use of mechanical and electrical equipment in performing the job.

11.4 ERGONOMICS OF HUMAN–MACHINE SYSTEMS

Ergonomics is concerned with the following aspects of the human–machine systems:

- the design of tools to match the physical characteristics of the worker with the functioning of the tools;
- the design of the workplace and work space to meet the physical characteristics of the worker;

the design of controls and displays to allow the worker to operate and monitor the work system processes efficiently with minimum errors;
- the development of job procedures that meet the worker's capabilities;
- the minimization of the effects of external factors (i.e., thermal conditions, illumination, noise, and vibration) on the worker in the workplace.

11.5 HUMAN–MACHINE SYSTEM ANALYSIS AND DESIGN

Since ergonomists are concerned with the human–machine interfaces in the work environment, they need to use a logical procedure to analyze the interfaces. A logical method for analysis of the human–machine interface in the work environment is the systems approach. This involves the analysis of the total system by dividing it into manageable subsystems, so that they can be individually analyzed for the factors of concern. The system is divided into its major components, that is, input, process, and output of the system. This method of the systems approach is called a **black box** method (Fig. 11.2). In the workplace, the raw materials, data, and instructions are the inputs that are necessary to perform the job. The process refers to all necessary steps taken to convert the inputs into the desired output. The process may involve activities such as fabrication and heat treating of metal, cutting, and painting, or just the information processing. The process results in the desired output.

11.5.1 STEPS IN SYSTEM DESIGNS

A system design involves the following seven steps:

1. **Objectives determination**: The desired output (i.e., the objectives of the system) must be defined first. What the system should accomplish is defined in this step. The characteristics of the output and tolerances or error level should also be defined in this initial step.
2. **Input specifications**: The required input to obtain the output must be

Fig. 11.2 The system black box.

defined and specified. The required materials, work skills, and resource constraints are specified in this step.

3. **Process description**: The process must be described to show how the input can be converted into the desired output. The required functions and tasks as well as the constraints must be defined.

4. **Functional allocation**: All specific functions or tasks to be performed to achieve the objectives must be identified. Functional flow diagrams are useful tools in the conceptualization of the interrelationships among system functions. The identified functions are allocated to the human operators or machines in this step. Human capabilities and limitations determine the functional allocation. If performance requirements of the function exceed human capabilities, the function should be allocated to machines.

5. **Interface design**: The interfaces between the human operator and other system components (e.g., machines, tools, workstations, and environmental factors) are so designed to optimize human performance and minimize risks of human errors and injuries. The following are of special consideration in this step (Mahone, 1993):

   • Workstations are so designed that the worker assumes natural postures during job performance.
   • Controls are designed based on worker capabilities and limitations to avoid contact stresses with the worker's body.
   • Displays should be user-friendly for the worker to understand them easily, so that the risk of errors is minimized.

6. **Facilitator design**: Facilitators include all materials and procedures that enhance operator performance, such as mechanical assists, instruction manuals, and training programs. Such materials should be carefully designed for the intended users.

7. **Testing and evaluation of the system**: The final step in a system design is to review and evaluate the system's performance. This means a complete review of the performance of the system's components, including individuals, workplace layout, equipment, and procedures. The system testing and evaluation either confirm that the system will function as intended or trace any detected problems back to where revisions are to be made (Mahone, 1993).

11.5.2 JOB ANALYSIS

Job analysis is used to describe the process. As the beginning point, the job is divided into five general stages (Byers *et al.*, 1978). These stages are:

1. **Preparation**: What steps are necessary to prepare the worker for performing the job?

2. **Observation**: What information and data must the worker have to perform the job?
3. **Control**: What steps (i.e., decisions and mental processes) must the worker take to control the job processes?
4. **Physical demands**: What physical tasks must the worker do to perform the job?
5. **Termination**: What steps must the worker take to terminate the job?

Stages 2 and 3 make up a feedback loop, in which the worker receives the information concerning the process and decides what control must be applied to the process to obtain the desired output. Once the control is applied, the process changes accordingly, then the information concerning the changes in the processes is fed back to workers to enable them to evaluate the result and take the next control action. This feedback (action–result) loop is graphically illustrated in Figure 11.3.

In job analysis, each job is divided into functions, such as shipping and receiving, production, and quality control. Each function can then be subdivided into its individual tasks or activities, such as manual materials handling, and hammering. Each task is, in turn, broken down into its elements. For example, lifting, carrying, and lowering are elements of manual materials handling. The elements of a packaging task for a given item may be as follows:

- Get 10 empty cartons and sheets of wrapping paper.
- Position a sheet of wrapping paper.
- Position an item on the wrapping paper.
- Wrap the item.
- Place the wrapped item in the carton.
- Close the carton and tape it.
- Label the carton.
- Set the packed carton aside.

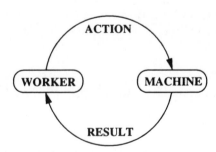

Fig. 11.3 Action–result feedback loop in the human–machine interface.

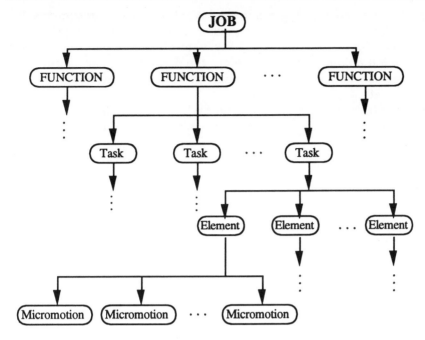

Fig. 11.4 Job breakdown for job analysis.

Finally, each element can also be analyzed in terms of micromotions that are involved in performing the element. Such an analysis follows the work done by Frank Gilbreath in motion study. Job analysis is illustrated in Figure 11.4.

How far a job should be divided depends upon the objective of the job analysis. If the objective is to determine the worker's duties, the task level seems to be sufficient. If the objective is to remove potential health hazards or to improve the worker's efficiency, it may be necessary to break the job down to the level of micromotions. Whatever the objective is, the job should be divided into small segments of manageable size such that the segments can be analyzed. It should be remembered that the initial five stages (preparation, observation, control, physical demands, and termination) may be found very helpful in identifying the groupings of elements within the task itself since it is likely that most of the stages will be present in the task.

Once the job breakdown is established, the analysis can begin. The analysis involves a number of questions about each task, element, or micromotion that is being analyzed. The following are some typical questions (Byers *et al.*, 1978):

- What are the initiating conditions? What is the cause or stimulus that results in the requirement of a job analysis in the first place?
- What actions must be taken? What steps must be performed by the worker and/or machine to achieve the desired result or output?
- What feedback is required to insure that the results of the action are as desired?
- What potential errors are possible? What is the cost of these errors in terms of damage to equipment, materials, and/or the worker's well-being?
- What potential hazards are present that can cause an injury or illness to the worker?
- What is the required reaction time that is necessary to initiate the task?
- What is the time frame in which the task must be completed?
- What tools and equipment are required to complete the task or element?
- Where is the task or element being performed? What is the physical location and structure of the workplace in which the task or element is performed?
- What physical demands are placed upon the worker to perform the task or element?
- What skills and knowledge are required of the worker to perform the task or element?

The objective of this analysis is to define each job task or element in order to determine how the task or element can be changed to meet the desired result while, at the same time, lowering the potential stress on the worker who performs the task. After an existing job is analyzed, it is desirable to answer certain questions that can help achieve the analysis objectives. The following are examples of such questions:

- Is it necessary for the worker to perform the job as it is currently done?
- Can equipment be used to remove some of the physical demands on the worker?
- Can the worker be seated rather than standing to perform the task?
- Does the worker have to perform the task or element in close proximity to the mechanical process?
- Can equipment be used to replace the worker totally?
- Can the entire process be automated to remove the worker from the workplace? (Such action will eliminate the current strain on the worker.)
- Should the equipment being used be redesigned?
- Are there any potential hazards associated with the equipment? If yes, can these potential hazards be designed out of the system by equipment modification or replacement?
- Can or should the workplace be modified?

- Can modifications be made to the work space, controls, displays, tools, and/or equipment being used to allow for more efficient operation?
- Can the task or elemental procedures be changed to reduce the potential for critical errors occurring? (Note that reducing critical errors will result in a more efficient and less costly operation, while providing a safer workplace for the worker.)

Using job analysis, the job, workplace, tools, and equipment can be modified to fit the worker's characteristics. Job analysis can also result in improved productivity, lower costs, improved morale, lower employee turnover, and improved labor utilization.

11.6 CONTROLS AND DISPLAYS

A proper design of the human–machine interface is of great importance for enhancing the operator's productivity, safety, and well-being. Errors are the major problems associated with displays and controls in human–machine systems. They are caused by poorly designed displays and controls and inappropriate layouts of the instruments.

11.6.1 DISPLAYS

A variety of displays are available to interact with the human sensory systems. The most common are visual displays and auditory displays. Visual displays can be either analog or digital. Analog displays include dials, graphs, and status indicators. Digital displays include mechanical counters and electronic alphanumeric displays. Other sensory displays (tactile, olfactory, and taste) are not common in industry. However, sensory input in the form of heat or burning odors often alerts workers to conditions needing further exploration. There are three common types of visual display instruments found in industry (Grether and Baker, 1972; Alexander, 1986; Grandjean, 1988; Tayyari, 1993):

- A window with digital read-out (counters) in which figures can be read directly (Fig. 11.5a).
- A (fixed) circular scale with moving pointer (Fig. 11.5b).
- A fixed pointer over a moving scale (Fig. 11.5c).

Each of these visual displays has its peculiar advantages over the others under certain conditions. For example, a window digital read-out (counter) is best for reading quantitative values. The reading can be made quickly without the requirement of making any approximation. However, when it is necessary to monitor how a process changes, a display with a moving pointer over a fixed scale may serve the purpose better than the other types. This is especially true when the process

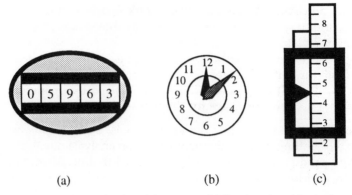

Fig. 11.5 Three types of display: (a) counter; (b) fixed scale with moving pointer; (c) fixed pointer with moving scale.

changes slowly and has to be checked periodically. A fixed pointer with moving scale may also be acceptable for such cases (Grandjean, 1988).

In general, a display with a moving pointer over a fixed scale is superior to the moving scale with a fixed pointer (Sanders and McCormick, 1993). The reason is that the moving pointer is more effective in attracting the operator's attention and is quick to read off. In contrast, if the display has to cover a wide range of values, the scale becomes too large and difficult to read and it is then better to see only a part of the scale moving against a fixed marker (Grandjean, 1988).

11.6.2 CONTROLS

Controls include valves, toggle switches, push-buttons, pedals, keyboards, levers, rotary knobs, thumb-wheels, hand-wheels, and cranks. Selecting a suitable control from a large number of types available is critical for the effectiveness of human–machine performance. The control should be selected based on its suitability for the particular task. A control that is suitable for a given task may be inappropriate for another. On the basis of the action required, controls are of the following two types:

- **Discrete-action controls**: This type of control can be set to any exact position from a limited number of positions. They are suitable for turning equipment on or off, changing modes of operations, and selecting meter scales. This type of control includes toggle switches, push-buttons, thumbwheels, legend switches, and rotary selector switches.
- **Continuous-action controls**: A continuous-action control can be set at any position within the range of its movement. This type of control is

used for opening or closing valves, adjusting displays, and varying potentiometers. Examples of these controls are thumb-wheels, hand-wheels, hand cranks, levers, knobs, joysticks, and pedals.

Based on the amount of muscular effort required for their use, manual controls are classified into the following two categories:

- **Nimble (light) controls**, which require little muscular effort. For example, finger push-buttons, rotary selector switches, joysticks, toggle switches, legend switches, rocker switches, rotating knobs which can be operated with fingers.
- **Heavy controls**, which require muscular effort. For example, the operation of hand-wheels, cranks, heavy levers, pedals and foot push-buttons require the use of the large muscle groups of arms and legs.

11.7 GUIDELINES FOR DESIGNING DISPLAYS AND CONTROLS

Displays and controls should be designed based on expectation in order for human–machine interfaces to be efficient. Stereotyping and standardization minimize human errors when interfaced with machines. A **stereotype** is what people typically expect to occur when an action is taken. For example, moving a lever backward is supposed to slow the system down, while moving it forward would make the system work faster. A **standard**, on the other hand, is used under circumstances where there is no clear stereotype. For example, a red button for stop, computer keyboards and typewriter key layouts, and telephone buttons have all been standardized.

Since controls and displays are usually interrelated, they should be designed to be compatible. They should be properly grouped together so that the relationship between each pair of controls and displays can be easily understood by the operator. The following can be used as general guidelines for design of controls and displays, their relationship, and their layouts (Grether and Baker, 1972; Kvålseth, 1985; Alexander, 1986; Grandjean, 1988).

11.7.1 GUIDELINES FOR DESIGNING DISPLAYS

The following principles should be used as general guidelines for designing or choosing displays:

- In general, a display should present the minimum amount of information necessary for the worker to make a decision based on the information displayed.

- Use digital read-out (counters) displays when quick and accurate readings are to be made, provided that the displayed data do not change too quickly to be missed. The fewer the number of digits displayed, the faster and more accurate the information will be presented to the user.
- Use circular or semicircular scales instead of vertical or horizontal scale displays where possible and if space permits.
- Increment scale numbering in a clockwise direction on a circular/semicircular scale, upward on a vertical, and to the right on a horizontal scale.
- Use a display with a moving pointer on a fixed, circular scale if it is necessary to monitor how a process is changing, or to note the amplitude of change. This type of display is the best for relative information.
- Avoid the use of displays with moving scales and fixed pointers since it is not easy to memorize the previous reading, nor to assess the extent of movement. Moving scale displays are poor in presenting qualitative data.
- The major scale markers should be numbered in a natural and easily recognizable fashion (e.g., (1, 2, 3, . . .), (5, 10, 15, . . .), (10, 20, 30, . . .), (100, 200, 300, . . .), etc. rather than (2, 4, 6, . . .) or (3, 6, 9, . . .)).
- Avoid multiscale, multiple-pointer, and non-linear-scale displays, whenever possible.
- Install displays in the direct line of sight of the operator and avoid any obstructions.
- Avoid glare on the displays.
- Avoid small-size or illegible displays.
- Use capital letters for single-word or short identification sentences since they are readable at a greater distance than lower-case letters.
- The display must be readable from the operator's normal location.
- The size of letters and figures, thickness of lines and their distance apart must be determined based on the viewing distance between the eye and the display. The height of the letter or number is determined based on the recommended values in Table 11.1. The stroke width (thickness) of a letter or number should be one-sixth of its height. The width of all letters and numbers should be three-fifths of the height, except for I, i and 1, which should be one-stroke-width wide, M, and W, which should be about one-fifth (20%) wider than the other letters, and 4, which should be one-stroke-width wider than the other numbers. These proportional values are illustrated in Figure 11.6.
- The heights and widths of major, intermediate, and minor graduation marks should not be less than the values given in Figure 11.7. A reading distance of 71 cm (28 in) is assumed for the dimensions

Table 11.1 Recommended minimum letter and numeral heights

Nature of markings	Low lighting	Adequate lighting
Critical markings, position variable (digital read-outs or moving scales)	0.20 in (5 mm)	0.12 in (3 mm)
Critical markings, position fixed (emergency instructions, numbers on fixed scales, and control and switch markings)	0.15 in (3.75 mm)	0.10 in (2.5 mm)
Non-critical markings (identification labels, instructions, non-emergency labels)	0.05 in (1.25 mm)	0.05 in (1.25 mm)

Notes

● The values are given for a 28-in (71-cm) viewing distance. The values will increase or decrease in proportion to the increase or decrease in the viewing distance.
● Luminance (brightness) up to 1 fL or 3.4 cd.m^{-2} is considered as low lighting.
● Adapted from Grether and Baker (1972).

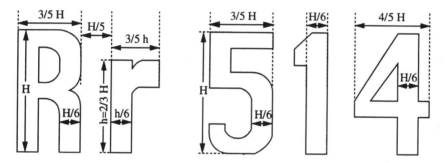

Fig. 11.6 Recommended proportions for letters and numerals. H = Height of capital letters or numerals; h = height of lower-case letters. Adapted from Grether and Baker (1972).

recommended in Figure 11.7; for other reading distances, the recommended minimum dimensions should be proportionally increased or decreased.

● Under normal illumination, graduation marks may be spaced as close as 0.9 mm (0.035 in), but the space should never be less than twice the stroke-width for white marks on black dial faces or less than one-stroke-width for black marks on white dial faces. As shown in Figure 11.7, this space for low illumination should be not less than 1.8 mm (0.07 in).

● Label displays clearly to prevent confusion. Labels are helpful during training and for use by new employees.

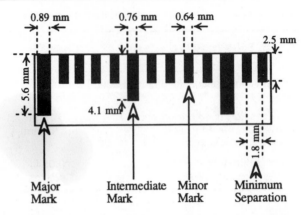

Fig. 11.7 Recommended minimum scale dimensions for low illumination and reading distance of 71 cm (28 in). Adapted from Grether and Baker (1972).

- Signs and labels should contrast with their background and be of proper size and color.
- Icons and symbolic signs are very quick (faster than text and/or color codes) in conveying information to the human operator. Since they do not use a written message, they require minimum time for perception and do not have to be translated into other languages for use by people from different nations. They are commonly used for warning messages, such as traffic signs.
- Provide adequate illumination and prevent shadows on the display surfaces.

11.7.2 GUIDELINES FOR DESIGNING WARNING LIGHTS

A special class of displays is warning lights. Warning lights are used to attract attention to actual or potentially dangerous situations. In spite of a lack of extensive studies of ergonomics of this type of signals, the following general principles have been developed to enhance their effectiveness.

- **Size**: Warning lights should subtend at least 1° of visual angle (Sanders and McCormick, 1993).
- **Color**: Colors in the order of eliciting fast human responses are first red; second, green; third, yellow; and fourth, white (Reynolds *et al.*, 1972). Since red normally indicates danger to most people, warning lights should generally be red.
- **Flash rate**: A flashing light is more effective than a steady light in attracting someone's attention. An optimum flash rate is 4 times per

second (Heglin, 1973). It can, however, be in the range of $3-10 \text{ s}^{-1}$ (Sanders and McCormick, 1993).

- **Contrast with background**: Proper brightness and color contrast enhance the effectiveness of a warning light. A warning light should be bright enough to get attention. Too much brightness will cause glare, which must be prevented. The light should be at least twice as bright as its immediate background (Sanders and McCormick, 1993). A warning light should also contrast with the background lighting. If there is a great deal of red in the background, then a red warning light would not be as effective as a blue (or other contrasting color) light.
- **Multimedia presentation**: An auditory signal can be used for getting extra attention or as a back-up system for the visual warning signal. Auditory signals are also effective for operators who move around the workplace and may be out of the visual field of a warning light.

11.7.3 GUIDELINES FOR DESIGNING AND SELECTING CONTROLS

The following recommendations may be considered as general guidelines for the design or selection of controls.

- Controls should be designed based on the anatomy and functional characteristics of the body limbs (e.g., fingers and hand should be used for quick, precise movements, arms and feet for operations requiring the application of force).
- The least capable operator should be able to operate the control. The force, speed, accuracy, or range of movement required by the control should not exceed the capabilities of the least capable operator.
- Use push-buttons, tumbler switches, and rotating knobs for operations requiring little movement or muscular effort, small travel and high precision, and for either continuous or stepped operation (click-stops). These controls can be easily operated with fingers.
- Use long-armed levers, cranks, hand-wheels, and pedals for operations requiring muscular effort over a long distance, and requiring relatively little precision. These controls involve the major muscle groups of arms and legs.
- The number of controls should be minimized, and their movements should be simple, natural, and easy to perform. Natural movements are more efficient and less fatiguing to operate than those that seem awkward or complex.
- Control movements should be as short as possible, consistent with the requirements of accuracy and feel.
- Controls should show a positive sign of activation so that malfunction can be immediately apparent to the operator.
- Control actions should result in a positive indication to the operator that there has been a system response.

- Controls should be resistant to abuse and tamper.
- Control surfaces should be designed to prevent the accidental activation of the control by a slipping finger, hand, or foot.
- Controls should have sufficient resistance so that their inadvertent activation by the weight of a hand or foot is prevented. However, the resistance should be minimized for controls operated continuously, or frequently.

11.7.4 GUIDELINES FOR INSTALLATION AND LAYOUT OF CONTROLS

The following recommendations may be used as general guidelines for installation and positioning of controls.

- Install controls near the operator, so that they are usable in the normal sitting or standing position.
- Install hand-operated controls so that they can easily be reached and grasped; that is, they should be located between the elbow and shoulder heights of the user.
- Install heavy levers, that require the use of a large muscle group of the arm, at slightly below shoulder height for a standing operator, but at elbow height for a seated operator.
- Install controls at either side of the center of the operator's position rather than mounting them in the center.
- Install cranks that require a high level of torque so that the turning handle is parallel to the frontal plane of the body.
- Install controls in full view of the user.
- The distance between neighboring controls must be sufficient to avoid accidental activation.
- Install foot controls in such a way that accidental activation, which may lead to potential hazards and undesirable consequences, can be avoided. Foot controls should be avoided for standing activities.
- Consider the human anatomy in specifying distance between controls, for example, finger-operated knobs or switches should be at least 1.5 cm (0.6 in) apart; and whole-hand operated controls should be at least 5 cm (2 in) apart.
- Use appropriate methods of control coding to enable the operator to identify each control correctly and quickly. Examples of common methods of coding are color, shape, size, texture, labeling and location.

11.7.5 GUIDELINES FOR CONTROL–DISPLAY COMPATIBILITY

Controls and displays affected by them should be compatible. There are two obvious problems associated with incompatible controls and displays. First, it is not certain that, even after long practice, the operators will achieve the same level of performance efficiency as they would with compatible controls and displays. Second, even if operators can establish new habits and become efficient in using incompatible controls and displays under normal conditions, they may not be able to have the same efficiency and may revert to their old expectations under stressful situations. There are certain natural or expected relationships between controls and displays that should be considered so that movement errors can be minimized. The following recommendations should be used as guidelines or a checklist for control–display compatibility:

- The direction of movement of controls should be compatible with the direction of movement of displays affected by them, as naturally expected by the users. The expected relationships between control and display movements are illustrated in Figure 11.8. When a control is moved or turned to the right, upwards, forwards, or clockwise, the corresponding display should show an increase, and its pointer must accordingly move to the right on a horizontal scale, clockwise over a circular scale, or upwards on a vertical scale. A right-handed or clockwise rotation implies an increase, so the display instrument should also record an increase.
- The moving scale with a fixed indicator should move to the left (counterclockwise) when the control is moved to the right (clockwise), while the scale values should increase from left to right (clockwise), so

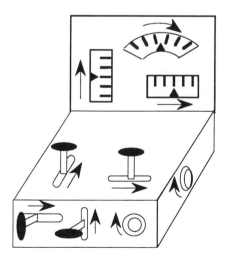

Fig. 11.8 Expected relationships between control and display movements to indicate increases.

that the rotation of the scale to the left (counterclockwise) shows increased readings.

• When a hand lever (control) is moved upwards, forwards, or to the right, the display reading should increase or a switch should move to the **on** position. To reduce the reading, or to switch to **off** position, the lever is moved downwards, or towards the body, or to the left.

• Control movements and location should be parallel to the axis of its corresponding display motion.

• Controls should be installed near the displays which they affect.

• The control should be installed either below the display it affects or, if necessary, to the right of it.

• The type, location, and orientation of the control should be compatible with the system response through displays. Examples of acceptable controls for various types of system responses are illustrated in Figure 11.9.

• Identification labels should be placed above the controls and identical labels above the corresponding display instruments.

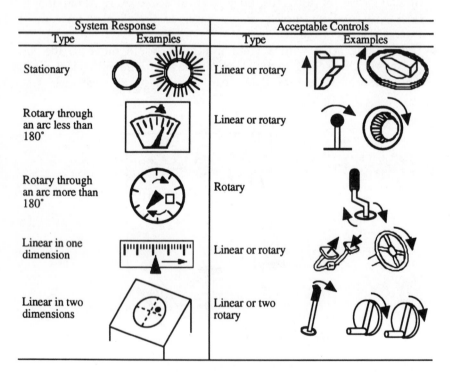

Fig. 11.9 Examples of acceptable controls for various types of system responses. Adapted from Chapanis and Kinkade (1972).

- If the controls have to be in one panel and the displays in another, the two sets should be laid out in the same order and arrangement.
- If a series of controls are normally operated in sequence, they and their corresponding display instruments should be arranged on the panel in the same order, from left to right.
- When arranging a group of controls and displays, the controls and displays that are used more often should be located directly in front of the operator, and others to the sides.

11.7.6 CONTROL–RESPONSE RATIO

The movement of the control will result in the machine response, which may be indicated by the movement of the corresponding display. Control movements should be proportional to the movements of the affected displays. This proportionality is expressed in a control–response (C/R) ratio, which is the ratio of movement of a control relative to that of the moving element (pointer, cursor, etc.) of the display. It may also be called the control–display (C/D) ratio. It applies to continuous controls only, not discrete controls. Movement of the control or display may be measured in linear distance (e.g., in lever controls with horizontal or vertical displays), or in angles or number of revolutions (e.g., in ball levers or knobs and circular displays). The lower the C/R ratio, the more **sensitive** is the control. A very sensitive control causes a major movement of the affected display with a slight movement; it, therefore, has a low C/R ratio. The reciprocal of the C/R ratio (i.e., the response–control, R/C) is called **gain**.

For **linear** controls (e.g., levers) that affect linear displays, the C/R ratio is defined as the ratio of the control's linear displacement (c) to the resulting display's linear displacement (d); that is:

$$C/R = c/d \qquad\qquad (11.1)$$

where c = the control's linear displacement and d = the display linear displacement.

For controls with significant **rotational** movement (e.g., ball-lever controls), that affect linear displays (Fig. 11.10), the C/R ratio should be determined by the following equation (Chapanis and Kinkade, 1972):

$$C/R = \left[\frac{\alpha}{360} \times 2\pi \cdot L\right] \div d \qquad\qquad (11.2)$$

where: α = the angular movement of the control (in degrees);
L = the length of the control lever arm;
d = the display movement (in the same unit as L).

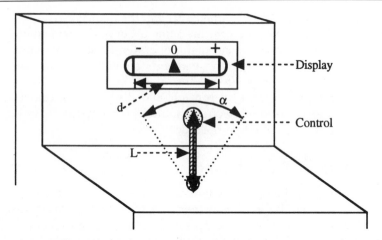

Fig. 11.10 An illustration of control–response (C/R) ratio for a ball-lever control affecting a linear display. Adapted from Chapanis and Kinkade (1972).

For rotary knobs, the C/R ratio should be the reciprocal of the display movement (*d*) for one complete revolution of the knob; that is:

$$C/R = 1/d \qquad (11.3)$$

where *d* is the linear movement (in inches) of the display element in one complete revolution of the knob. This is independent of knob diameter and hence independent of displacement around the circumference of the knob.

An appropriate (C/R) ratio is a critical factor in control design since the performance of the operator can be affected by it. This effect is a function of the following two types of human motions:

- A slewing (or sluing) movement: a slewing time (or travel time) is the time for gross adjustment of the control in which the operator rapidly moves the control close to the desired position. An increase in the C/R ratio will increase the slewing time because a longer control movement is required.
- A fine-adjustment movement: fine-adjustment movements follow the gross-adjustment movement by which the operator brings the controlled element (display) precisely to the final desired position. The fine-adjusting time is reduced by increasing the C/R ratio, which eases the adjustment.

Therefore, the optimum C/R ratio minimizes the total time (slewing plus fine adjusting) that is required to set the control at the final desired position (Figure 11.11). For ball-lever controls the optimum C/R ratio

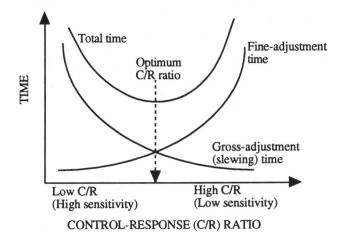

Fig. 11.11 An illustration of the control–response ratio and the control adjustment time.

ranges from 2.5 : 1 to 4 : 1, and for knobs it ranges from 0.2 to 0.8 (Chapanis and Kinkade, 1972). However, due to the many complexities involved, optimum C/R ratios should be determined experimentally where time and precision are critical.

11.8 APPLICATIONS AND DISCUSSION

Despite all the emphasis on the need for ergonomic designs, many instruments and products are still poorly designed, which results in undesirable consequences in human–machine interfaces. Designers and manufacturers seem to focus on the type of technology advancement that drastically increases productivity, as defined by the speed and volume of production. Therefore, they concentrate on the technology, and ignore the potential user needs, behavior, and preferences. For example, in order for electronic instruments to be cost-efficient, more functions are being integrated into them (Wiley, 1991). Such instruments are often difficult to understand and use without errors.

Another major problem with human–machine interfaces is the incompatibility between controls and displays. This causes operator errors or human errors. Most such errors are due to the poor ergonomic design of human–machine interfaces, and the human operator should not be blamed for the consequences of the poor design. The design or selection of appropriate controls and dislays and their compatibility and layout are not simple. Often designers use themselves as the human in the human–machine interface being designed. Designers consciously or unconsciously assume that if they can use the product, then everyone can. However, everyone might not be a right-handed male with a 90th-percentile upper-body strength, 20–20 uncorrected eyesight, and so familiar with the design that they see little necessity for written instructions. What is obvious to the designer who has spent a great deal of time with a system may be perplexing to a potential user.

The ergonomist should be a part of the design or evaluation team that assesses a product to determine its applicability in the workplace. The users of the product should also be consulted early in the design stages to incorporate any suggestions that they might have regarding the product or its use. Almost everyone can relate horror stories about products that have violated the ergonomics principles of human–machine interface design. Such stories can range from automotive replacement parts that are nearly impossible to replace, to water faucets that turn the wrong way, to poorly designed displays and controls on a variety of products.

A wide variety of controls and displays are commercially available, but selection of such instruments must be made based on the particular task requirements and the operator's physical and mental capabilities and limitations. The ergonomics principles and guidelines presented in this chapter can be applied to the design and selections of controls and displays, and their layout, to achieve these objectives to establish effective human–machine interfaces.

REVIEW QUESTIONS

1. What are the consequences of the complexity of a human–machine interface?
2. In a human–machine system, the human operator interacts with the machine. Explain why this sytem is called a closed-loop system and what media are used to facilitate the interaction.
3. What are the principal communication devices that interface the operator with machines? What is the role of ergonomics in this interface?
4. Describe the two modes of receiving information from the environment. Which mode is the main concern of ergonomists?
5. List three of the five characteristics of every workplace.
6. What are the three general types of functions the worker performs in a human–machine system?
7. With what aspects of the human–machine systems is ergonomics concerned?
8. Briefly explain what is meant by systems approach, as used for analysis of the human–machine interface in the work environment.
9. Describe the black box method of system analysis.

10. List and briefly describe the seven steps involved in system design.

11. List the five general stages into which the job is divided when a job analysis is performed to describe a process. What typical question is asked at each stage?

12. Once the job breakdown is established, the job analysis can begin. The analysis involves a number of questions about each task, element, or micromotion that is being analyzed. Give seven typical job analysis questions.

13. What is the objective of job analysis? What are the key questions whose answers might help achieve this objective? List at least seven questions.

14. What are the major problems associated with displays and controls in human–machine systems? What are their causes?

15. What are the three common types of display instruments? Which type is recommended when quick and accurate readings are to be made? Which type is recommended if it is necessary to monitor process changes, or to note the amplitude of some change? Which type should be used if the display has to cover a wide range of values?

16. Give five examples of controls.

17. Classify controls based on the action required by the particular task, and give some examples of such controls and their applications.

18. Classify manual controls based on the amount of muscular effort required for their use, and give some examples of each type.

19. Stereotyping and standardization of displays and controls minimize human errors when interfaced with machines. What does stereo-type mean? Under what circumstances is standardization used?

20. What principles are applied to digital read-out displays?

21. Between circular or semicircular scales and vertical or horizontal scale displays, which type is preferred?

22. When designing displays, in what direction should scale numbering increment on a circular/semicircular scale, on a vertical, and on a horizontal scale?

23. What type of display is the best for monitoring how a process is changing, or to note the amplitude of some change, where relative information is displayed?

24. What type of display should be avoided for monitoring qualitative data, and when is it necessary to memorize the previous reading, and to assess the extent of movement?

25. How can the major scale markers be numbered in order to be easily recognizable and in a natural fashion?

26. What type of lettering should be used to display single words or short identification sentences?

27. A non-critical instruction is to be read from a 5-m distance. The

illumination level is adequate. Determine the heights, stroke-width, and width of each letter and number on the display. (*Hint*: see Table 11.1 and Figure 11.6.)

28. List the general principles applied to the size, color, flash rate, contrast with background, and multimedia presentation of warning lights to enhance their effectiveness.

29. List seven recommendations for designing or selecting controls.

30. List six recommendations for the installation and positioning of controls.

31. What problems are associated with incompatible controls and displays?

32. When a control is moved or turned to the left, downwards, backwards, or counter-clockwise, should the display affected by it show an increase or decrease? In which direction must the display pointer move on a horizontal scale, over a circular scale, or on a vertical scale?

33. When a control is moved (or turned) to the right (or clockwise), in which direction should the moving scale of the affected display with a fixed indicator move, and what changes in the scale values (increase or decrease) should be expected?

34. To reduce the displayed reading, in which direction is the corresponding hand lever (control) moved:
    (a) if the lever moves horizontally?
    (b) if the lever moves vertically?
    (c) if the lever moves away (forwards) from the body or towards the body?

35. On which side of the display should an affecting control be installed?

36. On which sides of the controls and the display instruments affected by them should identification labels be placed?

37. What are your layout recommendations for the order and arrangement of a set of controls and the corresponding set of displays in relation to each other, if each set has to be on a different panel?

38. When arranging a group of controls and displays, how should the controls and displays be located in relation to the operator?

39. Assume that the optimum control–response (C/R) ratio for a 15-cm ball-lever control ranges from 2.5 : 1 to 4 : 1. What should be the range of angular movement of the lever if the display is to move 5 cm linearly?

40. Assume that the optimum C/R ratio for a 7.5-cm diameter knob ranges from 0.2 to 0.8. What should be the range of the linear movement of the display element in one complete revolution of the knob?

41. The effect of C/R ratios on human performance in operating controls

is a function of two types of human motions. Name and briefly describe these two types of motions.

REFERENCES

Alexander, D.C. (1986) *The Practice and Management of Industrial Egonomics.* Prentice-Hall, Englewood Cliffs, NJ.

Byers, B.B., Hritz, R.J. and McClintock, J.C. (1978) *Industrial Hygiene Engineering and Control: Ergonomics–Student Manual.* (Contract CDC–210–75–0076.) Division of Training and Manpower Development, National Institute for Occupational Safety and Health (NIOSH), US Department of Health and Human Services (formerly Health, Education, and Welfare–HEW), Cincinnati, Ohio.

Chapanis, A. and Kinkade, R.G. (1972) Design of controls. In: *Human Engineering Guide to Equipment Design* (Van Cott, H.P. and Kinkade, R.G., eds). US Government Printing Office, Washington, DC.

Grandjean, E. (1988) *Fitting the Task to the Man: A Textbook of Occupational Ergonomics,* 4th edn. Taylor & Francis, London.

Grether, W.F. and Baker, C.A. (1972) Visual presentation of information. In: *Human Engineering Guide to Equipment Design* (Van Cott, H.P. and Kinkade, R.G., eds). US Government Printing Office, Washington, DC.

Heglin, H.J. (1973) *NAVSHIPS Display Illumination Design Guide, II, Human Factors.* NELC/TD223. Naval Electronics Laboratory Center, San Diego, California.

Kvålseth, T.O. (1985) Work station design. In: *Industrial Ergonomics: A Practitioner's Guide* (Alexander, D.C. and Pulat, B.M., eds). Industrial Engineering and Management Press, Institute of Industrial Engineers, Norcross, Georgia.

Mahone, D.B. (1993) Review of system design employs ergonomics prior to work injuries. *Occupational Health and Safety* 62: 89–92, 105.

Reynolds, R.F., White, R.M. Jr and Hilgendorf, R.I. (1972) Detection and recognition of colored signal lights. *Human Factors* 14: 227–236.

Sanders, M.S. and McCormick, E.J. (1993) *Human Factors in Engineering and Design,* 7th edn. McGraw-Hill, New York.

Tayyari, F. (1993) Design for human factors. In: *Handbook of Concurrent Design and Manufacturing* (Parsaei, H.R. and Sullivan, W.G., eds). Chapman & Hall, London, pp. 297–325.

Wiley, R.C. (1991) Ergonomics: ticket to winning design. *Design News* 47: 82–83.

Thermal environments

12

12.1 INTRODUCTION

The human body, as a warm-blooded (homeothermic) animal, possesses an excellent thermoregulatory mechanism for maintaining an internal thermal balance despite a large range of variation in environmental temperatures. In addition, human ingenuity in controlling the conditions of the surrounding environment has led to a further enlargement in the range of tolerable temperatures. Advances in science and technology have made great contributions to the design of workplaces to improve workers' safety and comfort. However, the issues of thermal (both heat and cold) stress in many occupational environments have not yet been totally resolved. Heat and cold can cause illnesses and injuries in the exposed individuals. Thermal changes can also cause accidents (e.g., fire and explosion) and injury to people.

There are many occupational environments in which workers are exposed to excessive heat loads. Examples of such environments are given in Table 12.1 (National Institute for Occupational Safety and Health (NIOSH), 1986).

In such occupational environments workers are often exposed to high environmental heat loads. Heat stress, associated with these environments, is a major health hazard and reduces worker productivity.

In some other occupations workers are exposed to very low thermal conditions. Cold stress under such environmental conditions can present health hazards and reduce worker productivity. Examples of cold environments are outdoor construction sites, pipeline operations, meat packing, and power and utility activities during the cold season.

The effects of work under adverse thermal conditions on individuals have been extensively studied by many researchers. The results of these efforts have provided rational bases for the prevention of health impairment and physiological damage to workers due to heat stress

Table 12.1 Hot work environments

Iron and steel foundries
Non-ferrous foundries
Brick-firing and ceramics
Glass products manufacturing
Rubber products manufacturing
Boiler maintenance
Kitchens
Bakeries
Laundries
Mining
Steam tunnels
Roofing and home repair
Building construction
Road construction

(Stephenson *et al.*, 1974). Many industries have also become involved in heat stress investigations to provide safe workplaces for their workers and to comply with governmental standards for work in hot environments. Determination of heat effects on work quality and quantity has been another reason for industry's interest in heat stress studies.

12.2 HEAT STRESS TERMINOLOGY

To study the subject of heat stress better, one should know its terms. Some common terms are given below:

- **Heat stress**: Heat stress may be defined as a total load of all internal and external heat factors upon the body. Internal factors include metabolic heat, degree of acclimatization, and body temperature. External factors include ambient air temperature, radiant heat, air velocity, humidity, and clothing thermal resistance.
- **Heat strain**: Heat strain is the sum of all biochemical, physiological, and psychological adjustments in the person made in response to the heat stress. The severity of heat strain depends upon the level of heat stress, which can range from a feeling of discomfort to a serious heat disorder.
- **Heat stress threshold limits:** The body can tolerate heat only up to a certain level before signs of heat stress appear. Such a level of heat is called the threshold limit value, which depends on the workload and the duration of exposure. Threshold limit values (TLVs) in terms of wet-bulb globe temperature (WBGT) have been developed to assess the level of heat stress (American Conference of Governmental Industrial Hygienists (ACGIH) 1994).
- **Thermal comfort**: Thermal comfort for a person has been defined by

Fanger (1970) as "That condition of mind which expresses satisfaction with the thermal environment." As he has explained, under such a condition, the person does not know whether he or she would prefer a cooler or a warmer condition. The level of thermal comfort is not unique for all people.

According to ASHRAE standard 55–1981, *Thermal Environmental Conditions for Human Occupancy*, temperatures in the range from 22.8 to 26.1°C (73–79°F) new effective temperature (ET*), with an optimal temperature of 24.4°C (76°F) ET* are known as the summer comfort envelope for individuals with a clothing value of 0.5 clo. It has also been reported that with 0.9 clo, the winter comfort envelope consists of an optimal temperature of 21.7°C (71°F) ET* and boundary temperatures from 20 to 23.6°C (68–74.5°F) ET*. Holzle *et al.* (1983), in a study on 108 males and 108 females, validated the summer comfort envelope developed by ASHRAE and found a thermal comfort for 0.95 clo as 22.8°C (73°F) ET*. [The reader is referred to ASHRAE Standard 55-1981 for measurement and distinction between the (original) 'effective temperature (ET)' and the 'new effective temperature (ET*)'.]

- **Metabolic heat** is the body's heat generated internally, through metabolism, as it burns foodstuffs to release energy.
- **Evaporation** is the method of body heat transmission to the environment by evaporation of perspiration and respiration. Evaporation can only result in a transfer of heat from the body to the environment. Sweat must be evaporated from the skin surface for the heat transfer to occur. In saturated environments (relative humidity = 100%), no heat transfer can take place through evaporation.
- **Convection**: Convection is the method of heat exchange through a fluid medium (e.g., air or water). In general, the convection heat transfer is made by a gas (air) or fluid through movement of its molecules. The gas/fluid molecules are warmed up by touching (conducting) warmer objects in the environment; and colder objects in the environment, in turn, absorb heat from the warmer gas/fluid molecules. The circulation of the gas/fluid molecules is naturally caused by differences in density within the gas/fluid. This process is called **natural** or **free** convection. Sometimes circulation is created by the use of mechanical devices (e.g., pumps and blowers) and this is referred to as **forced** convection. When the air temperature is lower than the skin temperature, the body transmits heat to the air (called heat loss). If air temperature is higher than skin temperature, the heat transmission is reversed (called heat gain). A severe case of convective heat loss can occur if a person falls into very cold water and is not removed within a very short period of time.
- **Conduction**: Conduction is the method of heat exchange by direct contact of the body with other objects. In this method, heat is

transferred from one substance to another, or among molecules within a substance, without physical movement of the substance itself or its molecules. Since clothing insulates the body, the amount of heat exchange by conduction may become negligible. Conductive heat exchange in the body is usually minimal and is often ignored in the heat balance equation.

- **Radiation**: Generally, any substance with a temperature above absolute zero emits heat radiation in the form of electromagnetic waves (similar to light waves). Thermal radiation will travel through vacuum or transparent media. It can be reflected and its intensity will vary inversely with the square of the distance from the source, in the same way as light waves and other electromagnetic waves. The amount of heat radiated depends on the temperature and area of the radiating object, regardless of its mass. Radiant heat exchange between the body and other objects (e.g., walls, the sun, or furnaces) takes place without direct contact.
- **Clo** is the unit used for measuring the intrinsic thermal insulation value of clothing. One clo unit is equivalent to 5.55 kcal $\cdot$ m^{-2} $\cdot$ h^{-1} of heat exchange by radiation and convection for each degree Celsius of temperature difference between the skin and dry-bulb temperature (ACGIH, 1994).
- **Latent heat** is the amount of heat that must be absorbed when a substance changes phase from solid to liquid or from liquid to gas, or the amount of heat lost (emitted) during phase changes from gas to liquid or from liquid to solid. For example, at sea level (or where the barometric pressure is 760 mmHg, or 14.7 psi) approximately 540 kcal is needed to evaporate 1 kg (or 970 BTU to evaporate 1 lb) of water at 100°C (or 212°F). Also, 316 kcal of heat must be removed from 1 kg (or 143.5 BTU from 1 lb) of water in order for it to freeze at 0°C (32°F).

12.3 PHYSIOLOGICAL RESPONSES AND HEAT EXCHANGE

Maintenance of life demands a constant flow of energy from the environment through organisms. The intake of energy is in the form of chemical potential energy of foodstuffs. The potential energy of foodstuffs is released in a form usable for organs. The released energy is finally returned to the environment in the forms of mechanical work and heat. The mechanical work may be external (e.g., physical activities) or internal (e.g., heart beats, respiration, digestion, and brain activities). The energy used in internal work, finally, leaves the body in the form of heat. The heat produced by the body is called metabolic heat.

When people are exposed to a hot environment, they experience first, vasodilatation (expansion of blood capillaries near the skin surface), which facilitates increased heat transfer from the core to the shell of the body to be removed by evaporation; and second, an activation of the sweat glands (in the subcutaneous layer under the skin) to facilitate evaporative heat loss (Åstrand and Rodahl, 1986). These reactions are an adjustment by the thermoregulatory system to restore a new thermal balance between the body and environment. At some level of heat load the limit of heat dissipation will be reached. Beyond this physiological limit the establishment of a steady state for the core temperature will become impossible, and it may increase to dangerous levels. Furthermore, the physiological limit is usually reached before the body temperature rises much above 40°C (104°F) (Poulton, 1970).

To prevent the internal heat build-up, the body has to dissipate some of its metabolic heat. The body attempts to achieve thermal equilibrium with its surrounding environment through the following heat exchange methods: metabolism, evaporation, convection, conduction, and radiation. The heat exchange follows the second law of thermodynamics, according to which heat from a substance of a higher temperature is transferred to another object with a lower temperature. The process of heat exchange between the body and its surrounding environment can be expressed by the **heat balance equation**:

$$\pm S = M \pm CV \pm CD \pm R - E \tag{12.1}$$

where:

S = heat storage (positive sign indicates heat gain, while negative indicates heat loss. If the heat balance is achieved, $S = 0$);

M = metabolic heat (always positive);

CV = convective heat (positive sign indicates air temperature is higher than skin temperature, and negative indicates a reversed case);

CD = conductive heat (positive when the contacting objects are warmer than skin, and negative when the skin is warmer);

R = radiant heat (positive when surrounding objects are warmer than skin, and negative when the skin is warmer);

E = evaporative heat (always negative).

To be able to work safely and effectively in hot environments, the worker must maintain a body thermal equilibrium. The maintenance of this equilibrium depends upon the body's ability to dissipate metabolic heat. Under conditions of high ambient temperature and/or high radiant temperature, heat gain in body tissues can result. The problem arises when the environmental temperature approaches the skin temperature, and heat loss by means of convection and radiation gradually ends. At temperatures beyond skin temperature, convection and radiation reverse directions and increase the heat content of the body, and the only remaining means of heat loss is evaporation. In hot environments with high humidity, heat loss by evaporation also becomes impaired.

12.4 HEAT DISORDERS

Heat stress not only may reduce work performance, but also can lead to a heat disorder. At lower levels of heat stress there is no health damage risk, even though an individual may feel discomfort (Dukes-Dobos, 1981). However, when the heat stress exceeds the person's heat tolerance capacity, adverse health effects will occur.

Heat disorders and their causes have been well-described in detail in the literature (Lee, 1964; Leithead and Lind, 1964; Martinson, 1977; Kroemer et al., 1994). The most critical of heat disorders are heat stroke, heat exhaustion, heat cramps, and prickly heat. A brief explanation of each of these disorders and methods to prevent and/or treat them can be found in many articles on heat stress (American Red Cross, 1981).

- **Heat stroke**: Heat stroke or sunstroke is the most dangerous heat disorder. This is a state of acute collapse of the thermoregulatory system. It occurs when the body becomes so overheated that sweat glands and other organs cannot function normally. Heat stroke is a life-threatening emergency and is characterized by:

  - diminished or stopped sweating, and hot, dry, and flushed skin;
  - deep and rapid breathing;
  - rapid, but weak and possibly irregular pulse;
  - dizziness, confusion, or loss of consciousness;
  - hyperpyrexia (rectal temperature above 40.6°C);
  - disturbance of the central nervous system, which results in collapse and coma.

  Medical care is needed promptly. Heat stroke is fatal if not treated urgently. A heat stroke patient needs immediate medical care and hospitalization. If proper facilities are not readily available, the body temperature must be lowered as quickly as possible. Current recommendations for lowering body temperature include: the application of wet sheets and forced ventilation (the use of a fan), an alcohol sponge bath applied to the skin to increase evaporative heat loss, or pouring cool water on to the body to increase convective and evaporative heat loss.

- **Heat exhaustion**: Heat exhaustion (also called heat collapse or heat syncope) is a thermoregulatory failure which is caused by salt depletion (with or without dehydration) resulting in cerebral anoxia (insufficient supply of oxygen to the brain) due to peripheral vasodilatation, especially in the lower extremities as the result of prolonged standing, or by a reduction in blood volume due to dehydration. It may result in unconsciousness, vomiting, rapid pulse, fatigue, and headache. Skin is cool and sometimes pale and clammy with sweat. The victim must be placed in the horizontal position with the legs elevated and salt solution should be ingested (one-half teaspoon salt in one-half glass of water) every 15 min for three or four doses.

- **Heat cramps**: Heat cramps are painful intermittent spasms (contractions) of voluntary (skeletal) muscles due to depletion of salt in body fluid after excessive sweating, or drinking a large volume of water without replacing the lost salt.
- **Prickly heat**: Prickly heat or heat rash is a non-contagious skin disease. It is a form of damage to the sweat gland ducts, resulting in inflammation of the glands. Prickly heat is caused by reduced resistance of skin due to the continuous presence of unevaporated sweat. It appears as tiny red blisters in the affected area.

In addition to the above parameters, Fanger (1973) discussed the influence of other factors on the state of thermal comfort. These factors include age, sex, adaptation, season, physique, daily rhythm, menstrual cycle, ethnic differences, food consumption, and the color of the environment.

12.5 HEAT STRESS INDICES

A great number of heat stress indices (from simple to complex) have been developed. The following indices are most commonly used in monitoring heat stress:

- **Ambient air temperature** or **dry-bulb temperature** (T_{db}) is usually measured using a mercury-in-glass thermometer, or some other thermistors (thermally sensitive elements) exposed to the air. The thermometer must be shielded against the sun and other radiant heat sources without restricting the air flow around the bulb (i.e., the mercury reservoir or the thermally sensitive elements) of the thermometer.
- **Natural wet-bulb temperature** (T_{nwb}) is generally measured using a mercury-in-glass thermometer whose mercury reservoir is covered by a wetted cotton wick. The wick should extend over the bulk of the thermometer, covering the mercury reservoir and about an additional bulb length. The free end of the wick is immersed into a reservoir of distilled water. ACGIH (1994) recommends that:
  - the wick should also be wetted by direct application of distilled water from a syringe 30 min before each reading;
  - the wick should always be clean;
  - new wicks should be washed before using.

The ability of the air to remove heat from the wick-covered thermometer is a function of the vapor pressure or relative humidity of the air.

- **Wet-bulb temperature** or **forced wet-bulb temperature** (T_{wb}) can be measured by fanning a natural wet-bulb thermometer until there is no further drop in the wet-bulb temperature before taking a reading. It can also be measured using a sling thermometer.
- **Relative humidity** (*rh*) can be determined from a psychometric chart (see Appendix E) using the dry-bulb and wet-bulb temperature readings.
- **WBGT** is a heat stress index that combines the effects of humidity and air velocity (T_{nwb}), radiant heat (globe temperature, T_g), and ambient air temperature (T_a or T_{db}) into a single index. This is the simplest and most commonly used heat stress index. WBGT values are calculated based on basic thermal indices as follows:

Outdoors with solar load (under sunshine):

$$WBGT_{out} = 0.7T_{nwb} + 0.2T_g + 0.1T_{db} \qquad (12.2)$$

Indoors or outdoors without solar load (in shade):

$$WBGT_{in} = 0.7T_{nwb} + 0.3T_g \qquad (12.3)$$

where: WBGT = wet-bulb globe temperature index;
 T_{nwb} = natural wet-bulb temperature;
 T_g = globe temperature;
 T_{db} = dry-bulb temperature.

The globe temperature is read from a standard mercury-in-glass thermometer inserted into a 15-cm (6-in)-diameter copper globe, whose outside surface is painted with a matte black finish. The mercury reservoir or sensor of the thermometer is situated at the center of the globe.

An ideal index would account for the physiological response to heat stress. An international panel of experts convened by the World Health Organization (WHO) in 1969 recommended that workers' deep body temperature should not exceed 38°C (100.4°F) during their daily exposure to heavy work in hot environments. This is the best heat stress index because it is based on the physiological response to heat stress. The major drawback associated with the use of such an index is that the measurement of the deep body temperature is not practical for monitoring workers' heat strain.

It seems that WBGT is the simplest and most convenient index that has a reasonably close correlation with physiological responses.

12.6 ASSESSMENT OF HEAT STRESS EXPOSURE LIMITS

Although heat acclimatization improves the capacity of the person to work under hot thermal conditions, the dangers and problems associated with exposure to heat stress cannot be totally relieved. Even for an acclimatized person, the maximum tolerable levels of heat exposure depend on the combination of ambient temperature, radiant heat, humidity, air movement, metabolic heat during work, and the duration of exposure. The WBGT combines the thermal factors into one single index. This index is often used for heat stress evaluations.

Industrial hygienists and safety professionals determine the permissible exposure limits using the TLVs published by the ACGIH, which are based on the assumption that the WBGT values in both the workplace and resting place are the same or very close (ACGIH, 1994). The TLVs are given in Table 12.2, and graphically illustrated in Figure 12.1.

Table 12.2 Examples of permissible heat exposure threshold limit values (TLVs) in wet-bulb globe temperature (WBGT)

| | Equivalent metabolic rate (M) of workload | | |
| | Light | Moderate | Heavy |
Average hourly work–rest regimen	$M \leqslant 200$ kcal · h^{-1}	$200 < M \leqslant 350$ kcal · h^{-1}	$350 < M \leqslant 500$ kcal · h^{-1}
Continuous work	30.0°C (86°F)	26.7°C (80°F)	25.0°C (77°F)
75% work/25% rest	30.6°C (87°F)	28.0°C (82°F)	25.9°C (78°F)
50% work/50% rest	31.4°C (89°F)	29.4°C (85°F)	27.9°C (82°F)
25% work/75% rest	32.2°C (90°F)	31.1°C (88°F)	30.0°C (86°F)

Adapted from ACGIH (1994).

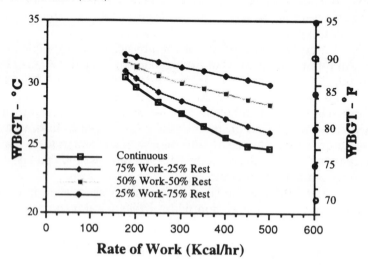

Fig. 12.1 Graphical presentation of the permissible heat exposure threshold limit values given in Table 12.2. WBGT = Wet-bulb globe temperature.

Since a work-shift may consist of various work activities (with varying metabolic rate), performed in different areas (each with a different thermal condition), ACGIH is recommending the use of a time-weighted average value for both environmental and metabolic heat.

The time-weighted-average (TWA) metabolic rate (M), and the TWA WBGT are determined by means of the following equations:

$$M_{TWA} = \frac{M_1 \times t_1 + M_2 \times t_2 + \ldots M_n \times t_n}{t_1 + t_2 + \ldots + t_n} \qquad (12.4)$$

$$WBGT_{TWA} = \frac{WBGT_1 \times t_1 + WBGT_2 \times t_2 + \ldots + WBGT_n \times t_n}{t_1 + t_2 + \ldots + t_n} \qquad (12.5)$$

where:

t_i = the number of minutes in period i (where $i = 1, 2, \ldots, n$), measured by a time study;

M_i = the measured or estimated metabolic rate for the activity or rest during period t_i (where $i = 1, 2, \ldots, n$);

$WBGT_i$ = the calculated WBGT for the activity-area or rest-area during the time period t_i (where $i = 1, 2, \ldots, n$).

12.7 FACTORS AFFECTING HEAT STRESS AND STRAIN

The following three general factors affect the level of heat stress of individuals who work in hot environments:

- **Environmental thermal conditions**: that is, ambient temperature, radiant heat, humidity, and air movement.
- **Physical workload**, which positively affects the body's heat generated internally, through metabolism.
- **Clothing**: Some clothes are permeable and allow heat exchange between the body and the surrounding environment. Others are impermeable, especially those used in personal protective garments for use in chemically hazardous environments, which constrict the heat exchange process. Holzle et al. (1983) found that at 79°F (26.1°C) ET*, individuals with a clothing value of 0.95 clo were significantly warmer than those with a clothing value of 0.54 clo, with no difference due to gender.

The threshold limit of heat stress should be modified to account for clothing. Clothing can be as unrestrictive as shorts and a T-shirt or as restrictive as a suit of armor. Restrictive clothing can interfere with the body's evaporation capability. As Ramsey (1978) stated, for unrestrictive clothing, a positive 2°C (3.6°F) is added to the threshold limit. As much as 10°C (18°F) may be subtracted from the heat stress TLVs to account for a suit of armor.

The level of heat strain experienced by an individual exposed to heat stress is influenced by the above three heat stress factors and the personal characteristics, such as gender, age, race, physical fitness, status of nutrition, medical history (health conditions), skill, body build, and heat acclimatization. These personal characteristics are described in the following section.

12.8 PERSONAL CHARACTERISTICS AFFECTING HEAT STRAIN

All individuals do not show the same level strain in response to the same level of thermal (heat or cold) stress. Personal characteristics contribute to the level of thermal strain experienced by the person under heat stress. Figure 12.2 illustrates the thermal strain of three hypothetical human subjects with different personal characteristics. The results of a survey of literature (Tayyari, 1995) are summarized in the following subsections.

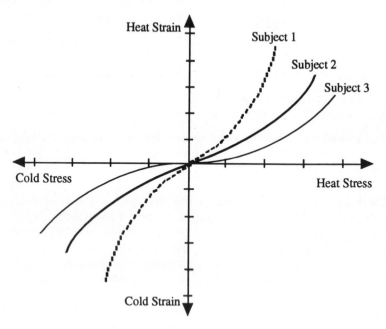

Fig. 12.2 An illustration of thermal (cold and heat) strains of three individuals with different personal characteristics; of the three subjects, Subject 1 is the most and Subject 3 is the least sensitive to thermal stress.

12.8.1 GENDER

Studies have shown that the heat strain response by women is higher than men. Therefore, women are less heat-tolerant than men. Investigators have explained the gender differences in physiological responses to heat stress as follows:

- Women thermoregulate less effectively than men.
- On the average, women have lower cardiorespiratory capacity than men.
- Women have a lower body weight and lower skin (body) surface area, but a higher body fat content and, more importantly, a higher ratio of body surface area to body mass than men.
- The hormonal levels of estrogen and progesterone in women fluctuate.
- In heat stress environments, women normally exhibit higher heart rates than men under similar conditions, yet women sweat less.
- Females tolerate hot-humid climates better than males, but males can tolerate hot-dry climates better than females. These phenomena can be explained by the fact that excessive sweat by men in a hot-humid climate cannot be evaporated, which only expedites the dehydration process, while women's lower sweating in a hot-dry climate interferes with the body's self-cooling system.
- Based on the core body temperature and skin temperature, differences between females and males are not significant; this is a disadvantage for females. This has been explained by the fact that since, on the average, women's body surface area is smaller, they have to transport more heat per square meter body surface area and are, therefore, under more cardiovascular strain than men.

12.8.2 AGE

Individuals of different ages respond to thermal conditions differently. The body systems deteriorate with aging (Ramsey, 1978). The cardiovascular and pulmonary systems degrade over time and the body's heat tolerance drops. This phenomenon is built into the heat stress thresholds by subtracting 1–2°C from the corresponding TLVs for the elderly. The following age effects reported in the literature attest to this process:

- Exposure to high-temperature environments causes greater thermoregulatory strain on the elderly than younger individuals.
- At temperatures about 21.1°C (70°F), the elderly express feeling more warmth than young, college-age individuals.
- The elderly prefer slightly warmer temperatures for comfort than do younger individuals.

- At about 38–40°C (100.4–104°F) ET, older males have demonstrated more heat tolerance than younger males, but the result has been the reverse for females.
- At 36°C (96.8°F) ET, older individuals have shown more heat tolerance than younger individuals for both genders.
- As can be demonstrated by increased heart rate, working in hot environments taxes the cardiovascular system. Because of such a physiological load, older people are at a higher risk of developing heart problems than younger people when working under high thermal conditions.
- Both the maximum aerobic capacity (maximum oxygen uptake) and cardiac output diminish with the aging process. These physiological changes are most likely the main factors responsible for making the elderly more vulnerable to hot-heavy jobs; that is, performing heavy workloads under high heat loads.

Although age is a factor that reduces heat stress tolerance, well-conditioned older individuals have often demonstrated performance capabilities as good as or exceeding those of younger individuals. For this reason the age factor should not be used as a criterion for excluding a person from working under heat stress, especially at moderate levels.

12.8.3 RACE

White and black skins absorb and radiate heat equally; however, in the visible range of short-wave radiant heat, white skin reflects considerably more radiation than black skin (Edholm, 1967). Hence, black skin may reach higher temperatures by absorbing an additional radiant load. However, the dark pigment (melanin) in the skin of dark-skinned individuals can represent a natural adaptation to the environmental heat.

12.8.4 PHYSICAL FITNESS

Training can increase muscle strength, cardiac output, and the amount of oxygen carried from the lungs to the tissues. Karpovich and Sinning (1971) postulate that individuals who do exercise develop better heart muscles, enabling them to postpone exhaustion. Such physiological improvements significantly help reduce the level of heat strain and, as a result, the risk of heat illnesses. Physical fitness is probably the most important factor that affects the level of heat stress a person can tolerate. A physically fit individual can handle higher work loads easier than less fit individuals. Under heat stress, a physically fit person demonstrates a lower body temperature, a lower heart rate, and a more efficient sweating mechanism.

12.8.5 STATE OF NUTRITION

The body's water loss through sweating increases with work intensity and environmental temperature, and such water loss varies among individuals. Sweat contains salt. Hence, excessive sweat can result in salt deficiency which must be replaced through increased salt intake in food; otherwise, it can cause serious heat illnesses, such as heat exhaustion (also called heat collapse or heat syncope) and heat cramps.

12.8.6 MEDICAL HISTORY

Health impairments can reduce the person's ability to cope with high levels of heat stress. When selecting workers for assignment to heavy tasks in hot environments, their medical histories must be taken into consideration to exclude heat-intolerant individuals. Heat-intolerant individuals include those with cardiovascular disease, recent heat disorders, poor work capacity, and obesity.

12.8.7 SKILL

The level of energy expenditure in performing a certain task may vary among individuals due to the degree of skill they possess. This may be best explained by the fact that a skilled person exhibits more economical movements and makes fewer mistakes and expends less energy in performing a skilled task compared to a trainee who is learning the task. Therefore, a skilled worker can tolerate working in a hot environment better than an unskilled worker.

12.8.8 BODY BUILD

Obese and stocky individuals have higher metabolic rates that produce heat which taxes the cardiovascular system. Fat and the body surface area are two major factors determining the amount of heat stress that can be tolerated. Passmore and Durnin (1955) reported that small individuals are at an advantage over large individuals in performing hard physical work in extremely hot environments. They believe that larger individuals have trouble dissipating metabolic heat because they have a smaller ratio of body surface area to metabolizing body mass as compared to smaller individuals.

12.8.9 HEAT ACCLIMATIZATION

When individuals are exposed to a hot environment for a few days, they will be able to tolerate the heat much better than prior to exposure. The

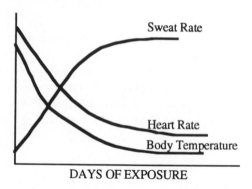

DAYS OF EXPOSURE

Fig. 12.3 An illustration of physiological adjustments during heat acclimatization. Adapted from Åstrand and Rodahl (1986).

process of improvement in heat tolerance is called heat acclimatization. Heat acclimatization can then be defined as "a process by which the body becomes tolerant or adapted to hot thermal conditions." The improvement in heat tolerance is associated with reduced heart rates, lowered skin and body temperatures, and improved sweat mechanism, facilitating more and sooner sweat production. Figure 12.3 illustrates these phenomena. According to Edholm (1967) acclimatization only develops if people sweat; and the more people sweat, the better they adapt to heat. Kamon *et al.* (1978) reported that the gender differences in tolerating heat stress will diminish with heat acclimatization.

Heat acclimatization usually requires about 7 days of exposure to the workload under the heat conditions in question. However, it may take 14 days or more to achieve a full acclimatization level. Acclimatization can be genetically inherited or occur as a result of nervous stimulus or a physiological process requiring time, supplemented by behavioral adjustments (Bell and Greene, 1982).

12.9 PROTECTION AGAINST HEAT STRESS

There are many ways to reduce the adverse effects of heat stress through engineering and administrative controls and personal protective equipment. To devise an appropriate solution for a heat stress problem, the heat sources must first be identified and the environmental factors measured. The following are some methods of reducing heat effects (Edholm, 1967; Kroemer *et al.*, 1994):

- Installing heat shields between the source of radiant heat (e.g., molten metal furnaces) and the worker.

- Installing an adequate ventilation system. Although radiant heat cannot be blown away, ventilation systems can reduce the air temperature and increase air movement, which facilitates heat loss through evaporation. However, excessive air movement can worsen the heat effects, especially if the air temperature exceeds the skin temperature. Edholm (1967) suggested an upper limit for air movement in the range 50–75 cm · s^{-1} (i.e., 1.8–2.7 km · h^{-1}, 0.5–0.75 m · s^{-1}, or 100–150 ft · min^{-1}) under hot conditions.
- Installing air condition or refrigeration systems to cool air and reduce the humidity.
- Providing an ample amount of cool water for workers in heat to drink. Workers must be encouraged to drink water frequently to prevent their dehydration through excessive sweating. Dehydration reduces the sweat rate and, consequently, the body temperature rises and the heat exposure becomes intolerable. People tend not to replace all their lost fluid through voluntary drinking, and should be encouraged to drink water or other low-sugar fluids.
- Replacing the body salt lost by sweating to prevent heat exhaustion and heat cramps. Usually dietary salt is adequate for ionic balance. However, under extremely hot working conditions, a lightly salted, low-sugar drinking solution may be recommended.
- Providing microclimate (personal) cooling systems (e.g., ice vests, air-ventilated garments, and water-circulated suits) when cooling the work environment is impractical or the worker has to wear protective clothing (e.g., in furnace cleaning operations, or in asbestos and chemical environments).

The risk of heat collapse or heat stroke increases rapidly as deep body temperatures rise above 39°C (Van Graan, 1975). WHO (1969) and ACGIH (1994) recommended that workers should not be permitted to continue work when their deep body temperature exceeds 38°C (100.4°F). As environmental temperature and metabolic workload increase, the body is less able to maintain its thermal equilibrium with the surrounding environment and deep body temperature may rise beyond the acceptable level.

In many situations, if a worker is to maintain thermoneutrality, minimize heat gain and the risk of heat collapse or heat stroke and/or, at least, be able to work for a reasonable time period before reaching a maximum allowable level of heat storage, protective devices must be provided. Focusing on a system which can provide some complete or partial body cooling to workers in hostile environments seems to be a promising solution to the problem.

12.10 COLD STRESS AND INJURIES

Cold stress is usually not as critical a problem as heat stress. However, its effects on exposed personnel should not be underestimated. As Malley (1992) has stated, under severe cold conditions, a major proportion of workers' time and energy is used in self-preservation, which reduces their efficiency. Winter clothing, which is rather bulky, decreases the person's efficiency. Wearing mittens or warm gloves will significantly reduce the sense of touch and the person's dexterity, which may lead to accidents.

Cold exposure causes vasoconstriction (contraction of capillaries near the skin surface) to direct warm blood to the body core to prevent excessive heat loss through the skin and preserve the heat for maintaining vital body functions (Ramsey, 1985). Extreme cold exposure can cause shivering, low body temperature, numbness, weakness, and drowsiness. The adverse effects of cold stress have been categorized by MacFarlane (1963) into three types of injuries: chilblains, wet-cold syndrome (also known as immersion foot and trench-foot), and frostbite.

- **Chilblains**: Chilblains can affect the hands and feet, and are a relatively mild form of tissue damage. The affected part becomes bluish-red in color, and is characterized by local itching and swelling (Hammer, 1989).
- **Wet-cold syndrome or hypothermia**: Wet-cold syndrome or hypo-thermia (meaning low heat) occurs when the rate of body's heat loss is greater than its heat replacement from metabolism and environment. It can cause a dangerous drop in the victim's core body temperature, which is crucial for the body's internal functions. Hypothermia is caused by exposure to extreme cold, prolonged exposure to mild cold (i.e., about 12°C and below), or immersion in cold water. The well-known immersion foot and trench-foot are examples of wet-cold syndrome (Hammer, 1989). Body wetness expedites its heat loss. Alcoholics and the elderly are more likely to develop hypothermia than others.
- **Frostbite**: Frostbite is an injury due to freezing of the soft tissues of a body part (usually, the extremities) exposed to extreme cold. Its symptoms include:

  - crystal formation on the skin;
  - numbness and intense cold sensation;
  - pale and glossy skin;
  - itching sensation;
  - pain;
  - blisters.

The part affected by frostbite should be warmed slowly, not rubbed.

- **Hypothermia**, or the lowering of body temperature, is also a very dangerous condition in the outdoor environment. Hypothermia occurs frequently in wet, windy weather. Most hypothermia cases occur in air temperatures between −1 and 10°C (30 and 50°F). The typical scenario is a person out for a hike, getting caught in a rain storm, getting wet, getting lost, going to sleep, and eventually dying due to loss of body heat. Symptoms of hypothermia include: uncontrolled shivering; vague, slow slurred speech; memory lapses; fumbling hands; frequent stumbling; drowsiness, and apparent exhaustion. To avoid hypothermia, use a buddy system, get out of the cold environment, get into dry clothing, and don't let the individual go to sleep until in a warm environment, or until body temperature has returned to normal.

12.11 MONITORING COLD STRESS

Windchill factor is widely used as a thermal index for cold environments. It combines the effects of temperature and air movement, and is expressed as the number of kilocalories of heat per hour (kcal $\cdot$ h^{-1}) that the atmosphere can remove from the body (Malley, 1992). It may also be measured in terms of kilocalories per hour per squared meter of body surface area (kcal $\cdot$ h^{-1} $\cdot$ m^{-2}) removed from the body, using the following equation (Siple and Passel, 1945):

$$\text{Windchill (kcal} \cdot \text{m}^{-2} \cdot \text{h}^{-1}) = (\sqrt{W_s \times 100} - W_s + 10.45) \times (33 - T_{db}) \qquad (12.6)$$

where: W_s = wind speed in meters per second;
 T_{db} = ambient (dry-bulb) temperature in degrees Celsius;
 33   = average skin temperature, in degrees Celsius.

The concept of windchill factor emphasizes that convection is the most important single avenue of heat loss to the air under windy, cold conditions. The windchill index has become popular in the reporting of climatic conditions during winter. Tables of windchill factors are readily available from a variety of sources. However, the effect of windchill is better felt by the equivalent temperature sensed by exposed flesh (nude skin). The US Army Research Institute of Environmental Medicine has developed a chart to measure the equivalent temperature on exposed flesh for various combinations of low temperatures and wind speeds (Table 12.3). Based on this data set, the authors have developed a line chart that can be used as an alternative instrument for the calculation of equivalent temperatures (Fig. 12.4).

Table 12.3 Equivalent temperatures on exposed flesh at various wind speeds*

Wind speed (km · h⁻¹)	Actual air temperature (°C)											
	10.0	4.4	−1.1	−6.7	−12.2	−17.8	−23.3	−28.9	−34.4	−40.0	−45.6	−51.1
	Equivalent temperature (°C)											
0 (Calm)	10.0	4.4	−1.1	−6.7	−12.2	−17.8	−23.3	−28.9	−34.4	−40.0	−45.6	−51.1
8	8.9	2.8	−2.8	−8.9	−14.4	−20.6	−26.1	−32.2	−37.8	−43.9	−49.4	−55.6
16	4.4	−2.2	−8.9	−15.6	−22.8	−31.1	−36.1	−43.3	−50.0	−56.7	−63.9	−65.0
24	2.2	−5.6	−12.8	−20.6	−27.8	−35.6	−42.8	−50.0	−57.8	−65.0	−72.8	−80.0
32	0.0	−7.8	−15.6	−23.3	−31.7	−39.4	−47.2	−55.0	−63.3	−71.1	−78.9	−85.0
40	−1.1	−8.9	−17.8	−26.1	−33.9	−42.2	−50.6	−58.9	−66.7	−75.6	−83.3	−91.7
48	−2.2	−10.6	−18.9	−27.8	−36.1	−44.4	−52.8	−61.7	−70.0	−78.3	−87.2	−95.6
56	−2.8	−11.7	−20.0	−28.9	−37.3	−46.1	−55.0	−63.3	−72.2	−80.6	−89.4	−98.3
64	−3.3	−12.2	−21.1	−29.4	−38.4	−47.2	−56.1	−65.0	−73.3	−82.2	−91.1	−100

Actual air temperature (°F)

Wind speed (mph)	50	40	30	20	10	0	−10	−20	−30	−40	−50	−60
	Equivalent temperature (°F)											
0 (Calm)	50	40	30	20	10	0	−10	−20	−30	−40	−50	−60
5	48	37	27	16	6	−5	−15	−26	−36	−47	−57	−68
10	40	28	16	4	−9	−24	−33	−46	−58	−70	−83	−85
15	36	22	9	−5	−18	−32	−45	−58	−72	−85	−99	−112
20	32	18	4	−10	−25	−39	−53	−67	−82	−96	−110	−121
25	30	16	0	−15	−29	−44	−59	−74	−88	−104	−118	−133
30	28	13	−2	−18	−33	−48	−63	−79	−94	−109	−125	−140
35	27	11	−4	−20	−35	−51	−67	−82	−98	−113	−129	−145
40	26	10	−6	−21	−37	−53	−69	−85	−100	−116	−132	−148

Wind speed greater than 64 km · h⁻¹ (40 mph) has little additive effect

Little danger zone: maximum danger is false sense of security within 5 h with dry skin

Danger zone: flesh may freeze within 1 min

Great danger zone: flesh may freeze within 30 s

*Developed by US Army Research Institute of Environmental Medicine, Natick, MA. Also cited by Ramsey (1983) and ACGIH (1994).

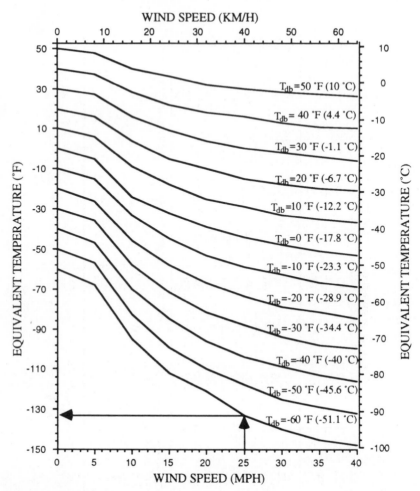

Fig. 12.4 Line chart for calculating equivalent temperature on exposed flesh at various combinations of actual air temperatures and wind speeds. To calculate the temperature equivalent to an actual air temperature (i.e., dry-bulb temperature, T_{db}) at a given wind speed, draw a vertical line from the wind speed to intercept the actual temperature curve (T_{db}), then draw a horizontal line from the point of interception to find the equivalent temperature from the desired vertical scale (°F on the left or °C on the right side). As shown, the temperature sensation by exposed flesh equivalent to −60°F when wind speed is 25 mph is −133°F.

12.12 PROTECTION AGAINST COLD STRESS

Humans' primary defense mechanism against the cold is to seek shelter, or to add more clothing. Physiological responses (vasoconstriction and shivering) are of limited use in maintaining body temperature. In order to provide protection against the cold:

- The worker should wear warm clothing and, especially, protect the neck, head, chest, groin area, hands, ears, nose, and feet.
- The worker should wear a waterproof, wind-resisting outer layer of clothing, preferably with a layer of wool inside.
- The worker should wear a warm hat because heat loss through the head takes place very rapidly.
- The worker should have spare dry clothing because the body loses heat more rapidly when it is wet.
- The worker should not work alone in cold.
- The worker should consume high-energy food and snacks that provide more calories to produce more body heat.
- The worker should refrain from consuming alcoholic beverages to prevent dehydration.
- The employer should provide a warm escape place for workers who are exposed to cold stress.
- The employer should provide non-alcoholic liquids such as soda, milk, water, and juice for the worker to maintain body fluid balance.
- Workers who take medications that make them more vulnerable to cold should not be exposed to cold.

12.13 APPLICATIONS AND DISCUSSION

Temperature regulation in the human is normally maintained in a very close range, around 37°C. As a worker is exposed to warm or cool environments, physiological changes take place to maintain the required body temperature. Through a process of acclimatization, the body adapts somewhat to the thermal load of the environment. However, as the thermal load exceeds the body's ability to adapt, engineering and administrative controls, or personal protective equipment must be incorporated to protect the worker.

There are many occupations in which workers are routinely exposed to hot and cold environments. If the worker is not properly protected, the consequences could become fatal. This chapter has provided tools for understanding the thermal environment and physiological responses to changes in temperature. The ergonomist should be able to assess a thermal environment, understand the exposure risks for workers, and to develop guidelines for employees exposed to such environments.

It should be pointed out that thermal stress during off-the-job activities (e.g., yard work and sports) performed under thermally hazardous conditions is as serious a problem as occupational thermal stress. It is, therefore, crucial for people to gain knowledge of the health hazards and protection strategies associated with exposure to potentially harmful thermal environments both at work and in their recreational activities.

REVIEW QUESTIONS

1. List five examples of hot environments.
2. List three examples of cold environments.
3. Define heat stress and heat strain, and note the difference between the two.
4. Describe heat stress threshold limits.
5. State Fanger's definition of thermal comfort.
6. Describe the evaporation method of metabolic heat dissipation to the surrounding environment. Also explain how this heat dissipation depends on the air humidity.
7. Describe the convection method of heat exchange, and free and forced types of convection. Also, explain under what conditions heat will be lost to and gained from the surrounding environment by this method.
8. Describe the conduction method of metabolic heat dissipation to the surrounding environment.
9. Describe the radiation method of heat exchange, and how this type of heat exchange takes place between the body and the surrounding environment.
10. What is clo and its equivalent amount of heat exchange?
11. What is latent heat?
12. Explain how the body's energy intake is returned to the environment.
13. Explain the reaction of the body's thermoregulatory system to heat exposure.
14. Can the body reach a steady-state internal heat level when the environmental conditions are such that no heat exchange can take place between the body and the environment? Why, or why not?
15. Write the heat balance equation that expresses the process of heat exchange between the body and its surrounding environment. Describe the components of the equation.
16. List the most critical heat disorders.
17. What is heat stroke? When does it occur and what are its characteristics?
18. What is heat exhaustion? How does it occur and what are its symptoms?
19. What are heat cramps and their causes?
20. What is prickly heat? How does it occur and what are its characteristics?
21. The three primary thermal measurements taken at a construction site were: 33°C dry-bulb, 27°C wet-bulb and 35°C globe. Calculate the wet-bulb globe temperature (WBGT) equivalent to these thermal conditions if the measurements were made:
    (a) under direct sunshine;
    (b) in the shade.

22. List the six factors on which the maximum tolerable levels of heat exposure depend.

23. The following environmental thermal and metabolic heat data were collected through a job analysis, in which a heat-acclimatized worker performed various job activities in various locations in the workplace. Evaluate the heat exposure to see whether the American Conference of Governmental Industrial Hygienists' (ACGIH's) recommended threshold limit values (TLVs) for heat exposure have been violated.

Location	Metabolic rate $(kcal \cdot h^{-1})$	WBGT (°C)	Exposure (h)
1	400	31.8	1
2	300	33.9	1
3	400	31.8	2
4	350	34.5	2
5 (rest)	100	22.8	2

24. What are the three general factors that affect the level of heat stress of individuals who work in hot environments?

25. How should the TLV of heat stress be modified to account for both unrestrictive clothing (such as shorts and a T-shirt) and restrictive clothing (such as a suit of armor)? Also state what is meant by restrictive clothing.

26. List the personal characteristics that can affect the level of heat strain in response to heat exposure.

27. Explain how females tolerate hot-humid climates better than males, and how males can better tolerate hot-dry climates than females.

28. Studies have shown that there are no significant differences between core body temperature and skin temperatures of females and males. Explain how this can be a disadvantage for females.

29. Explain how a physically fit individual under heat stress can handle higher workloads easier than less fit individuals.

30. Explain why a skilled worker can tolerate working in hot environment better than an unskilled worker.

31. Define the heat acclimatization process. How long does it usually take to become heat-acclimatized, and what are the signs of such acclimatization?

32. List four strategies for protection against heat stress.

33. What are the body's reactions to extreme cold exposure? Also, list the three common types of cold injuries.

34. What are chilblains?

35. What is frostbite? What are its symptoms?

36. Calculate the windchill factor for a climatic condition with a dry-bulb temperature of $-10°C$ and wind velocity of 9 km $\cdot$ h^{-1}. Also, find the equivalent temperature on exposed flesh for this condition.
37. List five strategies for protection against cold stress.

REFERENCES

ACGIH (1994) *Threshold Limit Values for Chemical Substances and Physical Agents and Biological Exposure Indices for 1994–1995*. American Conference of Governmental Industrial Hygienists, Cincinnati.

American Red Cross (1981) *Multimedia Standard First Aid Student Workbook*. American National Red Cross, Washington, DC.

ASHRAE standard 55–1981 (1981) *Thermal Environmental Conditions for Human Occupancy*. American Society of Heating, Refrigeration, and Air Conditioning Engineers (ASHRAE), Atlanta.

Åstrand, P-O. and Rodahl, K. (1986) *Textbook of Work Physiology: Physiological Bases of Exercise*, 3rd edn. McGraw-Hill, New York.

Bell, P.A. and Greene, T.C. (1982) Thermal stress: physiological comfort, performance, and social effects of hot and cold environments. In: *Environmental Stress* (Evans, G.W., ed.). Cambridge University Press, New York, NY.

Dukes-Dobos, F.N. (1981) Hazards of heat exposure: a review. *Scandinavian Journal of Work, Environment and Health* 7: 73–83.

Edholm, O.G. (1967) *The Biology of Work*. World University Library, McGraw-Hill, New York.

Fanger, P.O. (1970) Conditions for thermal comfort – introduction of a general comfort equation. In: *Physiological and Behavioral Temperature Regulation* (Hardy, J.D., Gagge, A.P. and Stolwijk, J.A.J., eds). Charles C. Thomas, Springfield, IL.

Fanger, P.O. (1973) The influence of age, sex, adaptation, season and circadian on thermal comfort criteria for man. In: *Annexe 1973–2 au Bulletin de l'Institut International du Froid, Proceedings from Meeting of IIR (Commission E1)* Vienna, pp. 91–97.

Hammer, W. (1989) *Occupational Safety Management and Engineering*, 4th edn. Prentice-Hall, Englewood Cliffs, NJ.

Holzle, A.M., Munson, D.M., McCullough, E.A. and Rohles, F.H. (1983) A validation study of the ASHRAE summer comfort envelope. *ASHRAE Transactions* 89: 126–138.

Kamon, E., Avellini, B. and Krajewski, J. (1978) Physiological and biological limits to work in the heat for clothed men and women. *Journal of Applied Physiology: Respiration, Environmental and Exercise Physiology* 44: 918–925.

Karpovich, P.V. and Sinning, W.E. (1971) *Physiology of Muscular Activity*, 7th edn. W.B. Saunders, Philadelphia.

Kroemer, K.H.E., Kroemer, H.B. and Kroemer-Elbert, K.E. (1994) *Ergonomics: How to Design for Ease and Efficiency*. Prentice Hall, Englewood Cliffs, New Jersey.

Lee, D.H.K. (1964) Terrestrial animal in dry heat: man in the desert. In: *Handbook of Physiology* (Dill, D.B., ed.). American Physiology Society, Washington, DC.

Leithead, C.S. and Lind, A.R. (1964) *Heat Stress and Heat Disorders*. Cassell, London.

MacFarlane, W.V. (1963) General physiology mechanism of acclimatization. In: *Medical Biometeorology* (Troup, S.W., ed.). Elsevier, New York, pp. 372–417.

Malley, C.B. (1992) Cold stress revisited. *Professional Safety* **37**: 21–23.

Martinson, M.J. (1977) Heat stress in Witwatersrand gold mines. *Journal of Occupational Accidents* **1**: 171–193.

NIOSH (1986) *Working in Hot Environments*. National Institute for Occupational Safety and Health, Center for Disease Control, Public Health Service, US Department of Health and Human Services, DHHS (NIOSH) publication no. 86–112.

Passmore, R. and Durnin, J.V.G.A. (1955) Human energy expenditure. *Physiological Reviews* **35**: 801–840.

Poulton, E.C. (1970) *Environment and Human Efficiency*. C.C. Thomas, Springfield, IL.

Ramsey, J.D. (1978) Abbreviated guidelines for heat stress exposure. *American Industrial Hygiene Association Journal* **39**: 491–495.

Ramsey, J.D. (1983) Heat and cold. In: *Stress and Fatigue in Human Performance* (Hockey, G.R.J., ed.). John Wiley, New York.

Ramsey, J.D. (1985) Environmental factors. In: *Industrial Ergonomics: A Practitioner's Guide* (Alexander, D.C. and Pulat, B.M. eds.). Industrial Engineering and Management Press, Institute of Industrial Engineers, Norcross, Georgia, pp. 85–94.

Siple, P.A. and Passel, C.F. (1945) Measurements of dry atmospheric cooling in subfreezng temperatures. *Proceedings of the American Philosophical Society* **89**: 177–199.

Stephenson, R.R., Colwell, M.O. and Dinman, B.D. (1974) Work in hot environments: II. Design of work patterns using net heat exchange calculations. *Journal of Occupational Medicine* **16**: 792–795.

Tayyari, F. (1995) Factors affecting the level of heat strain. In *The Proceedings of the Konz/Purswell Occupational Ergonomics Symposium*. Institute for Ergonomics Research, Department of Industrial Engineering, Texas Tech University, Lubbock, Texas, pp. 87–90.

Van Graan, C.H. (1975) A cooling suit for use in hot environments in gold mines. In: *The Proceedings of International Mine Ventilation Congress*. Johannesburg, South Africa.

WHO (1969) *Health Factors Involved in Working Under Conditions of Heat Stress*. World Health Organization Technical Report series no. 412, World Health Organization (WHO), Geneva.

Light and vision   13

13.1 INTRODUCTION

Light is the stimulus for vision. Thus, insufficient or too strong and, particularly, glaring illumination causes visual inefficiency, resulting in fatigue, headache, dizziness, and increased accident risk. Large amounts of time and money are lost due to visual inefficiency on jobs (Nakagawara, 1990). Good lighting, that enhances visual performance, has many advantages, including fewer mistakes, increased productivity, reduction of accidents, improved morale, and improved housekeeping. Ergonomics principles can be applied to achieve these goals through visual improvement.

This chapter offers a basic knowledge of light, methods for its control and measurement, and guidelines for visual improvement.

13.2 ELECTROMAGNETIC SPECTRUM

Heinrich R. Hertz (1857–1894), who first experimentally verified the existence of electromagnetic waves, found that all such waves (including light) travel at the same speed in a vacuum. These waves are different from one another in their frequency and wavelength. The relationship between the speed or velocity (V), frequency (f), and wavelength (λ) of an electromagnetic wave is as follows:

$$V = f \cdot \lambda \tag{13.1}$$

Since the speed (or velocity) of electromagnetic waves is constant, when the frequency changes, the wavelength will also change. The lower the frequency of the vibrating wave, the longer the wavelength and vice versa.

The classification of electromagnetic waves according to their frequency or wavelength is called the **electromagnetic spectrum**. Figure 13.1 gives a schematic illustration of the electromagnetic spectrum. The

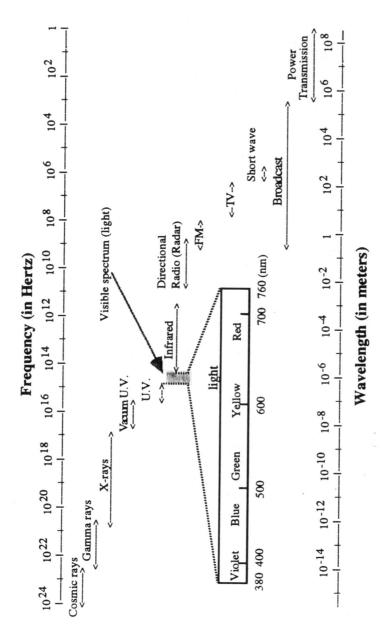

Fig. 13.1 The electromagnetic spectrum. Adapted from General Electric (1974).

Table 13.1 Primary colors of white light

Wave length (nm)	Color
380–435	Violet
435–480	Blue
500–560	Green
580–595	Yellow
595–650	Orange
650–670	Red

range of visible light is from 380 to 760 nanometers (nm). (The nanometer, formerly called millimicron, is a unit of wavelength equal to one-billionth (10^{-9}) of a meter.) The visible-light portion of the spectrum may be divided into the six principal colors, as listed in Table 13.1 (Bumsted, 1973). These colors can be seen when white light is diffracted into its primary colors upon passing through a prism.

Ultraviolet and infrared are the upper and lower limits of the visible-light frequencies. The absorption of ultraviolet energy causes energy changes involving ionization of atoms and molecules. The portion of the sun's energy that causes sunburn or other skin damages is ultraviolet. The infrared portion of the spectrum generates heat. The absorption of infrared energy causes molecular vibrations, such as bending and stretching of the interatomic bonds, and molecular rotation (Bumsted, 1973).

13.3 LIGHT AND ITS PHYSICAL CHARACTERISTICS

As defined by the Illumination Engineering Society (IES, Nomenclature Committee, 1979), light is "radiant energy that is capable of exciting the retina (in the eye) and producing a visual sensation." As previously mentioned and illustrated in Figure 13.1, the visible spectrum (light) ranges from about 380 to 760 nm. Light possesses the following three physical characteristics:

- Light travels in a straight line. It can be modified or redirected by means of a reflecting, refracting, or diffusing medium.
- Light waves pass through one another without alteration of either. That is, a ray (beam) of green light will pass directly through a ray (beam) of red light without changing the directions and colors of each other.
- Light is invisible in passing through space unless a particle or medium, such as dust, dirt, water, etc., scatters it.

13.4 LIGHT CONTROL

Light can be controlled in a number of ways by application of one or more of the following physical phenomena: reflection, refraction, polarization, diffraction, diffusion, absorption, and transmission. These phenomena are briefly described in this section. The reader is referred to basic physics books (Krauskopf and Beiser, 1966; Hudson and Nelson, 1990) for detailed explanations of these subjects.

13.4.1 REFLECTION

Reflection is a process by which a part of the light waves falling on an object (or medium) leaves that object from the incident side (Fig. 13.2). Reflection may be specular, spread, or diffused.

- **Specular reflection**: If a polished surface is perfectly smooth, it reflects specularly; that is, the angle between the incident ray and the normal to the surface will be equal to the angle between the reflected ray and the normal (Fig. 13.2).
- **Spread reflection**: If a polished surface is figured in any way (i.e., hammered, deeply etched, or engraved), it spreads out any rays it reflects (Fig. 13.3). Hence, this type of reflection is called spread reflection.

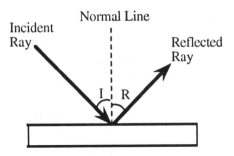

Fig. 13.2 Specular reflection of a light beam ($I = R$).

Fig. 13.3 Spread reflection of a light beam by a rough surface.

● **Diffused reflection**. If a medium has a rough surface or is composed of minute crystals or pigment particles, its reflection is diffused; that is, every incident ray is reflected at many angles (Fig. 13.4).

13.4.2 REFRACTION

Refraction is the process of bending of a light ray as it passes from one transparent medium to another of different optical density.

The velocity of the light (the speed of its propagation, not its frequency) changes when it leaves one material and enters another of greater or lesser optical density. The speed is reduced if the material entered is denser and increased if it is less dense. Except when the light enters at an angle perpendicular to the surface of the medium of different density, the change in speed is accompanied by a bending of the light from its original path at the point of entrance. This phenomenon is known as refraction. A ray passing from a rare to a denser medium is bent toward the normal, while a ray passing from a dense to a rare medium is bent away from the normal (Fig. 13.5).

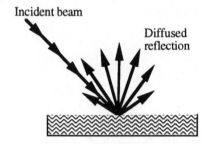

Fig. 13.4 Diffused reflection of a light beam by a rough surface.

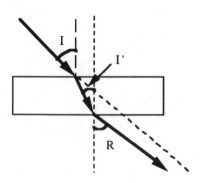

Fig. 13.5 Refraction of light rays.

13.4.3 POLARIZATION

Polarization is the process of modifying the vibration behavior of light waves. Light waves move away from the light source in a spherical form in all directions. A cross-section of such a sphere (Fig. 13.6) is analogous to circular water waves moving away from the point where a small object strikes the surface of a pool.

Light waves can be modified by polarization such that the vibrations are in one plane. The sky, water surfaces, transparent materials which are internally stressed, and multilayer glass or plastic are examples of polarizing media. Figure 13.7 illustrates how polarizing materials act as if they were screens with parallel slits which allow a light to vibrate only in the plane of the slits. When a second screen is turned 90° to the first one, then it blocks the light wave passed through the first screen so that no light will pass through the combination of the two.

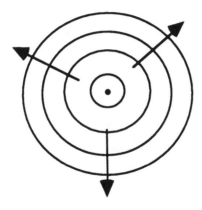

Fig. **13.6** A cross-section of spherical light waves moving away from its source.

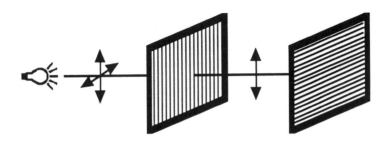

Fig. **13.7** A polarization process.

13.4.4 DIFFRACTION

Diffraction is the process by which light waves bend out of their straight lines and spread around the edges of an object, or light beams pass through an opaque object (e.g., a glass prism) and spread into a region behind the object (Fig. 13.8).

13.4.5 DIFFUSION

Diffusion is the scattering of light waves in many directions, or the breaking-up of a beam of light and spreading its rays in many directions (Fig. 13.9).

13.4.6 ABSORPTION

Absorption is the ability of a medium to soak up the light; that is, neither to reflect nor transmit light. How much light is absorbed depends on the surface, density, shape, and color of the medium.

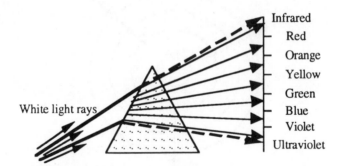

Fig. 13.8 Representation of a prism bending rays of white light.

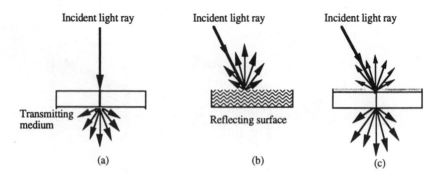

Fig. 13.9 Diffusion process. (a) Transmitted diffused rays; (b) reflected diffused rays; (c) partially reflected and transmitted diffused rays.

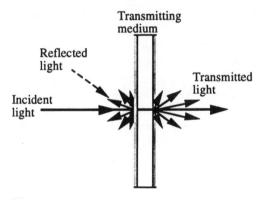

Fig. 13.10 Illustration of light transmission.

13.4.7 TRANSMISSION

Transmission is the ability of light waves to pass through a medium (Fig. 13.10).

13.5 PHOTOMETRY AND LIGHTING TERMINOLOGY

Photometry is a branch of metrology that is concerned with the measurement of light. The measurement is usually made with a sensing device that has a spectral sensitivity close to the sensitivity of the average human eye. The principal quantity in photometry is luminous flux, which is measured in lumens.

The terms often used in photometry (i.e., measurement of light) or in the design and evaluation of illuminated environments include: luminous intensity, illumination level, brightness (luminance), reflectance, contrast, and luminaries.

13.5.1 RADIANT POWER

Radiant power is the time rate of flow of radiant energy, which is called radiant flux. It is measured in watts (W).

13.5.2 LUMINOUS INTENSITY AND LUMINOUS FLUX

Luminous intensity, which is also called candlepower, is the amount of light output emitted from a source in a specific direction and is measured at the light source. The unit of luminance intensity is the **candela** or **candle** (cd). (Note that candlepower gives information about the luminous flux at its origin independent of the distance from the source at which it is measured.)

Luminance flux, which is also called luminous energy, is the time rate of flow of light output from a light source emitted initially in all directions. The lumen (lm) is the unit of light output from a source. Incandescent lamps have an efficiency of 17–23 lm · W^{-1}, while fluorescent lamps or tubes have an efficiency of 50–80 lm · W^{-1} (Lum-i-neering Associates, 1979; Cushman and Crist, 1987). For example, a 100-W incandescent lamp emits about 1700 lm initially, whereas a 40-W cool-white fluorescent lamp emits about 3200 lm initially (Kaufman, 1973). A light source of 1 cd produced 4π lm (i.e., 12.5664 lm).

13.5.3 ILLUMINATION LEVEL

Illumination level is the density or quantity of the flow of light falling on a surface and is measured in lux (lx) in the International System of Units (SI units) and in foot-candles (fc) in the British System (or the US Customary System, USCS) of measurement units. One lx is equal to 1 lm · m^{-2} and 1 fc is equal to 1 lm · ft^{-2}. One fc is the amount of light from a standard candle, which produces 4π lm or 12.5664 lm of luminous flux per square foot of the surface of a sphere of 1-ft radius surrounding the candle. Thus, the illumination level at a distance of d units from a light source, of some luminous intensity emitting a luminous flux in all directions, can be calculated as follows:

$$\text{Illumination} = \frac{\text{candlepower or luminous flux (in lumens)}}{4\pi \cdot d^2} \qquad (13.2)$$

where d is the radius of a sphere surrounding the light source and $4\pi \cdot d^2$ is the surface area of the sphere. Since 1 cd equals 4π lm, equation 13.2 can be reduced to the following equation:

$$\text{Illumination} = \frac{\text{candlepower or luminous flux (in cd)}}{d^2} \qquad (13.3)$$

where d is the distance of the illuminated surface from the light source. This equation shows that the level of illumination is inversely proportional to the square of the distance between the source and the surface. This is known as the inverse-square law, which is applicable only to point sources, and is illustrated in Figure 13.11.

Using the corresponding distance units, equation 13.3 can be rewritten for measurement of illumination in SI and USCS as follows:

$$\text{Illumination (lx)} = \frac{\text{candlepower (cd)}}{d^2 \text{ (in m}^2)} \quad \text{(SI units)} \qquad (13.4)$$

or

$$\text{Illumination (fc)} = \frac{\text{candlepower (cd)}}{d^2 \text{ (in ft}^2)} \quad \text{(USCS units)} \quad (13.5)$$

where d is the distance between light source and the illuminated surface.

For example, a light source emits 1000 lm on a surface of 1 m² and 1 m away from the source, the illumination level on the surface would be 79.6 lx (note that 1000 lm ÷ 4π = 79.6 cd). If the surface is located at 2 m away from the source, the illumination level is reduced to 19.9 lx (i.e., 79.6 ÷ 4).

The Lambert cosine law states that the illumination of any surface varies as the cosine of the angle of the incident light beam (Fig. 13.12). The angle of incidence (θ) is the angle between the normal to the surface and the direction of the incident light.

The inverse-square law and the cosine law may be combined and represented in the following general equation:

$$\text{Illumination (lx)} = \frac{\text{candlepower (cd)}}{d^2 \text{ (in m}^2)} \times \text{Cos } \theta \quad \text{(SI units)} \quad (13.6)$$

or

$$\text{Illumination (fc)} = \frac{\text{candlepower (cd)}}{d^2 \text{ (in ft}^2)} \times \text{Cos } \theta \quad \text{(USCS units)} \quad (13.7)$$

Conversion factors for the illumination level and other units of measurement useful in photometry are provided in Table 13.2.

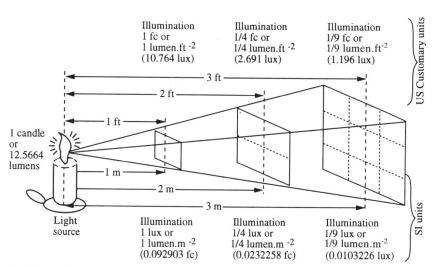

Fig. 13.11 Illustration of the inverse-square law of light distribution from a light source.

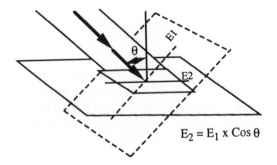

$$E_2 = E_1 \times \text{Cos } \theta$$

Fig. 13.12 The Lambert cosine law: E_1 and E_2 are respectively the illumination levels on the plane perpendicular to the light beam and on the surface whose normal line makes an angle (θ) with the direction of the incident light.

Table 13.2 Conversion factors for the units used in photometry

Measurement	Unit	System of measurement	Conversion factor*
Candlepower	Candela (cd)	USCS and SI	1 cd = 4π or 12.566371 lm
	Lumen (lm)	USCS and SI	1 lm = $1/4\pi$ or 0.0795775 cd
Illumination level (illuminance)	Foot-candle (fc)	USCS	1 fc = 1 lm·ft^{-2} = 10.76391 lx†
	Lux (lx)	SI	1 lx = 1 lm·m^{-2} = 0.092903 fc
Luminance	Foot-Lambert (fL)	USCS	1 fL = 3.4262591 cd·m^{-2}
	Candela per square meter (cd·m^{-2})	SI	1 cd·m^{-2} = 0.2918635 fL

*The conversion factors have not been rounded.
†1 fc ≈ 10.76391 lx since 1m^2 = 10.76391 ft^2.
USCS = US Customary System; SI = International System.

13.5.4 LUMINANCE (BRIGHTNESS)

Luminance is the amount of light per unit area reflected from or emitted by the surface. Brightness and luminance are frequently used as two equivalent terms, but brightness is actually the intensity of the subjective sensation that results from viewing a surface.

Luminance in SI units, where the unit of surface is square meters, is measured in candles per square meter (cd · m^{-2}; equation 13.8). In the British system (or the USCS) of measurement units, when the area of surface is in square feet, luminance is measured in foot-Lamberts (fL; equation 13.9).

$$\text{Luminance (cd} \cdot \text{m}^{-2}) = \frac{\text{illumination (lx)}}{\pi} \times \text{reflectance (\%)} \quad \text{(in SI units)}$$

$$(13.8)$$

$$\text{Luminance (fL)} = \text{illumination (fc)} \times \text{reflectance (\%)} \quad \text{(in USCS units)}$$
$$(13.9)$$

One fL equals the amount of luminance obtained from a perfect diffusing and reflecting surface illuminated by 1 fc. One fL is equal to $3.426 \text{ cd} \cdot \text{m}^{-2}$ since $1 \text{ fc} = 10.764 \text{ lx}$ and $10.764 \text{ lx} \div \pi = 3.426 \text{ cd} \cdot \text{m}^{-2}$.

13.5.5 REFLECTANCE

Reflectance is a measure of how much light is reflected on a surface. It is the ratio of the luminance (brightness) of a surface to the illumination level on the surface and is measured as a percentage. Reflectance is defined by the following equations:

$$\text{Reflectance (\%)} = \frac{\pi \times \text{luminance (cd} \cdot \text{m}^{-2})}{\text{illumination (lx)}} \times 100 \quad \text{(using SI units)}$$

$$(13.10)$$

$$\text{Reflectance (\%)} = \frac{\text{luminance (fL)}}{\text{illumination (fc)}} \times 100 \quad \text{(using USCS units)} \quad (13.11)$$

A perfect reflecting and diffusing (i.e., perfectly white) surface would not absorb light and, thus, would have a reflectance of 100%. A completely black surface would absorb light and, therefore, would have a reflectance of 0%. However, in reality, most surfaces have a reflectance of between 5 and 95%.

13.5.6 CONTRAST

Contrast is a measure of luminance difference between a target (task) and its background as defined by the following equation:

$$\text{Contrast (\%)} = \frac{L_b - L_t}{L_b} \times 100 \quad (13.12)$$

where L_t and L_b are the luminance of the target and its background, respectively.

Contrast can vary from 100% (positive) to zero for targets darker than their backgrounds, and from zero to minus infinity ($-\infty$) for targets brighter than their backgrounds (Grether and Baker, 1972).

13.5.7 LUMINAIRES

A luminaire refers to all hardware used in a light fixture consisting of one or more lamps together with all hardware (directing or diffusing elements, sockets, covers, and wires and cables) to distribute the light, to position and protect the lamps and to connect the lamps to the power supply.

13.6 LIGHTING SURVEYS

Evaluation of a workplace lighting condition depends on the quality of the lighting survey procedures. The quality of survey procedures, in turn, depends on the type and amount of information obtained in the survey, and on the surveyor's knowledge of operations and the corresponding illumination recommendations. During a typical lighting survey the following data are recorded (Kaufman, 1973):

- **Description of workplace**: This includes dimensions, colors, reflectance and conditions of surrounding surfaces in the illuminated area, and air temperature near luminaires.
- **Description of general lighting system**: This includes data regarding quantities, conditions, lamps, wattage outputs, spacing, mounting, and distribution. Description of any supplementary lighting system is also recorded.
- **Description of instruments used**: The information about manufacturers, models, and date of last calibration of instruments used in the lighting survey.
- **Light quantitative measurements**: Illumination and luminance at specific locations depending on type of lighting systems used and viewing directions. These and other essential measurements, and measurement instruments, are described in subsequent subsections.
- **Subjective evaluation of the lighting**: This includes answers to a series of questions regarding the light-servicing procedure and subjective evaluation of the lighting conditions.

13.6.1 QUANTITATIVE LIGHT MEASUREMENTS

A complete lighting survey should include the following measurable quantities:

- **Illumination levels of (fc or lx)**: illumination or lighting levels are measured at various task locations for task lighting, and at various locations throughout the workplace for general lighting, and on various surfaces for determining luminances and reflectances in the task surroundings.

- **Luminance (fL or cd · m^{-2})**: Luminance of luminaires, and tasks and their surrounding surfaces are measured.
- **Reflectance (%)**: Reflectance percentages of various surfaces are determined.
- **Temperature**: The air temperature near luminaires is measured to account for the heat generated by these heat sources.
- **Voltage**: The voltage input to the luminaires is measured.

13.6.2 LIGHT MEASUREMENT INSTRUMENTS

Since most illumination measurements and recommendations are given in foot-candles (or lux), it is desirable to have a device (meter) that measures foot-candles (or lux) directly. Illumination (light) meters are commercially available and can be purchased from a number of sources. The desired level of accuracy determines the type of meter to be used.

When light meters are not available, an ordinary photographic light meter can be used. Table 13.3 presents conversion factors to convert

Table 13.3 Converting exposure values (EVs) to foot candles

EV	Lux (lx)	Approximate foot-candles (fc)
−5	0.17	0.016
−4	0.35	0.032
−3	0.70	0.065
−2	1.40	0.13
−1	2.80	0.26
0	5.5	0.50
+1	11	1
+2	22	2
+3	44	4
+4	88	8
+5	175	16
+6	350	32
+7	700	65
+8	1400	130
+9	2800	260
+10	5500	500
+11	11000	1000
+12	22000	2000
+13	44000	4000
+14	88000	8000
+15	175000	16000
+16	350000	32000

exposure values (EV) to foot-candle readings (and lux readings). To determine the equivalent foot-candle reading, set the ISO (ASA) film speed scale to 50, and slide the spherical diffuser (white dome) over the cell window, as if taking incident light readings. Point the meter dome at the light source and null the meter needle as in normal light-meter readings. Equivalent foot-candle readings can be obtained from the EV values obtained from the meter.

Luminance meters are also available to measure luminance (brightness). If the reflectance of a surface is known, a light meter can be used to measure illumination and to determine the corresponding luminance (luminance = illumination × reflectance).

Reflectance can be measured using a Baumgartner reflectometer, or by visual comparison of the surveyed surface with color chips of known reflectance. However, if illumination and luminance are known, the reflectance is calculated by dividing luminance by illumination.

13.7 THE VISUAL APPARATUS

A schematic representation of the visual process is given in Figure 13.13. The visual apparatus consists of the following three elements: the eyes, eye muscles, and nerve pathways which take part in the seeing process.

Light passes through the cornea and lens. Its reception takes place in the photoreceptors of the retina. The retina consists of layers of photoreceptors and nerves to transform light into nerve signals that can

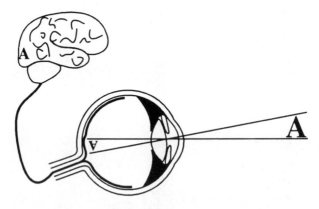

Fig. 13.13 A schematic representation of the visual process:
(1) Light passes through and is regulated in the cornea and lens.
(2) Light reception takes place in the retina.
(3) Information is transmitted along the optic nerve to the brain.
(4) Visual perception of the outside world takes place in the consciousness.
Adapted from Grandjean (1988).

be sent to the brain for processing. The majority of the photoreceptors in the retina are rods, which are sensitive to black and white vision and are used in night vision and to detect movement in the visual field. Color vision is initiated in the cones, which are the other photoreceptors in the retina. The cones are concentrated in the center of the retina, directly behind the lens (at the fovea), and provide the highest visual acuity and color vision. The rods are more widely spread and are important in peripheral vision. Once the photoreceptors transform the light into a neural signal, the information is transmitted along the optic nerve to the brain. Visual perception of the outside world occurs in the conscious sphere of the brain (not in the eye).

The human eye (Fig. 13.14) is similar to a photographic camera, as described below:

- The retina corresponds to the photo-sensitive film, on which the image of an object falls.
- The cornea, the lens, and the pupil together are similar to the focusing lens.
- The iris corresponds to the light-regulating diaphragm of a camera.

The eye lens is flexible, which allows the ciliary muscles to adjust its focusing power for forming sharp images of objects at varying distances. The ciliary muscles contract and elongate the lens to allow focusing on far-away objects. To focus on near objects, the ciliary muscles relax, allowing the lens to become more rounded. The ability of the eye to change its focal length is called accommodation. When the eye muscles are not sufficient, or the lens loses its ability to change shape, vision is corrected externally with glasses.

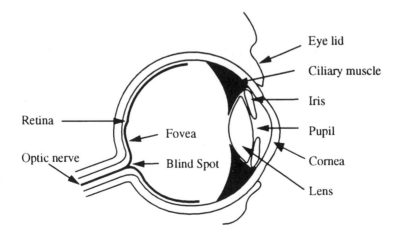

Fig. 13.14 The human eye.

13.8 VISUAL ABILITY

Visual ability or visual performance is the relationship between light and sight. Visual performance has a significant effect on psychomotor performance. Poor visual performance results in fatigue. The factors affecting visual ability can be grouped into two categories: physiological and physical factors.

13.8.1 PHYSIOLOGICAL FACTORS

Physiological factors affecting visual ability include: visual acuity, visual field, accommodation, adaptation, aging, and several other factors.

13.8.1.1 Visual acuity

Visual acuity refers to sharpness of vision. It is the ability of the eye to discriminate black and white detail. Visual acuity is evaluated by ophthalmologists using the Snellen letter chart. The chart consists of several rows of block letters of decreasing size (Fig. 13.15), placed at a distance of 20 ft from the test subject. The result of such evaluation is recorded as 20–X, an indication that the person whose visual acuity is being evaluated can see a detail from a 20-ft distance while a person with normal vision can see that detail from a distance of X ft. A normal vision has a 20/20 score. Therefore, a 20/15 vision is better than normal since a detail that is normally seen at a distance of 15 ft can be seen by the tested subject from a 20-ft distance. A 20/50 vision is poorer than normal since the subject can see a detail from a 20-ft distance, while it can be seen by a person with normal vision at a distance of 50 ft.

Another method of testing visual acuity is the minimum separable areas that can be distinguished. The result is expressed in terms of the inverse of the visual angle subtended at the eye by the smallest detail detected by the eye (Fig. 13.16). A visual angle of one minute arc is often considered as normal acuity (Grandjean, 1988) since its inverse is 1 (i.e., 100%). It should be noted that the ability to make such discriminations at near distances is relatively independent of the ability to do so at far distances (McCormick and Tiffin, 1974). Hence, if job performance is dependent on both near and far visual acuity, both should be tested.

13.8.1.2 Visual field

Visual field is the area which is visible when the head and eyes are kept still. The visual field is divided into:

- the area of maximum focus: vertical angle of 1°;
- the middle field: vertical angle of 40°;
- the outer field: vertical angle of 40–70°.

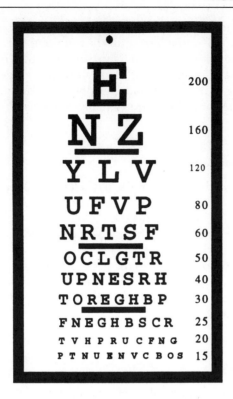

Fig. 13.15 Illustration of the general appearance of the Snellen letter chart. Adapted from McCormick and Tiffin (1974).

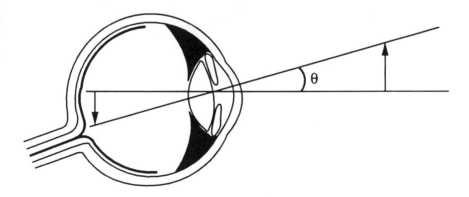

Fig. 13.16 The visual angle subtended at the eye by a small detail.

Within the middle field, sharp vision is not possible, but movement is noticed, and the eye shifts very quickly from one object to another so that visual perception remains undisturbed. The outer field corresponds to the extreme parts of the visual field which is bounded by eyebrows, nose and cheeks. In the outer field (peripheral vision) movement can be detected, but color and detail are virtually non-existent.

13.8.1.3 Accommodation

Accommodation is the ability of the eye's lens to focus on near and far objects. It is done by changing the thickness of the lens (from a relatively flat to a more convex shape). Accommodation also includes changes in pupil diameter or size. When the eye is focused on farther objects, the pupil is larger and vice versa.

As an object is brought close to the eyes, it reaches a certain point after which the object will be seen as two. This point is called the **near point of accommodation**. It is also the nearest point of accommodation without losing the details.

13.8.1.4 Adaptation

Adaptation is the ability of the eye to adjust to changing light conditions. The pupil regulates the amount of light entering the eye by changing its diameter, getting larger as the light level decreases and smaller as the light level increases. Dark adaptation takes several minutes to occur. When the lights are turned off, everything becomes very dark. However, after a few minutes you can begin to see objects and shapes. Initial dark adaptation occurs in 15–30 min, with total dark adaptation taking an hour or longer. It is important to recognize the performance limitations during dark adaptation and not expect workers to adapt immediately. Light adaptation is much faster and only takes a few minutes, as can be demonstrated when walking out of a dark movie theater into daylight, or turning on a room light at night.

13.8.1.5 Aging

The human eye deteriorates with increasing age; the lens loses elasticity, and as a result the focal point (the nearest point on which the eye can be focused sharply) moves farther away from the eye. The far focal point (the farthest point on which the eye can be focused sharply) generally remains the same.

13.8.1.6 Other factors

There are several other physiological factors that affect the effectiveness of visual performance. Examples of these factors include: depth perception, phorias (the ability to converge the two images of the same object on the two retinas into one image), color discrimination and color blindness, and night blindness.

13.8.2 PHYSICAL FACTORS

Physical or external factors that affect the visual performance include visual tasks and the visual environment.

13.8.2.1 Visual tasks

Visual tasks include all the things that have to be seen at a given moment. The visibility of the task is determined by the following four factors:

- size of the task (the object to be seen);
- time available for viewing the task;
- brightness of the task (the object);
- contrast (brightness and/or color) with other surfaces in the task background.

13.8.2.2 Visual environment

Visual environment includes not only the immediate task area, but also all other areas that will be seen continuously during a typical work day (or job performance), as shown in Figure 13.17.

13.9 QUANTITY OF ILLUMINATION

Quantity of illumination refers to the proper distribution of sufficient illuminance (lux or foot-candles) to provide the needed luminance (candelas per square meter or foot-Lamberts) for the visual task. In determining the amount of illumination, the four task factors (i.e., the size of the object, time to see the task, brightness of the task, and contrast) must be considered.

13.9.1 THE SIZE OF THE TASK

The larger the object to be seen, the easier it is to see. The visual size of any object depends on the size of the object and the distance from the object to the viewer. The visual size is expressed as the angle

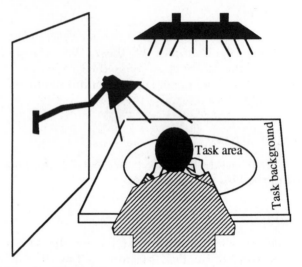

Fig. 13.17 Visual environment.

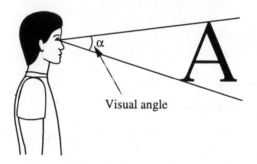

Fig. 13.18 Illustration of the visual angle.

subtended at the eye by the viewed object in degrees (Fig. 13.18). The larger the visual angle of an object, the easier it can be seen. For example, when you bring a small object closer to your eyes to be able to see it easier, you are increasing the visual angle by which the object seems to be larger to you.

13.9.2 THE TIME TO SEE THE TASK

The visual process requires some time to be completed. Therefore, if sufficient time is allowed, the person can see small details, even under low levels of illumination. However, the lower (or poorer) the light, the longer it takes to see the task, while increasing (improving) light increases the speed of vision.

13.9.3 THE BRIGHTNESS OF THE TASK

The brightness of the task depends on both the amount of light reaching it and its reflectivity. Increasing the amount of light falling on an object increases the brightness of the object.

13.9.4 THE CONTRAST

The contrast of the task refers to the relative differences between the target and its immediate surroundings. The differences in contrast generally refer to differences in reflection, color, or patterns between the target and the background. As shown in Figure 13.19, when the contrast is reduced, the effort required in seeing the target increases.

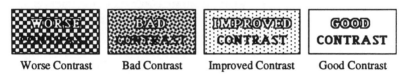

| Worse Contrast | Bad Contrast | Improved Contrast | Good Contrast |

Fig. 13.19 Poor contrast causes difficulties in seeing.

It should be remembered that size, viewing time, and contrast are the visual factors which may be inherent in the task, but brightness is a visual factor that can be easily controlled by changing the illumination level.

13.10 GLARE

Glare is the sensation produced by luminance (or brightness) within the visual field which is sufficiently greater than that to which the eyes can adapt (IES Committee on Industrial Lighting, 1991). Glare can cause annoyance, loss of visual performance and visibility, or discomfort. Glare can, therefore, be referred to as light noise since it is any light level in excess of what is required for performing the visual task under a comfortable visual condition.

An extreme example of industrial glare is the flash of a welding process that produces ultraviolet and infrared rays of energy. Ultraviolet rays can cause damage to the cornea. Infrared rays would pass through the cornea, but can damage the retina.

Glare may arise either from the light sources being very close to the line of sight or from reflections on the task. The IES Committee on Industrial Lighting (1991) has divided glare sources into two categories.

- **direct glare**, which results from high luminance of an insufficiently shielded light source in the visual field;

- **indirect** (or **reflected**) **glare** that results from images of high-luminance sources reflected by specular (i.e., shiny) surfaces in the visual field.

With respect to its effects on individuals, glare may be classified into three groups:

- **discomfort glare**, which causes a subjective feeling of annoyance (usually eye strain);
- **disability glare**, that can cause a diminished visual performance and ability to see;
- **photostress glare** from an intense light source that causes a delay in the recovery of the visual system after the macular region has been constantly stimulated with such a light source. This delay is normally 30 s or less (Nakagawara, 1990). (The macular region is the area in the eye near the center of the retina, which contains cone vision for detailed vision and color, and at which visual perception is most acute.)

13.11 VISUAL PROBLEMS

Most visual problems are very difficult to identify through review of the company's records of health and safety or productivity. They may not present obvious impacts on the operator's health, other than the operator's complaint of discomfort. Nevertheless, in addition to visual discomfort, inadequate lighting can lead to visual fatigue and mental weariness. Therefore, poor illumination can indirectly, through visual and mental fatigue, result in the following problems:

- irritation, fatigue, watering, and reddening of the eye;
- double vision;
- headache;
- decrease in the power of accommodation and convergence;
- decrease in visual acuity, contrast sensitivity, and speed of perception.

Although the human eye is able to adapt to a wide range of illumination levels, it will undergo discomfort and strain if exposed to too high (glare) or too little (insufficient) levels of light. The discomfort resulting from glare is due to the muscular effort to maintain the iris in a tighter (almost closed) position.

13.12 ILLUMINATION GUIDELINES

Appropriateness of illuminance for a given work environment depends on several factors. The visual difficulty (e.g., the size of the smallest details), importance and frequency of each task or job element, and time available for each task must all be considered in designing workplace lighting. Other important factors to be considered include worker comfort, age-dependent vision and expectations; the appearance of the workplace; and economics and feasibility.

Since the interrelationships among the factors affecting the illuminance appropriateness are complex, designers should not solely concentrate on lighting. They should rather consider various options. For example, tasks may be modified to become less difficult and require lower levels of general workplace illumination. Supplementary lighting, or special task lighting, is another option that can be used to illuminate the difficult tasks, rather than increasing general illumination.

As a general rule, the minimum level of illumination for the workplace, where multiple tasks are performed, must be determined based on the most difficult and critical tasks.

13.12.1 GENERAL RECOMMENDATIONS FOR VISUAL ENVIRONMENT

In order for the eyes to function efficiently and comfortably the luminances within the visual environment should not be too different (Kaufman, 1973). The eyes become adapted to the task luminance during task performance. However, the eyes frequently shift back and forth from the task to view the other objects of higher or lower luminance. In such a process, the task visibility can easily be disturbed. The following rules should be followed to achieve visual comfort:

- All objects and major surfaces in the visual field should be equally bright.
- The working field should be brightest in the middle and darker towards the edges.
- Surfaces in the middle of the visual field should not have a brightness–contrast ratio greater than 3 : 1.
- Contrast between middle and edge of the visual field should not be more than 10 : 1.
- Light sources should not contrast within their background by more than 20 : 1.
- Maximum permissible range of brightness within an entire room should not be more than 40 : 1.
- The use of polished surfaces or reflecting material on machines, table-tops, switch boards, or other apparatus should be avoided.
- To reduce direct visual glare, the angle between the horizontal line of sight and a line from the eye to the light source must be more than 30° (Fig. 13.20). If an angle of less than 30° cannot be avoided, the light must be screened on the sides.

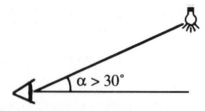

Fig. 13.20 The angle between the horizontal line of sight and a line from the eye to the light source should be greater than 30°.

Table 13.4 American National Standard Institute (ANSI)/Illuminating Engineering Society (IES) recommended maximum luminance ratios

	Environmental classification	
	A	B
Between tasks and adjacent darker surroundings	3 : 1	3 : 1
Between tasks and adjacent lighter surroundings	1 : 3	1 : 3
Between tasks and more remote darker surroundings	10 : 1	20 : 1
Between tasks and more remote lighter surroundings	1 : 10	1 : 20
Between luminaires (or windows, skylights, etc.) and surfaces adjacent to them	20 : 1	*
Anywhere within normal field of view	40 : 1	*

*Luminance ratio control is not practical.
A = Interior areas where reflectances of entire space can be controlled in line with recommendations for optimum seeing conditions.
B = Areas where reflectances of immediate work area can be controlled, but control of remote surround is limited.
ANSI/IES have recommended maximum luminance ratios for type C environmental classification, which includes areas (indoor and outdoor) where it is completely impractical to control reflectances and difficult to alter environmental conditions. It was, therefore, not listed in this table.
Adapted from IES Committee on Industrial Lighting (1991).

Table 13.4 presents maximum luminance ratios for various types of areas and situations which have been recommended by the IES of North America and the American National Standard Institute (ANSI) for industrial lighting.

13.12.2 RECOMMENDATIONS FOR GLARE REDUCTION

- Sunglasses should be used where discomfort glare prevails.
- Light sources should be positioned outside the operator's visual field.
- The workplace should be positioned so that the frequently used lines

of sight do not coincide with reflected light. Light sources should be by the sides, not in front of the operator.

- Naked light sources should be screened with glare shields (e.g., meshes and honeycomb crates).
- Windows and skylights should be tinted, or covered with blinds, curtains, or films.
- Special barriers (e.g., light shelves for high windows and skylights) should be used to redirect light towards the ceiling to provide glare-free softer light in the room.
- Luminosity of screened lights should not exceed 3000 cd·m^{-2} (~875 fL) for general lighting and 2000 cd·m^{-2} (~585 fL) for the workplace.
- As previously mentioned, the angle between the horizontal line of sight and a line from the eye to the unscreened (naked) light source should be greater than 30°.
- Partitions should be used to block light from ceiling light sources (Fig. 13.21).

13.12.3 MINIMUM ILLUMINATION FOR SAFETY OF EMPLOYEES

Table 13.5 gives the minimum illumination levels for safety of employees (Code of Federal Regulations, (CFR) 1926.56). For the recommended levels of illumination for areas and/or operations not shown in this table, or for illumination levels for efficient visual performance rather than for just safety alone, the reader is referred to the American National Standards A11.1-1983 (ANSI, 1983) and ANSI/IES-RP-7-1991 (IES Committee on Industrial Lighting, 1991). Examples of recommended

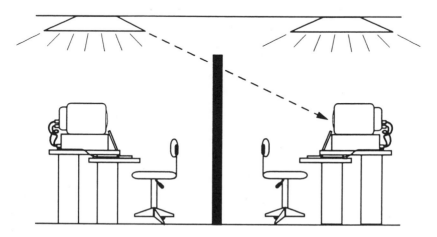

Fig. 13.21 A partition can block ceiling light from falling on computer screens.

Table 13.5 The minimum illumination levels for safety of employees

Area of operation	Foot-candles
General construction-area lighting	5
General construction areas, concrete placement, excavation and waste areas, access ways, active storage areas, loading platforms, refueling, and field maintenance areas	3
Indoors: warehouses, corridors, hallways, and exit ways	5
Tunnels, shafts and general underground work areas (exception: minimum of 10 fc is required at tunnel and shaft heading during drilling, mucking, and scaling. Bureau of Mines-approved cap lights shall be acceptable for use in the tunnel heading)	5
General construction plant and shops (e.g., batch plants, screening plants, mechanical and electric equipment rooms, carpenter shops, rigging lofts and active storerooms, barracks or living quarters, locker or dressing rooms, mess halls, indoor toilets, and workrooms)	10
First-aid stations, infirmaries, and offices	30

Adapted from CFR 1926.56 (1990).

Table 13.6 Examples of recommended illumination levels for efficient visual performance

Task/operation/location	Foot-candles
Fine assembly tasks	100
Accounting, proof-reading, design and drafting, and close work in laboratory or machine shops	80
Reading areas, medical examination areas, paint shops, storage for small parts, service and repair areas, and medium assembly tasks	50
Carpenter shops, conference rooms, rough work in machine shops, cafeteria dining rooms, kitchens, cashier areas, first-aid stations and infirmaries, X-ray and electrocardiogram rooms, and specimen rooms	30
Shipping and receiving areas, rough and easy-seeing assembly tasks	25
Lounges, rest areas, locker rooms, rest rooms, and storage for large items	20
Shelves and book rack areas in libraries, loading/unloading docks, boiler houses, and utilities	15
Stairways, walkways, corridors, traffic areas	10

Adapted from Kodak Park Division (1975).

illumination levels for efficient visual performance are given in Table 13.6.

13.13 APPLICATIONS AND DISCUSSION

Proper application of ergonomics principles of illumination requires knowledge and understanding of light characteristics, lighting measurements and lighting control techniques. Appropriate lighting surveys are also necessary to obtain the required information.

Lighting surveys are conducted in the workplace to determine illumination levels (e.g., for compliance with recommended or standard levels), luminance (e.g., for compliance with recommended or standard luminance ratio limits for visibility and safety), and assessment of the degree of comfort in the workplace. The results can be used to assess the adequacy of lighting maintenance and servicing procedures, or to investigate required changes for improving environmental visual conditions. The results of a lighting survey may indicate the existence of illumination deficiencies which can be resolved by application of higher or lower reflectance, better use of color, changing the type or location of luminaires to avoid shadow and glare, redirecting light towards the ceiling, or controlling day-lighting for better and more efficient workplace illumination.

REVIEW QUESTIONS

1. What are the adverse effects of insufficient or too strong and, particularly, glaring illumination? What are the advantages of good lighting?
2. Write an equation for expressing the relationship between the travel speed or velocity, frequency, and wavelength of an electromagnetic wave.
3. What is the range of wavelengths of visible light?
4. What type of electromagnetic waves are the upper and lower limits of the visible light frequencies?
5. What portions of the sun's electromagnetic spectrum cause sunburn and skin damage?
6. What portion of the electromagnetic spectrum generates heat?
7. List the three physical characteristics of light.
8. Describe the reflection process of light, and the three types of reflection.
9. Describe the refraction process of light.
10. What factor of the light wave changes when it leaves one material and enters another of greater or less optical density? Explain.
11. What is meant by light polarization?

12. Give four examples of polarizing media.
13. Describe the diffraction process of light.
14. Describe the diffusion process of light.
15. Explain light absorption.
16. What is light transmission?
17. What is photometry, and how is it made? What principal quantity is used in photometry?
18. What is radiant power and its units of measurement?
19. Define luminous intensity and luminous flux. What are their units of measurement?
20. How much luminance can a light source produce if the luminous intensity of the light source is 1 cd?
21. What is illumination, and what is its unit of measurement?
22. What is lux and its relationship with a foot-candle?
23. Calculate the illumination level at a distance of 3 m from a light source of 750-lm candle power emitting luminous flux in all directions.
24. Calculate the illumination level at a distance of 5 ft from a light source of 60-cd candle power emitting luminous flux in all directions.
25. A light source emits 1250 lm on a surface of 1 m^2, located 3 m away from the source. Calculate the illumination level on the surface if the angle between the surface and the direction of the incident light is 30°.
26. If a light source has an illumination reading of 4.5 fc at 10 ft away from the source, what is its illumination at 5 ft further away?
27. What is luminance, and how is it different from brightness? What are the units of measurement of luminance in both International System and US Customary System Units?
28. What is reflectance, and how is it measured?
29. What is contrast, and how is it measured?
30. What type of data are usually recorded during a lighting survey?
31. State at what part of the human visual apparatus light is regulated, light (or image of an object) reception takes place, and visual perception of the outside world occurs.
32. List five physiological factors that affect visual ability and provide a brief explanation of each.
33. What aspect of vision is evaluated by ophthalmologists using the Snellen letter chart?
34. Interpret what a 20/25 and 20/10 vision means.
35. For what purpose is the minimum separable areas testing used, and how are its results expressed?
36. List and briefly explain the three divisions of the visual field.
37. Explain what changes would take place by visual accommodation.

38. What is the near point of accommodation?
39. What part of the eye does adaptation change, and how?
40. What are the four factors that determine the visibility of the task?
41. What does the visual size depend on? How is visual size expressed?
42. What physical factors of the task visibility may be inherent in the task, and which factors can be controlled by changing the illumination level?
43. Define glare.
44. What part of the eye is acutely damaged by the flash of ultraviolet and infrared rays of the welding process?
45. List the two categories of glare sources.
46. List the three classifications of glare based on their effects on individuals.
47. List six visual problems caused by poor illumination, which can indirectly result from visual and mental fatigue.
48. What type of eye activity causes the discomfort resulting from glare?
49. List five factors on which appropriateness of illuminance for a given work environment depends.
50. When multiple tasks are performed in the workplace, the minimum level of illumination must be determined based on which type of task?
51. Why should the luminances within the visual environment not be too different?
52. What should be the contrast ratio between tasks and remote darker surroundings in interior areas where reflectances of entire space can be controlled?
53. What should be the contrast ratio between tasks and remote lighter surroundings in areas where reflectances of the immediate work area can be controlled, but control of remote surroundings is limited?
54. What is the limit of the contrast ratio between the middle and edge of the visual field?
55. What is the maximum permissible range of brightness within an entire room?
56. What should be the minimum angle between the horizontal line of sight and a line from the eye to the light source?

REFERENCES

ANSI (1983) *Practices for Industrial Lighting*. ANSI/IES RP-7-1983. American National Standards Institute, New York.

Bumstead, H.E. (1973) Spectrophotometry. In: *The Industrial Environment – Its Evaluation and Control*. National Institute for Occupational Safety and Health

(NIOSH), pp. 223–246. (For sale by the Superintendent of Documents, US Government Printing Office, Washington, DC 20402.)

CFR 1926.56 (1990) *Code of Federal Regulations, Title 29-LABOR, Part 1926 (Construction)*. Office of the Federal Register National Archives and Records Administration, US Government Printing Office, Washington.

Cushman, W.H. and Crist, B. (1987) Illumination. In: *Handbook of Human Factors* (Salvendy, G., ed.). John Wiley, New York, pp. 670–695.

GE (1974) *Light and Color (TP-119)*. Large Lamp Department, General Electric Company, Cleveland, OH.

Grandjean, E. (1988) *Fitting the Task to the Man: A Textbook of Occupational Ergonomics*. Taylor & Francis, London.

Grether, W.F. and Baker, C.A. (1972) Visual presentation of information. In: *Human Engineering Guide to Equipment Design* (Van Cott, H.P. and Kinkade, R.G., eds). The Superintendent of Documents, US Government Printing Office, Washington, DC, pp. 41–121.

Hudson, A. and Nelson, R. (1990) *University Physics*, 2nd edn. Saunders College Publishing, Philadelphia.

IES Committee on Industrial Lighting (1991) *Industrial Lighting*, ANSI/IES-7-1991. Illuminating Engineering Society of North America, 345 E 47th St, New York, NY 10017.

IES Nomenclature Committee (1979) Proposed American national standard nomenclature and definitions for illuminating engineering (revision of Z7.IR 1973). *Journal of Illuminating Engineering Society* 9: 2–46.

Kaufman, J.E. (1973) Illumination. In: *The Industrial Environment – Its Evaluation and Control*. National Institute for Occupational Safety and Health (NIOSH), pp. 349–356. (For sale by the Superintendent of Documents, US Government Printing Office, Washington, DC 20402.)

Kodak Park Division (1975) *Indoor Lighting*. Facilities Organization Standard (PEE-29).

Krauskopf, K. and Beiser, A. (1966) *Fundamental of Physical Science*, 5th edn. McGraw-Hill, New York.

Lum-i-neering Associates (1979) *Lighting Design Handbook*. Final report. US Government contract no. N68305-76-C-0017 (AD A074836). Civil Engineering Laboratory, Department of the Navy, Port Hueneme, California.

McCormick, E.J. and Tiffin, J. (1974) *Industrial Psychology*, 6th edn. Prentice-Hall, Englewood Cliffs, New Jersey.

Nakagawara, V.B. (1990) Glare vision testing: application in occupational health and safety programs. *Professional Safety* 35: 25–27.

Occupational noise environment

14

14.1 INTRODUCTION

Workers are exposed to various kinds of noise in their everyday jobs. Some kinds of noise may not be harmful but are annoying. However, exposures to extremely loud noises, even for short periods of time, or prolonged exposures to loud noises cause temporary or permanent hearing loss. Noise, in general, is annoying sound that can inversely affect safety and performance. It can also affect physiological responses. In the long run noise may result in health impairment and, at least, contribute to stress at the workplace.

The following are two primary attributes of sound which are of great concern to ergonomists and safety professionals: intensity (or loudness) of the sound energy (or sound pressure) and frequency of the sound pressure waves.

These attributes are described in the following section.

14.2 BASIC SOUND TERMINOLOGY

The following are descriptive terms used to study the characteristics of sound or noise (Michael, 1973):

- **Amplitude**: The sound amplitude refers to either the amount of sound produced at a certain location away from its source that is described by the sound intensity or sound pressure, or the ability of the source to emit sound which is described by the sound power of the source.
- **Frequency**: Sound frequency describes the rate at which complete cycles of high- and low-pressure regions are produced by the source of sound. Frequency is expressed in cycles per second (cps), also called hertz (Hz). The range of sound frequencies audible by a normal human ear (in young individuals) is about 20–20000 Hz. However, the frequency detected by the human ear will decrease or increase as

the distance from the sound source is increased or decreased, respectively, during listening. This phenomenon is known as Doppler effect (Halliday and Resnick, 1967). When the distance remains constant, the frequency of the sound heard will be the same as the frequency of the vibrating source.

- **Loudness**: The loudness of a sound is the person's impression of the sound amplitude which depends on the ear's characteristics.
- **Noise and sound**: The terms noise and sound are often used interchangeably. However, the term sound is used to describe useful communication or pleasant sounds, while noise is used to describe distracting or unwanted sound.
- **Pitch**: Pitch is a measure of auditory sensation which depends on the frequency, pressure and waveform of the sound.
- **Intensity (I):** The sound intensity at a given location is the average rate at which sound energy is transmitted through a unit area perpendicular to the direction of the sound propagation and is measured in joules per square meter per second ($J \cdot m^{-2} \cdot s^{-1}$). It is often expressed in terms of a level (sound intensity level, L_I), which is measured in decibels referenced to 10^{-12} W $\cdot$ m^{-2}.
- **Sound power (E):** The sound power of a source is the amount of sound energy emitted by the source per unit time and is usually expressed in watts. The sound power is also expressed in terms of a level (sound power level, L_E) and measured in decibels referenced to 10^{-12} W.
- **Sound pressure (P):** The sound pressure refers to the root-mean-square (rms) value of the pressure changes above or below atmospheric pressure when used to measure steady-state (continuous) noise. Short-term or impulse-type noise is described by peak pressure values. The sound pressure is measured in Newtons per square meter ($N \cdot m^{-2}$), dynes per square centimeter ($d \cdot cm^{-2}$), or microbars. It is, however, commonly expressed in terms of a level (sound pressure level, L_p) in decibels referenced to 2×10^{-5} N $\cdot$ m^{-2}, as $L_p(dB) = 20 \log (p/p_o)$. The relationship between sound pressure and time at a given location is shown in Figure 14.1.
- **Decibel** (dB): The decibel is defined as 10 times the logarithm of the ratio of a measured intensity to the reference intensity (bels). That is,

$$L_i \text{ (bels)} = \log \frac{\text{measured intensity}}{\text{reference intensity}} \tag{14.1}$$

$$L_i \text{ (dB)} = 10 \log \frac{\text{measured intensity}}{\text{reference intensity}} \tag{14.2}$$

where the reference intensity corresponds to the hearing threshold.

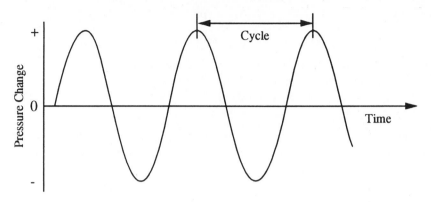

Fig. 14.1 Relationship between sound pressure and time at a given location.

Sound intensity is expressed as the average sound power passing through a unit area perpendicular to the direction of the sound propagation. The sound power of reference intensity is $I_o = 10^{-12}$ W · m^{-2} (hearing threshold). We can set up the following logarithmic equation for sound intensity level:

$$\text{bels} = \log \frac{I}{I_o} \qquad (14.3)$$

where: bels: = sound intensity level;
I = actual sound intensity (in W · m^{-2});
I_o = reference sound intensity (10^{-12} W · m^{-2}).

When the calculated bels is multiplied by 10, the common unit of sound intensity level (L_i), that is, decibels (dB), is found. (*Note*: 0.1 bels = 1 dB.) Thus,

$$L_i \text{ (dB)} = 10 \log \frac{I}{I_o} \qquad (14.4)$$

where: L_i = sound intensity level (dB);
I = reference sound intensity (W · m^{-2});
I_0 = reference sound intensity (10^{-12} W · m^{-2}).

14.3 RELATIONSHIP BETWEEN SOUND INTENSITY AND PRESSURE

Sound intensity describes, in part, characteristics of sound in the medium through which it passes, not at its source. Its relationshp with sound pressure can be expressed as shown in equation 14.5:

$$I = \frac{P^2}{d \cdot v} \tag{14.5}$$

where: I = sound intensity (W $\cdot$ m^{-2}) in the medium;
P = sound pressure (N $\cdot$ m^{-2}),
d = density of the medium;
v = speed of sound in the medium.

We can replace I and I_o, in equation 14.4, by $P^2/(d \cdot v)$ and $P_o^2/(d \cdot v)$, respectively, to derive an equation for calculating sound pressure level, as follows:

$$L_I = 10 \log \frac{I}{I_o} = 10 \log \left[\frac{\dfrac{P^2}{d \cdot v}}{\dfrac{P_o^2}{d \cdot v}} \right]$$

Simplification of the right side of this equation gives an equation in terms of P and P_o by which the decibel level (i.e., sound pressure level) can be calculated:

$$L_P = 10 \log \frac{P^2}{P_o^2}$$

or

$$L_P(\text{dB}) = 20 \log \frac{P}{P_o} \tag{14.6}$$

where: L_P = sound pressure level (dB);
P = actual sound pressure (N $\cdot$ m^{-2});
P_o = 0.00002 or 2 $\times$ 10^{-5} N $\cdot$ m^{-2} = 0.0002 μbar (microbars) (in air, this reference pressure corresponds to the reference intensity, $I_0 = 10^{-12}$ W $\cdot$ m^{-2}.)

14.4 RELATIONSHIP BETWEEN SOUND INTENSITY AND POWER

The relationship between sound intensity and sound power is shown in the following equation:

$$I_{\text{average}} = \frac{E}{4 \cdot \pi \cdot r^2} \tag{14.7}$$

where: I_{average} = the average sound intensity at distance r from its source, whose acoustic power is E (W);
E = the acoustic power of the sound source;

$4 \cdot \pi \cdot r^2$ = the surface area of a sphere of $2r$ diameter, surrounding the sound source, at which the intensity is averaged.

Now, we can write:

$$L_I = 10 \log \frac{I}{I_o} = 10 \log \left[\frac{\frac{E}{4 \cdot \pi \cdot r^2}}{\frac{E_o}{4 \cdot \pi \cdot r^2}} \right]$$

Simplification of the right side of this equation gives an equation in terms of E and E_o by which the decibel level (i.e., acoustic power level) can be calculated:

$$L_E = 10 \log \frac{E}{E_o} \qquad (14.8)$$

where: L_E = acoustic power level (in dB);
E = actual acoustic power (in W) at the source;
E_0 = 10^{-12} W, the reference sound power that corresponds to reference intensity
I_o = 10^{-12} W $\cdot$ m^{-2} and reference sound pressure
P_o = 0.00002 N $\cdot$ m^{-2} or 0.0002 µbar.

From equation 14.7, it becomes obvious that the sound intensity decreases with the square of the distance from the source. This phenomenon is known as the **inverse-square law**. Hence, the amount of sound (intensity) level is reduced by moving from a location to another location further away from the sound source. The noise (sound) reduction can be measured as follows:

$$NR = L_{I_1} - L_{I_2} = 10 \log \frac{I_1}{I_0} - 10 \log \frac{I_2}{I_0}$$
$$= 10 \, [\log I_1 - \log I_0 - \log I_2 + \log I_0] = 10 \, [\log I_1 - \log I_2]$$
$$= 10 \log \frac{I_i}{I_2}$$

where: $I_1 = E/(4 \cdot \pi \cdot (r_1)^2)$;
$I_2 = E/(4 \cdot \pi \cdot (r_2)^2)$

Hence,

$$NR = 10 \log \frac{E/(4 \cdot \pi \cdot r_1^2)}{E/(4 \cdot \pi \cdot r_2^2)} = 10 \log \frac{r_2^2}{r_1^2}$$

Then, the amount of noise reduction (in dB) can be expressed as:

$$NR = 20 \log r_2/r_1 \qquad (14.9)$$

where: NR = noise (sound) intensity level (dB reduced by
moving from location 1 to location 2);
r_1 = the distance from location 1 to the sound source;
r_2 = the distance from location 2 to the sound source
$(r_2 > r_1)$.

14.5 HEARING PHENOMENON

The sound pressure intensity is sensed as loudness or volume, whereas its pressure frequency is sensed as pitch. A high-pitch sound refers to a high-frequency sound (Fig. 14.2a and 14.2c). Figure 14.2 shows sound characteristics.

In occupational environments, the pressure intensity of the sound wave is more critical than its pitch (sensed frequency) in reference to the safety of employees. High-peak pressure waves can cause permanent hearing loss. The human ear is a very delicate hearing system that can detect the tiny pressure of the faintest audible sounds, while it can withstand without damage a sound pressure 10 million times as great as the faintest sound it can hear (Asfahl, 1990). Asfahl concludes from this ear characteristic that as sounds get into the upper part of this pressure range, the human ear becomes less sensitive in detecting changes in the intensity of the pressure, even if the increase in intensity is very large (e.g., doubled or tripled).

Hearing is a complicated process. It depends on the following three factors:

• the intensity of the sound wave;
• the frequency of the sound wave;
• the hearing ability of the person.

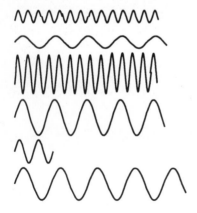

(a) High-pitch, soft sound

(b) Low-pitch, soft sound

(c) High-pitch, loud sound

(d) Low-pitch, loud sound

(e) Short-duration sound

(f) Long-duration (sustained) sound

Fig. 14.2 Sound characteristics.

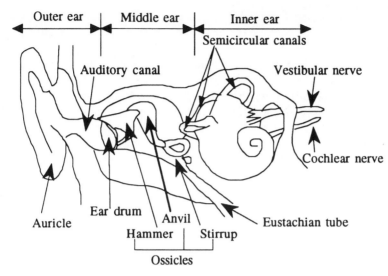

Fig. 14.3 Cross-section of the human ear.

Figure 14.3 show a sketch of the cross-section of the human ear. The sound waves are collected by the auricle (the portion of the ear seen outside the head) and directed into the auditory canal to the ear drum, which vibrates in response to the sound pressure and frequency. The vibrations generated by the ear drum, in turn, cause vibrations in the ossicles (i.e., the hammer, anvil, and stirrup) in the middle ear, from which they are transmitted to the cochlea in the inner ear. Cochlear fluid moves in response to the stirrup vibration and generates nerve impulses which are transmitted to the brain by the cochlear (acoustic) nerve. The semicircular canals and the vestibular nerve, the other major systems located in the ear, provide feedback regarding balance and the sense of motion. Pressures on both sides of the ear drum are balanced by the eustachian tube, which connects the middle ear to the pharynx in the nasal cavities.

The human ear does not respond to the same sound levels for all frequencies. The differences are due to the fact that the cochlear nerve cells nearer to the middle ear are stimulated by higher frequencies while those farther away from the middle ear are stimulated by lower frequencies. Thus, sound with a low frequency must be louder to have the same hearing response as that received by a high-frequency sound.

The audible frequency range for humans is 20–20 000 Hz. The human ear is able to respond to sound pressure ranging from 0.0002 to 2000 dynes cm^{-2} (microbars). A sound level of 0 dB is the human hearing perception threshold, and 120–140 dB is the pain threshold. (Note: 1 bar = 10^5 N ·m^{-2} and 1 N · m^{-2} = 1 Pa (pascal) = 10^{-5} bar = 10 µbar.)

Noise is described by its two components: intensity (or loudness) of the sound energy (or sound pressure), and frequency. Standard sound-level meters usually have three weighting scales (A, B, and C). Noise pressure intensity (or loudness) readings are indicated by the letter representing the scale used. For example, 78 dB(A) (or dBA) is the noise level measured using the A scale. Since the human ear is less sensitive to low frequencies, the A-scale circuit of the sound-level meter attenuates very low frequencies to approximate the human ear's response. It is represented by a logarithm scale ranging from about zero (0) dBA (hearing perception threshold) to about 120–130 dBA (pain threshold); Ramsey, 1985). Noise frequency is measured in cps or Hz. Different frequencies are sensed as different tones or pitches.

Octave-band analyzers are used to determine the frequency bands of the sound source, or to detect the noise source, based on its frequency characteristics. When the switch of an octave-band analyzer is set for a certain frequency band, it will block out all other noise frequencies which are outside the octave range in question. Octave-band sound pressure levels can be converted into their equivalent A-weighted sound levels (i.e., dBA) using Figure 14.4.

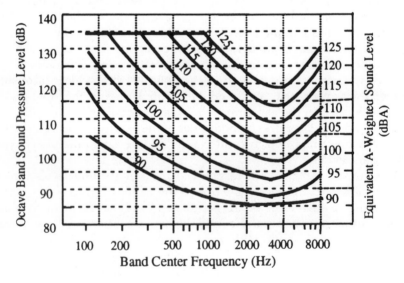

Fig. 14.4 The Occupational Safety and Health Administration's (OSHA's) standard dBA, equivalent to octave band sound pressure in dB.

14.6 COMBINATION OF NOISE SOURCES

If a machine (noise source) is very loud (e.g., X dB) and a second machine of exactly the same loudness is put beside it, the total loudness will not be simply doubled. The human ear senses only a slight increase in the loudness, although the actual sound pressure has been doubled by adding the second noise source. The noise level is rather increased by 3 dB, that is;

$$X \text{ dB} + X \text{ dB} = (X + 3) \text{ dB} \qquad (14.10)$$

This can be proven by a simple mathematical calculation on the logarithmic relationship between the decibel level and the sound intensity. Table 14.1 gives a scale for combining decibels to find a total noise level from two noise sources.

If there are more than two noise sources, two sources are combined first and then the result is treated as one source and combined with a third source, and so on, until all sources have been combined in a single total.

Example 14.1

Assume that the following machines with given noise levels are operating together in a shop:

Machine 1   80 dB
Machine 2   80 dB

Table 14.1 Scale for combining decibels

Difference between two decibel levels to be added (dB)	Amount to be added to the higher decibel level to find the sum (dB)
0	3.0
1	2.6
2	2.1
3	1.8
4	1.4
5	1.2
6	1.0
7	0.8
8	0.6
9	0.5
10	0.4
11	0.3
12	0.2

Source: NIOSH, 1978.

Machine 3   87 dB
Machine 4   82 dB

The combination of machine 1 and machine 2 produces 80 + 3 = 83 dB. Then machine 3 is added as follows:

$$\text{Difference dB} = 87 - 83 = 4 \text{ dB}$$

From Table 14.1, a difference of 4 dB between two sources is equivalent to adding 1.4 dB to the higher noise level of the two. Thus, the combination of machines 1, 2, and 3 produces a noise level of 87 + 1.4 = 88.4 dB.

Finally, adding machine 4 to the combination results in a noise level calculated as follows:

$$\text{Difference dB} = 88.4 - 82 = 6.4 \text{ dB}$$

From Table 14.1, a difference of 6.4 dB between two sources is equivalent to adding approximately 0.9 dB to the higher noise level of the two, that is, to the 88.4 level. Thus, the overall noise level by machines 1, 2, 3 and 4 is 89.3 dB (i.e., 88.4 + 0.9).

14.7 NOISE-EXPOSURE STANDARDS

14.7.1 OSHA'S STANDARDS

The Occupational Safety and Health Administration (OSHA) of the US Department of Labor has set a (maximum) permissible exposure limit (PEL) and an action limit (AL) to prevent noise hazards. The PEL is 90 dBA and AL is 85 dBA time-weighted average (TWA) for an 8-h exposure. At the AL of 85 dBA, OSHA requires that a hearing conservation program be implemented. Such a program would require record keeping, noise surveys, and audiometric testing of employees. The AL is a threshold trigger to alert the organization of potential noise exposure problems.

It is widely accepted that a worker can tolerate noise levels higher than the PEL for shorter periods of time. The PEL exposure duration can be calculated using the following equation:

$$T = \frac{8}{2^{(L-90)/5}} \text{ (PEL duration in h)} \tag{14.11}$$

where L is the noise intensity in dBA. For example, a 92-dBA environment would yield a PEL of $T = 8/2^{(L-90)/5}$ or $T = 8/2^{0.4}$, or $T = 6.04$ h. Similarly a 100-dBA environment would lead to a PEL of $T = 2$ h. In short, for every 5-dBA increase in the noise exposure to the worker, the PEL is cut in half.

14.7.2 ACGIH STANDARDS

The American Conference of Governmental Industrial Hygienists (ACGIH, 1994) has determined threshold limit values (TLVs) for noise (sound) pressure levels to represent conditions to which workers may be repeatedly exposed without adverse effect on their ability to hear and understand normal speech. The ACGIH uses an 8-h criteria level of 85 dBA, as opposed to the OSHA's PEL of 90 dBA. The TLV duration for any given noise level can be calculated using the following equation:

$$T = \frac{8}{2^{(L-85)/3}} \text{ (TLV duration in h)} \tag{14.12}$$

where L is the noise intensity in dBA.

14.7.3 STANDARD FOR ASSESSING INTERMITTENT NOISE EXPOSURE

Noise level and exposure duration are two critical factors with which safety professionals must be concerned. However, the noise level varies from time to time during the work-shift since workers move about the workplace. Tasks often change during the shift too, and noise-generating machines are not always working. Hence, to assess the noise exposure under such variable conditions, the worker exposure duration (C_i) to each noise level (L_i) during the work-shift must be measured, and then the **exposure dosage** (D_i) for each noise level is calculated using the following equation:

$$D_i = \frac{C_i}{T_i} = \frac{\text{actual time of exposure to the noise level } i}{\text{calculated PEL time for the noise level } i} \tag{14.13}$$

where $T_i = 8/2^{(L-90)/5}$ if the OSHA's PEL dosage is of concern, and $T_i = 8/2^{(L-85)/3}$ if the ACGIH's TLV dosage is used.

The total exposure dose, D for the entire work-shift is calculated by summing all partial doses, as shown in the following equation:

$$D = 100 \sum_{i=1}^{n} \frac{C_i}{T_i} = 100 \frac{C_1}{T_1} + \frac{C_2}{T_2} + \ldots + \frac{C_n}{T_n} \tag{14.14}$$

where: D = dose, total shift noise exposure as a percentage of PEL;
C_i = time duration of exposure at noise level i, that is, L_i;
T_i = maximum PEL or TLV time at noise level i, calculated by equations 14.11 or 14.12, respectively;
n = number of all noise levels observed.

The total dose, D must not exceed 100%. If it exceeds 100%, appropriate corrective measures must be used to reduce the dose to acceptable levels; that is, to 100% or below.

Using equation 14.13, it can be shown that the OSHA's AL of 85 dBA for an 8 h exposure is equivalent to a dose equal to 50% of PEL over an 8-h period, since:

$$D = 100 \sum_{i=1}^{n} \frac{C_i}{T_i} = 100\left(\frac{8}{16}\right) = 50\%$$

where 16 is PEL duration for 85 dBA. When the action level is reached, action must be taken. The action includes first, a hearing conservation program; second, noise surveys; third, training, and fourth, record-keeping.

Table 14.2 provides calculated values for the OSHA's PELs and ALs, and ACGIH's TLVs, which can be useful for assessing intermittent noise exposure.

Example 14.2

Assume that a worker is exposed to the following noise levels during an 8-h work shift:

08:00–09:30	95 dBA
09:30–11:30	90 dBA
11.30:13:00	75 dBA
13:00–15:00	95 dBA
15:00–16:00	90 dBA

Noise duration for each level is found as follows:

	Duration C_i	$T_i = PEL_i$
At 75 dBA	1.5 h	64 h
At 90 dBA	3.0 h	8 h
At 95 dBA	3.5 h	4 h
Total	8.0 h	

Calculation of PELs using equation 14.11:

$$T_i = \frac{8}{2^{(L_i-90)/5}}$$

$$T_1 = \frac{8}{2^{(75-90)/5}} = 64 \text{ h}, \quad T_2 = 8 \text{ h}, \quad T_3 = \frac{8}{2^{(95-90)/5}} = 4 \text{ h}$$

$$D = 100 \sum_{i=1}^{n} \frac{C_i}{T_i} = 100\left[\frac{1.5}{64} + \frac{3}{8} + \frac{3.5}{4}\right] = 127\%$$

Since $D = 127\% > 100\%$, the PEL is exceeded, and the worker is at risk. Engineering and administrative controls should be sought.

Table 14.2 Calculated permissible exposure limit (PEL), action limit (AL) time, and threshold limit values (TLVs) for a range of sound levels

Noise level (dBA)	OSHA PEL (h)*	OSHA AL (h)†	ACGIH TLV‡	Noise level (dBA)	OSHA PEL (h)*	OSHA AL (h)H†	ACGIH TLV‡
80	32.0	16.00	24.00 h	107	0.76	0.380	2.98 min
81	27.9	13.95	20.16 h	108	0.66	0.330	2.36 min
82	24.3	12.15	16.00 h	109	0.57	0.285	1.88 min
83	21.1	10.55	12.70 h	110	0.50	0.250	1.49 min
84	18.4	9.20	10.08 h	111	0.44	0.220	1.18 min
85	16.0	8.00	8.00 h	112	0.38	0.190	0.94 min
86	13.9	6.95	6.35 h	113	0.33	0.165	44.65 s
87	12.1	6.05	5.04 h	114	0.29	0.145	35.44 s
88	10.6	5.30	4.00 h	115§	0.25	0.125	28.12 s
89	9.2	4.60	3.17 h	116	0.22	0.110	22.32 s
90	8.0	4.00	2.52 h	117	0.19	0.095	17.72 s
91	7.0	3.50	2.00 h	118	0.165	0.082	14.06 s
92	6.1	3.05	1.59 h	119	0.144	0.072	11.16 s
93	5.3	2.65	1.26 h	120	0.125	0.0625	8.86 s
94	4.6	2.30	1.00 h	121	0.109	0.0544	7.03 s
95	4.0	2.00	47.62 min	122	0.095	0.0474	5.58 s
96	3.5	1.75	37.80 min	123	0.0825	0.0412	4.43 s
97	3.0	1.50	30.00 min	124	0.0718	0.0359	3.52 s
98	2.6	1.30	23.81 min	125	0.0625	0.0313	2.79 s
99	2.3	1.15	18.90 min	126	0.0544	0.0272	2.21 s
100	2.0	1.00	15.00 min	127	0.0474	0.0237	1.76 s
101	1.7	0.85	11.91 min	128	0.0412	0.0206	1.40 s
102	1.5	0.75	9.45 min	129	0.0359	0.0179	1.11 s
103	1.3	0.65	7.50 min	130	0.0313	0.0156	0.88 s
104	1.1	0.55	5.95 min	135	0.0156	0.0078	0.27 s
105	1.0	0.50	4.72 min	136	0.0136	0.0068	0.22 s
106	0.87	0.435	3.75 min	139	0.0090	0.0045	0.11 s

*Occupational Safety and Health Administration's (OSHA's) PEL duration is calculated using equation 14.11.
†OSHA's AL duration is one-half of PEL.
‡American Conference of Governmental Industrial Hygienists' (ACGIH's) TLV duration is calculated using equation 14.12.
§OSHA states that workers should not be subjected to noise levels greater than 115 dBA for even short periods of time without hearing protection.

14.7.4 IMPULSIVE NOISE

OSHA has specified 140 dBA as a ceiling or C value that should be considered as a limit for acute exposures only and is a safety hazard. According to ACGIH (1994), no exposure of an unprotected ear in excess of a C-weighted peak sound pressure level of 140 dB should be allowed.

14.7.5 HEARING IMPAIRMENT

The American Academy of Ophthalmology and Otolaryngology (AAOO) has defined hearing impairment as an average hearing threshold level in excess of 25 dB at 500, 1000, 2000, and 3000 Hz (ACGIH, 1993). The ACGIH's noise TLVs have been established to prevent hearing loss at 500, 1000, 2000, 3000, and 4000 Hz (ACGIH, 1994). ACGIH also warns that a hearing conservation program including audiometric testing is necessary when workers are exposed to noise at or above the noise TLVs.

14.8 NOISE SURVEY

Noise is usually measured with a sound-level meter, that should be used in accordance with American National Standards Institute (ANSI) S1.4, *Specifications for Sound Level Meters* (ANSI, 1983). The sound-level meter should be calibrated on a regular basis. The factors to be considered in a noise survey include first, the layout of the work area; second, the height at which a worker is usually stationed; and third, the number of working machines. Sometimes, especially during the design phases of a project, the sound pressure cannot be measured. However, the manufacturer (or other sources) can most likely provide noise level information for the equipment being considered.

The noise level at a point r ft away from the noise source can be calculated using the following equation:

$$L_p = L_s - 10 \log (4 \pi r^2) - 10 \tag{14.15}$$

where:

L_p = the noise level at a point away from the noise source in decibels;
L_s = the noise level at the source in decibels;
r = the distance between the point and the noise source, in feet.

Noise surveying is an important part of safety professionals' duties and should be performed systematically. The following guidelines are recommended for performing noise surveying (Botsford, 1973):

- Use only a sound-level meter that conforms to the requirements of the American National Standard Institute S1.4 (1983) type S2A, *Specification for Sound Level Meters* (ANSI, 1983).
- Read and understand thoroughly the instructions in the instrument operation manual provided by the manufacturers before making a noise survey.

- Check the instrument batteries and calibration periodically during the noise survey.
- Use the A-weighted scale (network) with slow meter response.
- Protect the instruments against shock and vibration during transportation and use.
- Protect the instruments against temperature extremes. Overheating (e.g., leaving the instruments in a car parked in sunshine) can damage circuit components. Water vapor condensation in very cold conditions (e.g., leaving the instruments in a car overnight in the winter) can cause electrical shortages, resulting in low readings when the instruments are used in a warmer place.
- When making noise measurements, make sure the meter indication is due to noise, not to other factors (e.g., wind, electric and magnetic fields.) Use a wind screen to prevent wind from blowing across the instrument microphone.
- Reorient the instrument when the movement of its needle is ceased by electromagnetic interferences around welding on large assemblies or electric furnaces. Minimum electrical and magnetic field interferences are indicated by minimum meter reading.
- Make sure that recorded readings represent data. The microphone should be moved about the workplace to obtain a spatial average, not just the standing waves.
- Make sure the noise measurements are made at the auditor's location. Do not measure noise right at the ear of the auditor since diffraction around the head can alter the noise field. Select a location a foot away from the auditor at levels the same as his or her location with an equal distance from the noise source.
- Record all significant factors (i.e., sound level, duration, intermittence, etc.) which are necessary to evaluate the noise exposure properly.
- When noise levels are highly variable to evaluate exposure with a sound-level meter and stop-watch, use a noise monitor (dosimeter) that is capable of automatically computing the fractional exposure.
- When the objective is to locate a major noise source in the workplace or in a machine, probe the sound field with a sound-level meter. The noise level increases as its source is approached. Connect the microphone to the instrument using its extension cable for more convenience. Octave-band analyzers are also available to isolate the noise source.

14.9 EFFECTS OF NOISE

Noise can have two undesirable effects on individuals working in noisy environments: disturbed communication, and hearing loss. Unheard or misinterpreted vocal instructional messages at the workplace can lead to human errors and accidents. In noisy situations where proper vocal communication is not possible, other modes of communication should be adopted.

Long-term exposure to excessive noise, especially at levels of 90 dBA and higher, can cause hearing loss. Such hearing loss is of two types. If hearing loss is caused by noise-damaged nerves in the inner ear or brain, it is called **sensorineural hearing loss**. Hearing loss due to damage to the outer or middle ear is called **conductive hearing loss**. The latter is diagnosed by physicians by the means of a tuning fork placed on the bone behind the auricle of the ear (mastoid bone) for testing bone conduction.

Hearing losses due to non-occupational factors are classified as **sociocusis** and **presbycusis**. Sociocusis is the hearing loss caused by exposure to non-occupational noise sources, such as household noises (i.e., radio, television, etc.), and traffic and aircraft noises. Presbycusis, which is sensorineural hearing loss, refers to the gradual hearing loss due to physiological changes that occur during the normal aging process. Some physiological changes, such as atrophy of the brain cells and cochlear nerves, and decreased blood supply, cause hearing loss in the elderly.

14.10 NOISE CONTROL

There are three distinguished noise control techniques: engineering controls, administrative controls, and personal protective equipment.

Wherever noise levels exceed the PEL, the US federal regulations require that feasible engineering or administrative solutions be applied. If these remedies are found to be insufficient to reduce the noise exposure to within the PEL, personal protective equipment must be provided and used to reduce noise levels to within the PEL. However, engineering controls should always be considered as a more adequate and permanent solution to the noise problem. Examples of engineering and administrative controls (Woodson and Conover, 1964; NIOSH, 1978) are given below.

14.10.1 EXAMPLES OF ENGINEERING CONTROLS

- Maintain and lubricate machines and equipment to eliminate rattles and squeaks.
- Replace loose and worn parts of machines.
- Place equipment on rubber mountings to reduce its vibration.
- Acquire new equipment with low noise levels.
- Use larger, low-speed fans instead of smaller, high-speed fans. The

fan casing should be rigid and damped. Fans with silent motors should be selected.

- Use presses instead of drop hammers.
- Use large, slow machines instead of small, high-speed ones which have the same output capacity.
- Mount vibrating machines on firm, solid foundations. Foundation bolts must be kept tight.
- Isolate noisy equipment, by moving the operator away from the noise source. The intensity of noise from a source varies inversely with the square distance from the source, given that the walls are not reflective.
- Modify or eliminate the process.
- Instal padding inside sound booths or enclosures to prevent reflected noise from adding to the overall noise level.
- Use plastic and wooden materials instead of noisy metal materials.
- Modify the plant layout to spread noisy machines out over the plant.
- Construct irregular wall patterns.
- Stagger doors along a hallway instead of locating doors directly opposite each other.
- Avoid domed ceilings in small rooms and enclosures which concentrate noises above the head of the occupant.
- Install acoustic materials (sound-absorbing barriers) between machines.
- Cover walls with perforated acoustic tiles, rough plaster, or heavy curtains.
- Use asphalt, cork, or carpets on floors made of bare boards and concrete, that magnify impact noise.
- Use boards suspended from the ceiling at different points
- Use double-pane windows with air space between panels.
- Avoid sharp directional and velocity changes in pipelines and ducts.
- Minimize the velocity of air and liquids flowing through ducts and pipes.
- Secure pipes firmly to prevent rattling actions.

14.10.2 EXAMPLES OF ADMINISTRATIVE CONTROLS

- Rotate workers to maintain noise exposure dosage in an acceptable range.
- Schedule production runs to split between shifts.
- Perform high-noise maintenance after the regular shift hours to minimize the number of exposed people.
- Warn workers of noise hazards and the necessity for using protective devices by posting proper posters where such devices must be worn.
- Make sure that workers wear their protective devices at all times when in high-noise areas.

- Periodically check the noise levels to identify areas where the noise levels exceed the OSHA's TWA 85 dBA action level.

14.10.3 PERSONAL PROTECTIVE EQUIPMENT

Feasible engineering or administrative solutions must be applied first. If these remedies are found to be insufficient to reduce the noise exposure to an acceptable level, effective personal protective equipment must then be provided and used to reduce noise levels. Personal protective equipment includes ear muffs, ear-canal caps and ear plugs. Noise reduction ratings (NRR) should be available from the manufacturer of the ear muffs or plugs. The NRR value should lower the sound level in the employee's ear to a level below the 90 dBA limit (or 85 dBA limit, depending on the standard being used). Personal protective equipment is the most common – but least reliable – method of noise reduction. The major problem with the use of hearing-protective devices is that they are not comfortable, and the workers may not use them. They also make communication more difficult. They should be the last resort for noise reduction.

14.11 APPLICATIONS AND DISCUSSION

Exposure to excessive noise levels can result in temporary or permanent threshold shifts or hearing loss. In many cases there is legislation regulating worker exposure to noise. A sound-level meter can be utilized to conduct a noise survey to document sound levels in the workplace. If excessive noise levels are found, the ergonomist should recommend engineering or administrative controls, or personal protective equipment to limit the worker's exposure to the noise. Engineering controls are the preferred solution, with personal protective equipment considered as a temporary, last-resort solution to the noise problem.

The ergonomist should be involved in equipment selection and design to understand the impact of the new machine in the workplace. There are usually equipment vendors that make efforts to reduce the noise levels of their machines. Understanding the work environment and the noise characteristics of the new equipment can help the ergonomist protect those workers exposed to the noisy environment. In some cases, a frequency analysis of the noise may provide clues to the modifications or reductions of the noise level.

REVIEW QUESTIONS

1. What are the two primary attributes of sound which are of great concern to ergonomists and safety professionals?

2. What is the sound amplitude?
3. Describe sound frequency, and state the range of sound frequencies audible by a normal hearing human ear.
4. Explain the Doppler effect of sound.
5. Describe sound loudness and pitch.
6. What is sound intensity and its unit of measurement?
7. What is sound power and its unit of measurement?
8. What is sound pressure and its unit of measurement?
9. Write an equation that shows the relationship between sound intensity and sound power.
10. What aspect of sound can cause permanent hearing loss?
11. On what factors does hearing depend?
12. Describe why the human ear does not respond to the same sound levels for all frequencies.
13. What are the human ear's hearing perception and pain thresholds?
14. Explain why the A-scale circuit of sound-level meters is used in assessment of human noise exposure.
15. If a noise level reading of 110 dBA is made between two identical machines working together, what reading could be expected if one of the machines were shut down?
16. Assume that in a shop three machines with the following noise levels are working together:

Machine 1	80 dB
Machine 2	85 dB
Machine 3	85 dB

Determine the resulting noise level.
17. Assume that in a shop three machines, each with a noise level of 90 dBA, are operated simultaneously. Determine the resulting noise level.
18. What are the Occupational Safety and Health Administration's (OSHA's) time-weighted-average (TWA) permissible exposure limit (PEL) and action limit (AL), and ceiling for acute exposure to noise?
19. At what level of noise exposure does OSHA require a hearing conservation program? What are the components of such a program?
20. What is the PEL duration for exposure to a continuous noise level of 105 dBA?
21. What is the American Conference of Governmental Industrial Hygienists' (ACGIH's) 8-h TWA threshold limit value (TLV) for noise exposure? Then, what is the TLV duration for noise exposure at 100 dBA?
22. To what dose of PEL is the OSHA action limit equivalent? Explain why.

23. Assume that a worker is exposed to a dose of noise equal to 70% of PEL. Is the employer required to take any action? If yes, what action or actions?

24. Assume that a worker is exposed to a dose of noise equal to 70% of PEL. What reading should be expected from a sound-level meter in dBA?

25. A worker is exposed to noise at different levels during a work-shift in a given plant. The following are readings for various time periods during an 8 h shift:

> 08:00–09:00   95 dBA
> 09:00–11:30   90 dBA
> 11:30–12:30   95 dBA
> 12:30–13:00   70 dBA
> 13:00–16:00   85 dBA

Calculate the dose of noise to which the worker is exposed over the work-shift.

26. In question 25, assume that an engineering control could be applied which would cut the noise pressure levels in half either in the morning or in the afternoon, but not both. Which alternative would you select?

27. What is the OSHA's noise ceiling (C) value that should be considered as a limit for acute noise exposures only? What is the ACGIH recommendation for such exposures?

28. What is the American Academy of Ophthalmology and Otolaryng-ology's definition of hearing impairment?

29. The noise level at 15 ft away from a noise source is 80 dBA. What is the noise level at 5 ft away from the source?

30. When the distance from the noise source is doubled, by how many decibels is the noise reduced?

31. Name and describe the two types of hearing loss due to long-term exposure to occupational noise environments.

32. Name and describe the two classes of hearing losses due to non-occupational factors.

33. What are the three ergonomic noise control techniques? Which one is the preferred method of noise reduction?

34. List five methods for reducing noise produced by a machine.

35. List five examples of administrative controls for reducing noise exposure.

REFERENCES

ACGIH (1993) *1993–1994 Threshold Limit Values for Chemical Substances and Physical Agents and Biological Exposure Indices*. American Conference of Governmental Industrial Hygienists, Cincinnati, Ohio.

ACGIH (1994) *1994–1995 Threshold Limit Values for Chemical Substances and Physical Agents and Biological Exposure Indices.* American Conference Governmental Industrial Hygienists, Cincinnati, Ohio.

ANSI (1983) *ANSI 51.4-1983, Specification for Sound Level Meters.* American National Standard Institute (ANSI), New York.

Asfahl, C.R. (1990) *Industrial Safety and Health Management*, 2nd edn. Prentice-Hall, Englewood Cliffs, New Jersey.

Botsford, J.H. (1973) Noise measurement and acceptability criteria. In: *The Industrial Environment – Its Evaluation and Control.* US Government Printing Office, Washington, DC.

Halliday, D. and Resnick, R. (1967) *Physics.* John Wiley, New York.

Michael, P.L. (1973) Physics of sound. In: *The Industrial Environment – Its Evaluation and Control.* US Government Printing Office, Washington, DC.

NIOSH (1978) *Industrial Noise Control Manual.* NIOSH 79-117, National Institute for Occupational Safety and Health, US Department of Health and Human Services (formerly Department of Health, Education and Welfare: DHEW), Cincinnati, Ohio.

Ramsey, J.D. (1985) Environmental factors. In: *Industrial Ergonomics: A Practitioner's Guide* (Alexander, D.C. and Pulat, B.M., eds). Industrial Engineering and Management Press, Norcross, Georgia.

Woodson, W.E. and Conover, D.W. (1964) *Human Engineering Guide for Equipment Designers.* 2nd Edn. University of California Press, Berkeley, California.

Occupational vibration 15

15.1 INTRODUCTION

A large number of people cope with some forms of vibration in performing their jobs. The use of vibrating tools, especially in conjunction with rapid repetitive motion and force, can cause various forms of cumulative trauma disorders (e.g., vibration white finger, carpal tunnel syndrome, tenosynovitis, and trigger finger). The most common sources of vibration are vehicles and powered (pneumatic, electric, and gasoline-driven) tools. These include cars, trucks, trains, industrial movers, earth-moving equipment, chain saws, pneumatic hammers, and wrenches. Heavy machinery such as large presses and compressors are other major sources of vibration.

The subject of vibration in general, and its assessment and adverse effects on the exposed workers, as well as the ergonomics measures for safeguarding workers, are offered in this chapter. The associated cumulative trauma disorders were presented in Chapter 8.

15.2 VIBRATION TERMINOLOGY

Vibration or oscillation is a periodic back-and-forth or up-and-down motion, such as the motion of a pendulum or spring. The vibratory motion reverses itself twice in every complete cycle. Vibration is inherent in any machine because of the motion of its moving parts that oscillate, rotate, and/or reciprocate. It can be characterized by frequency and intensity. Vibratory frequency is the rate at which vibration oscillations occur. This quantity is expressed in hertz (Hz) which is the number of cycles per second (cps), where each cycle represents the motion of the device from a mean position to one extreme, and return to the mean position. Vibration intensity is measured in terms of a variety of parameters such as amplitude, displacement, velocity, acceleration, and jerk.

In general there are two types of vibration: sinusoidal vibration and random vibration. **Sinusoidal** vibration may be a single sine wave of a certain frequency or multiple sine waves of different frequencies. The particular characteristic of this type of vibration is the repetition of the vibration sine wave at regular intervals. **Random** vibration, which is most commonly encountered in the real world, is an irregular and unpredictable type of vibration.

The following are several terms which are used by analysts and researchers in vibration studies and assessment of its effects on workers who are exposed to vibration in the workplace (Ayoub and Ramsey, 1975; Key *et al.*, 1977; Sanders and McCormick, 1993):

- **Amplitude**: Unless otherwise stated, the amplitude normally refers to the distance the device travels from the mean position to either extreme. Peak-to-peak amplitude is the distance from the maximum positive to the maximum negative displacement, and is referred to as double amplitude.
- **Displacement**: Vibratory displacement is the distance between the normal resting position of an object and its position at a given time in its vibratory cycle.
- **Velocity**: Vibration velocity is the rate of change in displacement. This is the first derivative of displacement and is expressed in meters (or feet) per second (i.e., $m \cdot s^{-1}$ or $ft \cdot s^{-1}$).
- **Acceleration**: Vibration acceleration is the time rate of change of velocity during a cycle (i.e., the rate at which the velocity of the vibratory motion changes in direction); which is the second derivative of displacement, or the first derivative of velocity. It is usually measured in meters (or feet) per second per second (i.e., $m \cdot s^{-2}$ or $ft \cdot s^{-2}$). It may also be expressed in a number of gravities (shown by **g**s, where 1 **g** is equivalent to $9.8 \ m \cdot s^{-2}$ or $32.2 \ ft \cdot s^{-2}$).
- **Jerk**: Jerk is the rate of change in acceleration and expressed in meters (or feet) per second per second per second (i.e., $m \cdot s^{-3}$ or $ft \cdot s^{-3}$). It is the third derivative of displacement.
- **Resonance**: Every object has a natural vibration frequency, known as the resonance frequency. When an object (or the body) is excited at a frequency beyond its resonant frequency, it will vibrate at a higher intensity. This increase in intensity is known as amplification.
- **Attenuation**: At frequencies lower than the resonant frequency the object (body) vibrates at a lower intensity than that applied to it; that is, the object (body) absorbs a part of the applied force. This is known as attenuation.

Due to resonance, body movements can become difficult to control, and attenuation (vibration absorption) may cause structural damages (Osborne, 1987).

15.3 TYPES OF VIBRATION WITH RESPECT TO ITS SOURCE

Depending on the behavior of its source, vibration may be classified as forced vibration or free (or natural) vibration. When an object is forced to vibrate under the action of a periodically reversing force or under the action of reversing displacement of a support, its vibration is called **forced vibration**. Vibrations occurring in machines are typically forced vibrations.

When the periodic forcing action is absent and vibration is induced by displacing an object (e.g., a spring) and then releasing it, permitting it to vibrate freely, the vibration is called **free vibration** (or **natural vibration**). Under such conditions the object will vibrate naturally between the two extremes (plus and minus amplitudes). In this case, if damping factors which resist the vibration motion are absent, the free vibration would continue indefinitely.

15.4 TYPES OF VIBRATION WITH RESPECT TO THE HUMAN BODY

With reference to the human body, vibration is commonly classified into whole-body (or head-to-toe) or segmental vibration (Key *et al.*, 1977).

15.4.1 WHOLE-BODY VIBRATION

Whole-body vibration is vibration transmitted to the entire human body via some supporting structure. Common modes of transmitting vibration are through the building floor to the feet, and vibration through a vehicle seat to the buttocks. Vibration with a frequency in the 2–100 Hz range is of interest for analysis of whole-body vibration (Wasserman *et al.*, 1974).

Whole-body vibration is a major concern when considering human reactions to vibration in the workplace. As the name implies, whole-body vibration is applied from head to foot. Transportation and farming vehicles present common whole-body vibration problems. Vibration sources are commonly associated with heavy construction vehicles, trucks, buses, and farming equipment (e.g., tractors). Walk-through tours in 45 different plants by a National Institute for Occupational Safety and Health (NIOSH) research team (Wasserman *et al.*, 1974) revealed that operating heavy equipment exposes probably the greatest number of workers to industrial vibration. Other occupations experiencing whole-body vibration include foundry operations, mining, forestry, lumber and wood production, printing and publishing, steel mills, blast furnace work, metal stamping, and can manufacturing. However, exposure to vibration in these occupations is relatively small in numbers compared to agriculture and transportation.

Chronic effects of whole-body vibration are not completely known. Vibration should at least be regarded as a generalized stressor and may affect multiple body parts and organs, depending on the characteristics of the vibration (Wasserman and Badger, 1973). However, research has shown that workers exposed to whole-body vibration can exhibit increased musculoskeletal, digestive, and circulatory strains (Cope, 1960). Musculoskeletal difficulties can lead to increases in oxygen consumption and pulmonary ventilation and may even cause deformation of the spine and bone structure. Such difficulties occur because the vibration entering the body through the seat or feet tends to be directed up the spine. Also, when vibrating the whole body, muscle tissue tends to become unstable or shaky, which causes a greater demand for oxygen. It should be noted that digestive strain can lead to abdominal and gastrointestinal problems. As Cope has stated, this affects the intestinal area the most, and the kidneys. Furthermore, circulatory disorders may cause dizziness and irregularity of blood flow into the extremities. Some initial effects of whole-body vibration are nausea, weight loss, decline of visual acuity, insomnia (sleeplessness), and cramps. The reported effects of whole-body vibration are summarized below (Cope, 1960; Key et al., 1977; Thompson, 1990):

- development of nausea, vomiting, and/or motion sickness (mostly resulting from stimulation of the vestibular apparatus of the inner ear);
- lumbar spinal disorders;
- changes in bone structure;
- severe reduction in manual tracking capability resulting from vibration at the 5-Hz level;
- chest pain in human beings after 25 s exposure to vertical vibration of about 0.15-in-amplitude at frequencies of 8–15 Hz. Lung damage (observed in mice, cats, and monkeys).
- changes in gastric secretions in the gastrointestinal tract, causing digestive problems;
- development of urinary tract problems;
- development of hemorrhoids and hernias;
- increased oxygen consumption and pulmonary ventilation;
- occurrence of the lowest subjective-discomfort tolerance level at about 5 Hz;
- difficulty in maintaining steady body posture during intense whole-body vibration;
- significant impairment in visual acuity caused by vibration in the 1–25 Hz range.

However, performance of tasks such as those involving reaction time, monitoring, and pattern recognition seems not to be affected by exposure to vibration (Grether, 1971; Shenberger, 1972).

Most experiments in the past have been performed in such a way that the subject is either sitting or standing while frequencies of vibration vary from 1 to 30 Hz (Weaver, 1979). Results of these experiments lead to the conclusion that the human body is most sensitive to frequencies between 4 and 8 Hz. Also, resonance (a decrease in response of the system due to a small change in the frequency of excitation) in the human body has been found to occur at about 5 Hz. At low frequencies, the organs begin to vibrate also, causing them to "bounce off" each other. Resonance in the human body can lead to serious trauma and even hemorrhage if it is not considered and controlled.

15.4.2 SEGMENTAL VIBRATION

Segmental vibration is vibration entering the human body through specific body parts, such as the hands and feet. This type of vibration, which is of greater concern to most ergonomists and safety professionals than whole-body vibration, is also known as upper-extremities or hand–arm vibration. The users of the following hand tools are exposed to segmental vibration:

- pneumatic tools: hammers, chisels, chippers, etc.;
- electrical tools: drills, saws, grinders, etc.;
- gasoline-powered tools: chain saws, vacuum cleaners, lawn mowers, etc.

Compared to whole-body vibration, hand–arm (segmental) vibration is harder to define and measure. Segmental vibration with a frequency in the range of about 8–1400 Hz is of interest (Wasserman et al., 1974). Segmental vibration, unlike whole-body vibration, is usually a localized stressor which can create injury to the fingers, hands, and wrists of workers who use vibrating tools. There is a wide array of harmful effects produced by segmental vibration. Extensive use of such tools, especially in cold environments, has been associated with the development of a hand–finger disease, called vibration white finger or dead hand, often referred to as Raynaud's phenomenon, which was discussed in Chapter 8.

15.5 ASSESSMENT OF VIBRATION

Recommendations, standards and guidelines for a proper assessment of vibration have been prepared by various organizations, including the International Organization for Standardization (ISO, 1986), American

National Standards Institute (ANSI, 1986) and American Conference of Governmental Industrial Hygienists (ACGIH, 1994). Based on ISO and ANSI standards and guidelines, ACGIH has adapted a method for assessing vibration exposure. Since vibration is a vector quantity, it is necessary to measure vibration in the three mutually orthogonal directions (x, y, and z) at a reference point close to where vibration enters the body (e.g., the hand in the case of segmental vibration). The direction x is horizontally in the sagittal plane (from back to front and vice versa), y is horizontally in the frontal plane (laterally from side to side), and the direction of z is vertically in the sagittal plane or perpendicular to the transverse plane (up-and-down or head–feet) in the biodynamic (anatomical) position. The coordinate systems for the whole body and the hand are illustrated in Figures 15.1 and 15.2, respectively.

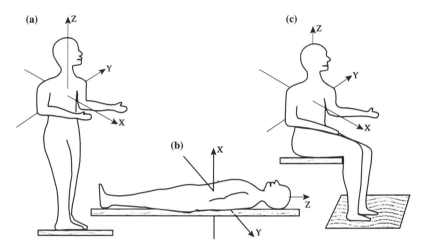

Fig. 15.1 The coordinate system for whole-body vibration in (a) standing, (b) lying and (c) seated positions.

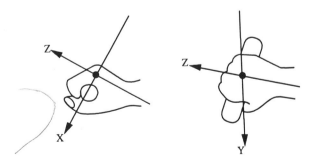

Fig. 15.2 The coordinate system for hand vibration.

As shown in Figure 15.2, the direction of x is across the hand, the direction of y is perpendicular to the palm of the hand, and the direction of z is along the hand–arm long axis.

The ACGIH method assesses the magnitude of vibration during normal operation of the vibrating tool or equipment in terms of the root-mean-square (RMS) value of the frequency-weighted (or time-weighted average) acceleration in each direction.

The ACGIH guidelines use the largest component acceleration (of the three x, y, and z components) as the basis for vibration exposure assessment. They recommend that if the total daily vibration exposure of the hand in any direction includes several exposures at different RMS accelerations, then the equivalent RMS, frequency-weighted (time-weighted) acceleration in that direction should be calculated by the following equation:

$$^aK_{eq} = \sqrt{\sum_{i=1}^{n} (^aK_i)^2 \frac{T_i}{T}}$$

$$= \sqrt{(^aK_1)^2 \frac{T_1}{T} + (^aK_2)^2 \frac{T_2}{T} + \ldots + (^aK_n)^2 \frac{T_n}{T}} \qquad (15.1)$$

where: T $= \Sigma T_i = T_1 + T_2 + \ldots + T_n =$ total daily exposure duration;

   ^aK_i $=$ the ith frequency-weighted, RMS acceleration component with duration T_i;

   K $=$ the component (direction x, y, or z).

Accelerometers are commercially available that can be used to measure vibration. The vibration-measuring system, including the accelerometer and electronics, is commonly called a human response vibration-measuring instrument.

Example 15.1

Assume that the following vibration measurements have been made using a human response vibration-measuring instrument:

Vibration exposure duration (h)	Component acceleration (m·s^{-2})		
	x	y	z
2	5	8.7	7.7
2	4	2.5	13.0
3	3	4.0	8.5
1	0	0	0

Using equation 15.1, the RMS values of the frequency-weighted component accelerations are calculated as follow:

$$a_{X_{eq}} = [25(2/7) + 16(2/7) + 9(3/7)]^{1/2} = 3.95 \text{ m} \cdot \text{s}^{-2} \text{ or } 0.402 \text{ g}$$

$$a_{Y_{eq}} = [75.69(2/7) + 6.25(2/7) + 16(3/7)]^{1/2} = 5.50 \text{ m} \cdot \text{sec}^{-2} \text{ or } 0.561 \text{ g}$$

$$a_{Z_{eq}} = [59.29(2/7) + 169(2/7) + 72.25(3/7)]^{1/2} = 9.81 \text{ m} \cdot \text{sec}^{-2} \text{ or } 1 \text{ g}$$

15.6 ACCEPTABLE EXPOSURE LIMITS TO SEGMENTAL VIBRATION

The ACGIH (1994) has developed threshold limit values (TLVs) for hand–arm (segmental) vibration. The TLVs refer to acceleration levels and duration of exposures that represent conditions to which most workers may be repeatedly exposed without progressing beyond stage 1 of vibration white finger in the Taylor–Pelmear classification (see Chapter 8). It is, however, crucial to know that susceptibility to vibration white finger varies among individuals and, thus, these TLVs should be used as general guidelines in the assessment of vibration exposure. The TLVs are given in Table 15.1.

For compliance with the TLVs given in Table 15.1, it is necessary to measure the vibration entering the hand of the exposed worker. This measurement is usually made using an accelerometer, placed at the point where the worker grasps the vibrating tool. Such a device measures vibration traveling in each of the three perpendicular

Table 15.1. The American Conference of Governmental Industrial Hygienists (ACGIH) threshold limit values (TLVs) for exposure to segmental vibration

| Total daily exposure duration* | Values of dominant† frequency-weighted, root-mean-square component acceleration that shall not be exceeded‡ a_k ($a_{k_{eq}}$) | |
	m s^{-2}	g§
Exposure < 1 h	12	1.223
1 h ⩽ exposure < 2 h	8	0.815
2 h ⩽ exposure < 4 h	6	0.612
4 h ⩽ exposure < 8 h	4	0.408

*The total time vibration enters the hand per day, whether continuously or intermittently.
†Usually one axis of vibration is dominant over the remaining two axes. If one or more vibration axis exceeds the total daily exposure, then the TLV has been exceeded.
‡Any worker can be exposed to segmental (hand–arm vibration) at or below the TLVs without progressing beyond stage 3 vibration white finger in the Taylor–Pelmear classification (see Chapter 8).
§g = 9.81 m · s^{-2}.

directions x, y, and z. If the total daily exposure value measured along with any of these three axes exceeds the TLVs given in Table 15.1, then corrective actions must be taken to reduce the worker's exposure to vibration. In the previous example, the y and z components exceeded the TLVs for an 8-h exposure, and ergonomics interventions would be recommended.

15.7 VIBRATION CONTROL AND PREVENTION OF VIBRATION WHITE FINGER

Since vibration from hand-held tools is a primary cause of developing vibration white finger, any provision for reducing vibration exposure can lessen the risk of this disease. The following are some possible methods for reducing exposure to vibration and the risk of vibration white finger (AFS Segmental Vibration Task Force, 1983):

- **Work modification**: Automation (e.g., using robots), where practical, eliminates all unnecessary chipping and grinding processes and reduces vibration exposure.
- **Tools with less vibration**: Vibration exposure can be decreased by choosing tools which vibrate less. Tool manufacturers should be able to provide vibration information for their equipment.
- **Tool maintenance**: Tool maintenance is a crucial practice in vibration exposure reduction. Tools vibrate much less when well-maintained than when not maintained. Chisel bits should be kept sharp and grinding wheels balanced.
- **Employee training**: Workers need to be trained in how to perform their jobs to minimize exposure to vibration. Tight gripping and leaning on to the tool increase vibration exposure. Hence, workers should be instructed to let the tool do the work, and not to force it by gripping it tighter or leaning on to it.
- **Tool modification**: The exhaust on air tools should be redirected away from the region of the handles of the tool, where the hands come into contact with the tool. Exposure to both the vibration and cold air, together, will increase the risk of developing vibration white finger or triggering its attack.
- **Padding tool handles**: Padded tool handles may be effective. Putting shock-absorbing material on the tool grips minimizes vibration entering the hands. Where practical, suspending and counterbalancing the tool should also be used.
- **Warm and dry the hands**: Hands should be kept warm and dry. Employees should be instructed to dry their sweaty or wet hands periodically when continuing to use vibrating tools. Dry gloves should be provided to employees using vibrating tools.

- **Padded gloves**: Padded gloves should be worn by employees using vibrating tools. Gloves are now commercially available which have special pads in the palm and fingers that absorb vibration. It should be noted that gloves should have external seams; otherwise internal glove seams press into the palm and fingers, cutting off some blood circulation during tight gripping. The latter case restricts blood flow and may aggravate the vibration injury or trigger an attack of vibration white finger.

15.8 APPLICATIONS AND DISCUSSION

The effects of vibration on the well-being of the worker have been extensively studied by ergonomists (especially biomechanists) and health specialists. Studies have shown that exposure to vibration can affect not only the productivity and efficiency of workers, but also adversely affect their physiological and psychological health conditions. Workers encounter occupational vibrations (both segmental and whole-body) through the use of powered tools or the operation of vehicles or heavy equipment. The vibration transmitted from such equipment is usually complex, however regular, which makes it possible for professionals to analyze and understand the various aspects of vibration. This chapter provides a methodology for assessing vibration exposure, the ACGIH TLVs for vibration, and recommendations for minimizing vibration exposure.

REVIEW QUESTIONS

1. Define vibration.
2. Describe vibration frequency and its units of measurement.
3. List and explain the five parameters that characterize vibration intensity.
4. Describe vibration resonance and attenuation. How do they affect worker performance and health?
5. Explain both forced vibration and free vibration.
6. What is whole-body vibration? What are its common modes of entering the body? What range of frequencies of this type of vibration is of interest for analysis of human exposure?
7. What are the adverse effects of vibration on the musculoskeletal system of the body?
8. List eight adverse effects of whole-body vibration.
9. To what frequency range of whole-body vibration is the human body most sensitive?
10. At what frequency does resonance occur in the human body?

11. What are the four basic effects of whole-body vibration imposed on the human being?
12. What is meant by segmental vibration? What are other names for this type of vibration?
13. What range of frequency of segmental vibration is of interest for analysis of human exposure?
14. During the work-shift a worker is exposed to the following segmental vibration from hand-held power tools:

Vibration exposure duration (h)	Component acceleration ($m \cdot s^{-2}$)		
	x	y	z
3.0	4.5	2.9	10.7
3.0	8.2	2.5	9.5
1.5	2.5	3.8	8.5
0.5	0	0	0

(a) Calculate the root-mean-square values of the frequency-weighted acceleration for each component in number of **g**s.
(b) Does the worker's total daily exposure to vibration exceed the American Conference of Governmental Industrial Hygienists recommended threshold limit values for segmental vibration? If yes, which components?
15. List eight methods of vibration control and vibration white finger prevention.

REFERENCES

ACGIH (1994) *1994–1995 Threshold Limit Values for Chemical Substances and Physical Agents and Biological Exposure Indices.* American Conference of Governmental Industrial Hygienists, Cincinnati, Ohio.

AFS Segmental Vibration Task Force (1983) Vibration white finger. *Modern Casting* **73**:38.

ANSI (1986) *Guide for the Measurement and Evaluation of Human Exposure to Vibration Transmitted to the Hand.* ANSI S3.34. American National Standards Institute (ANSI), New York.

Ayoub, M.M. and Ramsey, J.D. (1975) The hazards of vibration and light. *Industrial Engineering* **7**: 40–44.

Cope, F.W. (1960) Problems in human vibration engineering. *Ergonomics* **3**: 35–43.

Grether, W.F. (1971) Vibration and human performance. *Human Factors* **13**: 203–216.

ISO (1986) *Mechanical Vibration: Guide for the Measurement and the Assessment of Human Exposure to Hand Transmitted Vibration.* ISO 5349. International Organization for Standardization (ISO), Geneva.

Key, M.M., Henschel, A.F., Butler, J., Ligo, R.N. and Tabershaw, I.R. (eds) (1977) *Occupational Diseases: A Guide to their Recognition*. Publication no. 77–181. US Dept. of Health, Education, and Welfare (DHEW), National Institute for Occupational Safety and Health (NIOSH), Washington, DC, pp. 515–520.

Osborne, D.J. (1987) The physical environment. In: *Ergonomics at Work*. John Wiley, New York, pp. 222–245.

Sanders, M.S. and McCormick, E.J. (1993) *Human Factors in Engineering and Design*, 7th edn. McGraw-Hill, New York.

Shenberger, R.W. (1972) Human response to wholebody vibration. *Perception Motor Skills* **34** (suppl. 1):127. (Cited in Key *et al.*, 1977.)

Thompson, D. (1990) Ergonomics and the prevention of occupational injuries. In: *Occupational Medicine*, (LaDou, J., ed.), Appleton & Lange, Norwalk, Connecticut, pp. 38–57.

Wasserman, D.E. and Badger, D.W. (1973) *Vibration and the Worker's Health and Safety*. Technical report no. 77. National Institute for Occupational Safety and Health, Cincinnati, Ohio.

Wasserman, D.E., Badger, D.W., Doyle, T.E. and Margolies, L. (1974) Industrial vibration – an overview. *Professional Safety* **19**:38–43.

Weaver, L.A. III. (1979) Vibration: an overview of documented effects on humans. *Professional Safety* **24**: 29–37.

Shift work

16

16.1 INTRODUCTION

Many firms operate more than 8 hours per day. Indeed, some are in operation around the clock, and their employees have to work in one of two or three shifts. Some firms assign employees to a shift on a permanent basis, while others rotate them periodically. Usually shift-work employees who work in evening or night shifts receive extra pay as compensation for the inconvenience of their working time.

Shift work (and overtime) is established to maximize the utilization of available resources (e.g., plant, equipment, and staff). In the late 1970s, about 20% of the work-force in the USA and Europe was employed on some sort of shift work (Folkard and Monk, 1979). This figure reached 25% by the mid 1980s (Schultz and Schultz, 1986). Shift work facilitates an around-the-clock operation resulting in more output (production). Factory owners run their machinery 24 hours a day to spread their capital cost over more units of production to reduce their unit costs. Many industries (e.g., chemical, fertilizer, food-processing, power generation, and nuclear plants) and services (e.g., nursing, police, fire fighters, computer services), however, have no other choice than adopting shift work. Technological advances and increased automation provide for continuous operations in industries.

Often, shift work or shift workers become dangerous. The dangers can occur to the shift workers themselves, their fellow workers, or to their community (such as the incident that happened at Three Mile Island). The purpose of this chapter is to provide the reader with information to understand why shift work can be such a liability, and to develop knowledge for the implementation of programs to suppress these risks.

16.2 SHIFT-WORK TERMINOLOGY AND NOTATIONS

There are a tremendous number of different work schedules. For example, Tasto and Colligan (1977) have reported 150 different schedules for fire fighters in the USA and Knauth *et al.* (1983) have analyzed 120 shift systems practiced by the police in the Federal Republic of Germany. Industry has also tried many types of work schedules. However, the variety in types of work schedules is not the problem of concern; they rather render a great range of options to choose from for scheduling shift work or, even, normal day work. Since some schedules are more complex than others, it is difficult to compare them. There is also some inconsistency in the use of work schedule terminology and notations (Teaps and Monk, 1987).

Table 16.1 presents the terms and notations that have been proposed by the pioneers in this field (International Labour Office (ILO), 1978; Adler and Roll, 1981; Teaps and Monk, 1987). The reader is encouraged to study these terms and notations to understand clearly the subjects presented in the remainder of this chapter.

Example 16.1

Consider Figure 16.1 illustrating sequences of morning (M) or first (1), evening (E) or second (2) and night (N) or third (3) shifts, and off days (O). The first elemental cycle consists of the first 9 days, one complete sequence of the three shifts and one free period. The second 9 days and the last 10 days make up, respectively, the second and third elemental cycles. The whole 28-day period constitutes a minor cycle, in which an employee will perform the three different shifts and takes free periods, each of an equal number of days. Since the length of the minor cycle is a multiple of 7 days (i.e. 28 = 4 × 7), the length of the major cycle of this rotation is also 28 days.

Example 16.2

In a given shift pattern, the arrangement of first (1) or morning (M), second (2) or evening (E) and third (3) or night (N) shifts, and off days (O) is as follows:

[M, M, E, E, N, N, O, O] or [1, 1, 2, 2, 3, 3, O, O]

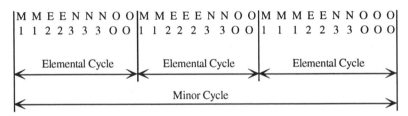

Fig. **16.1** Illustration of elemental and minor cycles in a shift rotation.

Table 16.1 Work system terminology and notations

Term/notation	Description
Shift	The time period of the day in which a group of one or more employees is scheduled to be at work in the workplace
Off time, off day or free period	The time period of the day in which an employee is off duty and normally not required to be at the workplace
Work schedule	The sequence of consecutive shifts and off time to be carried out by an employee
Work system	The collection of all work schedules implemented in a given workplace to carry out the required work activities of a given organization
Elemental cycle or basic sequence	The minimum number of days required by a work schedule to repeat itself. In this cycle, a group of one or more employees goes through one complete sequence of all different shifts and at least one off-time (free) period
Minor cycle	The minimum number of days required for the full sequence of shifts and free periods to repeat continuously until the given group of employees performs each and all shift and off days as an equal number of days. The number of elemental cycles in a minor cycle is an indication of the shift pattern complexity
Major cycle	The minimum number of days required for the sequence of shifts and free periods (one or more minor cycles) to be repeated until it begins to repeat on the same day of the week as the first sequence started. When the minor cycle is a multiple of 7 days, the lengths of the major cycle and the minor cycle are the same; otherwise, the length of the major cycle in weeks equals the number of days in the minor cycle
Permanent or fixed shifts	The work schedule is fixed; that is, employees always work the same shift and are not required to work more than one shift
Rotating or alternating shifts	The work schedule requires employees to work more than one shift. That is, employees switch from one to another according to a certain cyclic pattern of shift rotation
Continuous work weeks	A work schedule that requires the employee to work on some weekends
Discontinuous work weeks	A work schedule that does not require the employee to work on weekends
Shift pattern	A certain sequence of shifts and free periods within the elemental cycle

The length of the minor cycle in this pattern is 8 days, which is made up of only one elemental cycle. The length of the major cycle is 56 days [i.e. (7 days/week) × (8 days/minor cycle)].

16.3 SHIFT ORGANIZATION AND PATTERNS

Different companies have different natures of business, working conditions, products, number of workers, and set-ups. This is the main reason for the development of so many different shift-work systems. However, the systems pose many similar problems, such as discordance between the workers' work and rest periods, and the rhythms of social and family life (ILO, 1978).

The issue of shift-work scheduling is to insure that negative aspects of the schedule do not dramatically affect workers' morale, safety, health, and productivity (Monk, 1986). In planning the rotation of shift schedules for 24-h coverage, a key issue is how many consecutive days or nights should be worked before a worker moves on to a different shift. As Monk pointed out, expert opinions are divided between the benefits and drawbacks associated with rapid shift rotations of less than 5 days, comparing them with slowly rotating 21 or 28 days with intervening rest days and fixed shift systems. Some companies prefer rapidly rotating schedules as they may have more benefits than drawbacks, whereas for others they may not be suitable. Hence, there is no fixed rule of thumb, but some guidelines are available for the selection of proper shift schedules.

Based on the needs and regulations concerning the duration of work, rest period and leave, the shift organization becomes relatively complex and requires a number of crews that is greater than the number of shifts. This becomes more complicated when the operation is continuous for 24 h a day, all year round, regardless of weekends and holidays. ILO (1978) has distinguished the following three types of shift-work organizations:

- **discontinuous** systems, in which the duration of daily operation is less than 24 h and work stops every day and usually at the end of the week (after 5 or 6 consecutive working days);
- **semicontinuous** systems, which operate 24 h a day without daily interruption but stop at the end of the week (after 5 or 6 consecutive working days);
- **continuous** systems, which operate 24 h a day, 7 days a week, and where the work goes on even during public or national holidays.

Shifts are divided into two general classes: permanent or fixed shifts, and rotating or alternating shifts. As defined in Table 16.1, in permanent or fixed shifts workers always remain in the same shift, and in rotating or alternating shifts they switch from one shift to another according to a certain pattern or schedule of shift rotation.

The shift rotation can be forward, backward, or mixed. In **forward rotation**, the direction of shift rotation is from first (morning) to second (evening) shift, from second (evening) to third (night) shift, and from third (night) shift to off days. In **backward rotation**, the direction of shift rotation is reversed. A **mixed rotation** allows shift changes in both directions.

As will be explained later in this chapter, from the physiological point of view, fixed shifts, when the body is adapted to the circadian rhythm, are preferred over frequent-rotating shifts. However, the social problems induced by a fixed night shift make fixed-shift patterns undesirable. Based on the combined effects of physiological and social factors, many different patterns of shift rotation have been developed.

16.4 A CONVENTIONAL FORMAT FOR SHIFT PATTERN NOTATION

Adler and Roll (1981) proposed a conventional form of describing the pattern of shift sequence. The format was based on a number of consecutive morning, evening, and night shifts, as well as off days, in each elemental shift cycle. This conventional notation is written as: m-e-n (f), where m, e and n, respectively, denote the numbers: morning or first; evening or second; and night or third shifts; and f is the number of free (off) days, shown in parentheses in their corresponding locations within the elemental cycle. Whenever the elemental cycles are of different arrangements, all elemental cycles are written and separated by a bar (|) or slash (/). The notation may be extended to include the number of hours worked in each shift where they vary from shift to shift. If this feature is necessary, the shift lengths are written in their respective orders and in braces{ }. This notation format is followed in this chapter.

Example 16.3

Two commonly practiced methods of rotating three 8-h shifts, especially in European countries, are the **continental rota** and the **metropolitan rota**. The conventional formats for these two systems of shift work are as follow:

The continental rota: 2–2–3 (2)/2–3–2 (2)/3–2–2 (3)
The metropolitan rota: 2–2–2 (2)

noting that since the number of hours in each of the three shifts is 8 h, the shift lengths are not written in the notations. For more convenience, most authors refer to the continental rota as the 2–2–3 pattern, and to the metropolitan rota as the 2–2–2 (or 2–2–2–2) pattern.

Since there are three shifts and one off-time period, the shifts in either shift pattern must rotate among four crews of employees, i.e., there are three crews working and one resting. In the continental rota, one crew works 2 days on the morning, 2 days on the evening and 3 days on the night shifts during a week. During the next week, the crew first takes 2 days off, then works 2 days on the morning and 3 days on the evening shifts. During the third week, this crew works 2 days on the night shift, takes 2 days off and then works 3 days on the morning shift. During the following shift it works 2 days on the evening and 2 days on the night shift, then takes 3 days off. By the end of the fourth week this crew has worked an equal number of days on each shift and taken the same number of days off. This means that the crew has completed one minor cycle of shift work. A second crew follows the same pattern, but starting with working 2 days on the evening and 2 days on the night shifts and taking 3 days off. Following the same shift pattern, a third crew starts off with working 2 days on the night shift, and a fourth crew with having 2 days off first.

In the metropolitan rota, one crew works 2 days on each of the morning, evening and night shifts, in their respective order, then takes 2 days off. Another crew works 2 days on the evening and 2 days on the night shifts, takes 2 days off and, then works 2 days on the morning shift. A third crew starts off with working 2 days on the night shift, then takes 2 days off and works 2 days on the morning and 2 days on the evening shifts. A fourth crew starts the same pattern with having 2 days off first, then works 2 days on each of morning, evening, and night shifts, respectively. Every 8 days, one minor cycle is completed in this pattern of shift work.

A 4-week schedule for shift rotation using the 2–2–2 and the 2–2–3 patterns among four crews of employees is illustrated in Table 16.2 and Table 16.3, respectively.

16.5 EFFECTS OF SHIFT WORK ON WORKERS

Various effects of shift work on workers have been well-documented. It has been proven that shift work disturbs the length and quality of sleep, causes nervous and digestive disorders, and affects workers' behaviors and reactions depending on the level and type of familial and social pressures and other factors (Carpentier and Cazamian, 1977; Wojtczak-Juroszowa, 1977; Levi, 1984; Monk, 1988, 1989). The effects of shift work on circadian rhythms, physiological responses and health, psychological behaviors, work performance, and familial and social conditions of shift workers are presented in the following subsections. It is outside the

Table 16.2 Shift schedule using forward-rotating, continuous shift work – the continental rota

Crew	Week 1							Week 2							Week 3							Week 4						
	M	T	W	Th	F	Sa	Su	M	T	W	Th	F	Sa	Su	M	T	W	Th	F	Sa	Su	M	T	W	Th	F	Sa	Su
1	M	M	E	E	N	N	O	O	O	M	M	E	E	N	N	N	O	O	M	M	E	E	E	N	N	O	O	M
2	E	E	N	N	O	O	M	M	M	E	E	N	N	O	O	O	M	M	E	E	N	N	N	O	O	M	M	E
3	N	N	O	O	M	M	E	E	E	N	N	O	O	M	M	M	E	E	N	N	O	O	O	M	M	E	E	N
4	O	O	M	M	E	E	N	N	N	O	O	M	M	E	E	E	N	N	O	O	M	M	M	E	E	N	N	O

Notes
- M = morning shift; E = evening shift; N = night shift; O = off (rest) day.
- Elemental cycle = 9 days for the first two cycles, and 10 days for the third.
- Minor cycle = major cycle = 28 days.

Table 16.3 Shift schedule using forward-rotating, continuous shift work – the metropolitan rota

Crew	Week 1							Week 2							Week 3							Week 4						
	M	T	W	Th	F	Sa	Su	M	T	W	Th	F	Sa	Su	M	T	W	Th	F	Sa	Su	M	T	W	Th	F	Sa	Su
1	M	M	E	E	N	N	O	O	M	M	E	E	N	N	O	O	M	M	E	E	N	N	O	O	M	M	E	E
2	E	E	N	N	O	O	M	M	E	E	N	N	O	O	M	M	E	E	N	N	O	O	M	M	E	E	N	N
3	N	N	O	O	M	M	E	E	N	N	O	O	M	M	E	E	N	N	O	O	M	M	E	E	N	N	O	O
4	O	O	M	M	E	E	N	N	O	O	M	M	E	E	N	N	O	O	M	M	E	E	N	N	O	O	M	M

Notes:
- M = morning shift; E = evening shift; N = night shift; O = off (rest) day.
- Elemental cycle = minor cycle = 8 days.
- Major cycle = 8 weeks = 56 days.

scope of this book to discuss the economical justification and sociological aspects of shift work. However, some social factors in relation to workers are pointed out.

16.5.1 EFFECTS OF WORK SHIFT ON CIRCADIAN RHYTHMS

The circadian or diurnal rhythms are the body's interrelated internal multiprocesses which are to a great extent tuned to the time changes over the 24 h of the day. The circadian rhythms provide the physiological and psychological foundations for regular daily cycles of sleep and wakefulness. The rhythms follow a diurnal pattern of restful sleep during the night and an active life during the day.

Many physiological and psychological functions and states of the body have a regular circadian rhythm related to the 24 h of the day (Edholm, 1967; Seward, 1990). However, Seward argues that many of the human circadian rhythms follow a 25-h cycle, and hence the usual 24-h day requires a backward, 1-h-daily adjustment. This means that in changing our daily schedules, it is generally much easier to delay awakening than to arise progressively earlier. This theory explains why the shift rotation should proceed from morning to evening to night shifts rather than the reverse order. Such forward rotation put less stress on the adaptability of the internal biological clock.

Variations in the body temperature, metabolic rate, heart rate, blood pressure, and chemical composition are related to the circadian rhythms. The rhythms have shown an association with changes in environmental factors, such as daylight and darkness, and climatic temperature changes. The following are three typical examples of circadian rhythms in the heart rate, body temperature, and chemical composition during the course of a day:

- The heart rate is lowest during the night when the person is in the deepest sleep; it rises rapidly in the morning period, and peaks in the late afternoon or early evening; it slows down in the night again. This rhythm is consistent from one 24-h cycle to the next.
- The average normal body temperature curve of a healthy individual, who works during the day, is at its lowest early in the morning (36.1°C or 97.0°F about 4:00 a.m.), gradually rising during the day to reach 37°C (or 98.6°F) by mid-morning (between 9:00 and 10:00 a.m.) and to its maximum (37.2°C or 98.9°F) late in the afternoon (between 5:00 and 8:00 p.m.), and then falls gradually during the night until early in the next morning (Emerson and Taylor, 1946). The normal body temperature change of permant night shift workers is the reverse of day workers.
- Adrenocorticotropic hormone (ACTH), also known as corticotropin, a secretion of the adrenal cortex, varies according to a circadian rhythm,

and thus plasma cortisol tends to rise and fall in response to this variation (Ganong, 1987). The secretion of this hormone is most frequent in the morning and least in the evening. If the day's activities are spread over a period of longer than 24 h, the adrenal cycle also lengthens, but the secretion of ACTH still increases during the period of sleep.

Circadian rhythms of physiological (biological) functions determine the worker's proneness or susceptibility to environmental stressors. For example, worker vigilance (alertness) is very low in the early morning. When the circadian rhythms are disrupted, the body undergoes dramatic changes and strain to adapt to the new condition. However, some people can adapt easier and faster than others to shift changes (Zedeck et al., 1983).

Workers permanently assigned to one shift are usually able to adjust to a new circadian rhythm and, therefore, encounter fewer problems than those working in a rotating-shift system (Jamal, 1981). In a rotating shift-work system, however, workers have to make a new adjustment every time their shift is changed (e.g., every 2 or 3 days, or week). In fact, they may not have fully adjusted to one schedule before starting the next. This can have severe ill-effects on their health.

The effects of night-work on circadian rhythms can be summarized as follow:

- Night work requires the reversal of the normal sleep/wake cycle.
- At the beginning of night-shift work the body temperature is higher; it falls gradually throughout the night regardless of the physical activities of the shift. Also, the worker will develop a higher body temperature during day sleep and daytime hours when compared with that during night working hours. This is the opposite of what occurs in individuals working on day-shift work.
- The systolic blood pressure is lower during the night-shift when compared with day-shift work.

16.5.2 PHYSIOLOGICAL EFFECTS OF WORK SHIFT

Shift work, especially the night shift, has been shown to have adverse effects on the well-being of shift workers. These effects may be due, however, to the disturbed circadian rhythms. The following are examples of the physiological effects of shift work on workers:

- Shift workers encounter more gastrointestinal problems than non-shift workers.
- Shift work disturbs circadian rhythms or the biological clock of human beings. This disturbed schedule causes stomach upsets, sleep loss,

nervous and gastric disorders. However, studies report that workers who perform shift work have a similar overall rate of incidence of sicknesses compared to non-shift workers.

- The incidence of certain illnesses may not differ between shift workers and non-shift workers, but the duration of sick periods can be longer among shift workers.
- Shift workers have been found to have longer sleeping periods and longer naps when compared with non-shift workers. This increase in sleeping time is a compensation for the sleep debt incurred during the shift work.
- No significant difference in mortality rate has been found for shift workers when compared to non-shift workers.

16.5.3 PSYCHOLOGICAL EFFECTS OF WORK SHIFT

Shift-work stress results in fatigue which, in turn, can cause other psychological problems in workers, such as dissatisfaction and irritation. Shift work may not be particularly dangerous; however some level of risk is always involved. The rate of accidents can increase by elevation of stress, fatigue, and dissatisfaction resulting from shift work. The worker's complaints usually concern difficulties with falling asleep, difficulties with retaining sleep and lowered spirits, in addition to gastrointestinal problems (Åkertedt, 1976). The psychological effects of shift work are summarized below:

- Difficult and uneasy hours may cause uneasiness, tiredness, irritation, and a disturbed social life.
- Shift work can cause stress which is fatiguing. Fatigue, in turn, can result in dissatisfaction, turnover of employees, decrease in productivity, and increased risks of accidents.
- Shift work causes social and/or family problems, as opposed to medical disturbances, mostly due to psychological disturbances.

16.5.4 EFFECTS OF WORK SHIFT ON DOMESTIC AND SOCIAL FACTORS

Domestic and social factors have not been carefully studied. Most reported investigations are rather subjective, which may not be suitable for drawing general conclusions; they usually deal with personal perceptions of the advantages and disadvantages of shift work and workers' attitudes towards their own situations. The study deficiency is also due to the fact that it is extremely difficult to obtain the relevant information for drawing valid conclusions. For example, it may be impossible to obtain complete and honest information about familial problems from the majority of workers. Nevertheless, the following effects of shift work on the worker's domestic and social life can be expected:

- Disrupted family life, such as marriage problems caused by increased tension and conflict, problems with taking children to school or other activities and games. Family gatherings and interactions are disturbed by shift changes, or even by fixed night shifts.
- Disrupted interaction with friends and relatives.
- Disrupted hobbies, group activities, and community involvement. Such activities usually require regular participation, which is interrupted by shift rotation. A shift worker who is not able to attend such activities on a regular basis can easily develop the feeling of being dissociated from groups and community.
- Commuting problems, including traffic accidents, especially on the way home from work when the night worker can be sleepy and exhausted. This is also a problem for day-shift workers in crowded urban areas who must report to work early in the mornings. Workers may not be able to go to bed long before midnight, but may have to get up by 2 or 3 a.m. in order to get ready and be at work on time. Such sleep deprivation causes mental fatigue and other disturbances, resulting in traffic and job accidents and health problems.

16.5.5 SHIFT-WORK EFFECTS ON WORKER PERFORMANCE

Research conducted in the USA and Europe has revealed that shift work can affect worker performance in a variety of ways (Kabaj, 1978; Tilley *et al.*, 1982; Schultz and Schultz, 1986). However, these effects may be secondary to other effects of shift work, where the primary effects can be physiological, social, personal, etc. The effects of shift work on worker performance can be summarized as follows:

- Generally, shift-work performance is affected by a combination of the following factors:
  - type of work. Mentally demanding jobs (e.g., quality control and inspection) require patience and alertness. Shift workers may develop a deficiency in both;
  - type of shift system. Disturbed circadian rhythms can be a detriment to the mental and physical abilities of shift workers. This is particularly a problem in frequently rotating and night shifts;
  - type of worker. For example, older workers have less ability to stabilize their circadian rhythms in shift changes, particularly in frequently rotating shifts.
- Poor night shift performance can be related to:
  - impaired circadian rhythms;
  - slow adaptation to the night-shift requirements.

- Workers are more productive on the day shift than the night shift.
- Workers make less errors and have fewer accidents on the day shift than the night shift.
- The alertness (vigilance) of workers declines during night-shift work, especially in the early morning. This may be critical in tasks which require continuous monitoring (e.g., control console operations).
- If the worker gets an insufficient amount of sleep prior to the work shift, job performance can be adversely affected, especially when the job requires a high level of alertness.

Permanent night shifts should be preferred over rotating shifts where serious consequences may exist as the result of an error (e.g., nuclear or chemical power plant operations). Workers on permanent shift assignments should be encouraged to maintain their routines of sleep/wake on the weekends. This prevents any possible readjustment to a different sleep/wake schedule.

16.6 COMPARISON OF THE THREE SHIFT SCHEDULES

Each shift has some advantages and some disadvantages. The following comparisons were adapted from Carpentier and Cazamian (1977).

- The **morning shift** enables workers to have the evening meal with their family and to participate in family and social activities. Although it allows for more free time, the morning shift starting very early in the morning (e.g., at 4 a.m.) is often stated to be very tiring. However, the common 8 a.m. to 4 p.m. shift schedule practiced in the USA is the least disadvantageous to the shift worker.
- The **afternoon (evening) shift** seems to be the most disadvantageous from the family and social points of view. Workers will be home in the morning when their children are not, and the children will be home in the evening while workers are not. Participation in all evening activities is interrupted. However, the afternoon shift allows plenty of time for sleep and rest.
- The **night shift** is regarded as the least desirable shift and raises most problems because of its effects on the family atmosphere. If it begins at 9 p.m. or 10 p.m., it interrupts most social activities in the evening. The night work schedule in use in the USA, which begins at midnight, cuts off less of the evening. However, the night shift allows a great deal of free time during the day. Management's special attention to the atmosphere of the night shift and extra compensation make the night shift more desirable for workers.

Due to the succession of these advantages and disadvantages and the difficulties with periodic changes, a system of fixed shifts is preferred over rotating shifts. Also, a system of fixed shifts enables the worker to move from one shift to another, which is very difficult in the system of rotating shifts. However, the system of rotating shifts is widely practiced owing to its compromise between advantages and inconveniences (Carpentier and Cazamian, 1977).

16.7 GUIDELINES FOR SHIFT WORK

There are volumes of conflicting research findings (Urban and Dutta, 1995) that make it difficult to develop a comprehensive strategy for shift work. None the less, the following guidelines should be consulted to help shift workers ease their coping with shift-work problems.

16.7.1 GUIDELINES FOR SHIFT WORKERS

- Night-shift workers should take a hot meal during the night shift well before the end of the shift in the morning.
- The shift worker should set aside a specific block of time to sleep. This time should be regular and predictable, and free from social or other commitments (Monk, 1989). This time block for night-shift workers should be as soon as possible after the shift, rather than waiting for the afternoon sleep.
- Shift workers should spend about seven continuous hours in bed (even if sleep may not occur for the whole period). Night-shift workers usually need more sleep than day-shift workers for the following two reasons:
  - It takes longer for night-shift workers to fall asleep.
  - Night-shift workers usually sleep more lightly.
- The following symptoms indicate the lack of adaptation to shift work and require special attention:
  - loss of body weight;
  - nervous disorders;
  - gastrointestinal problems;
  - respiratory disorders;
  - fatigue after short work exposures.
- Shift workers should perform some off-the-job physical exercise at least twice a week.
- Shift workers should refrain from smoking, or at least reduce it if they do smoke.
- Shift workers should refrain from taking sleeping pills, the overuse of which by some shift workers has been reported (Fossey, 1990). Studies have shown that sleeping pills (triazolam) do not have circadian resetting properties (Monk, 1989).
- Shift workers should not consume alcohol as a sedative to help them

to fall asleep. Some researchers have found that alcohol causes further sleep disturbances (Wedderburn and Scholarios, 1993).
- Night-shift workers should be encouraged to consume coffee or tea, if it does not induce stomach troubles, to stay awake and alert during the shift and on their way home. However, care should be taken as caffeine consumption within a few hours of sleep time is not recommended.

16.7.2 GUIDELINES FOR MANAGING SHIFT WORK

- If possible, the night-shift duration should be reduced without reducing the amount of compensation and other benefits.
- The number of required night-shift workers should be reduced to allow more flexibility for reducing the number of nights each shift worker must work.
- A work-shift duration should not exceed 8 h.
- Each day or evening shift should be followed by at least 24 h free time, and each night shift by at least two consecutive free days, so that workers can regulate their sleep habits.
- Social interactions with other fellow workers should be allowed.
- Physical activity facilities, such as basketball hoops and ball, should be provided, especially for night-shift workers.
- Non-monotone and rhythmic music may be useful, especially during the night shift.

16.8 CRITERIA FOR SELECTION OF SHIFT WORKERS

The effects of shift work on individuals vary significantly. For this reason, determination of suitability of individuals for shift work is difficult. Development of a systematic technique for selecting shift workers is necessary. This technique should be comprehensive and based on certain criteria, such as physical and mental abilities, age, experience, motivations, subjective ratings, family structure and needs, and community involvement. However, no technique seems to insure that selected shift-workers would be able to adapt to shift work. Therefore, worker tolerance and acceptance need to be evaluated continually (Urban and Dutta, 1995). The following general criteria should be considered in selection of shift workers:

The individuals with the following characteristics are better candidates for shift work:

- independent;
- more adjustable to changes in working conditions;
- high self-esteem;

- young;
- healthy.

The individuals with the following characteristics are considered less suitable for shift work (especially for night shift):

- having a history of gastrointestinal problems;
- having a history of respiratory disorders;
- having a history of nervous disorders;
- developing abnormal fatigue after a short work exposure;
- past age 45, especially with health problems;
- pregnant.

Since shift work has many social disadvantages, the solution for measures which make shift work more acceptable for workers (optimum pattern) is of vital importance. Social, economic, and technical conditions differ from each other, and vary from country to country and industry to industry.

Table 16.4 shows the economic impacts of three basic shift-work strategies (A, B, and C). Strategy A offers minimum capacity utilization and strategy C maximum. With the same fixed assets, the theoretical production and employment opportunity level varies from 100 to 365 (Kabaj, 1978).

The question that now arises from the practical point of view is which of these three strategies is closest to optimal in modern conditions? How can the optimum be defined and achieved by accommodating both economic and social concerns? Obviously, the optimal shift-work strategy corresponds to the situation that guarantees rational utilization of equipment and human resources. However, it would not be possible to suggest any single shift that can universally be efficient for all industries, as the shift strategy might change under various economic conditions. As stated earlier, shift work and the shift-work strategy are influenced by both economic and social phenomena. Hence, they should be based upon a compromise between economic advantages and social disadvantages of multiple shift work.

16.9 ALTERNATIVE WORK SCHEDULES

An alternative to a fixed schedule is the concept of flexible time (or flexitime) in which all employees are allowed to select their own 8-h working time within a 12-h period. In this plan, all employees would be working during the middle 4 h of the day. Such a schedule gives tremendous flexibility to the worker, but creates scheduling and time-keeping problems for the employer. A flexitime scheme would obviously not work for multishift operations or production operations where work

Table 16.4 The three basic shift-work strategies

Shift-work strategy	Number of working days per year (approx.)	Number of hours per working day	Shift coefficient	Number of working hours per year	Index of production and employment opportunities (Strategy A = 100)
Semicontinuous processes					
A One shift	300	8	1	2400	100
B Two shifts	300	8	2	4800	200
C Three shifts	300	8	3	7200	300
Continuous processes					
A One shift	365	8	1	2920	122
B Two shifts	365	8	2	5840	243
C Three shifts	365	8	3	8760	365

Source: Kabaj (1978).

flow is continuous or is dependent on other employees completing their portion of the task before it is moved along to the next worker.

Another alternative to traditional shift work is shared work, where two employees share a work shift. Such a sharing might be on a 50/50 basis for each shift (each person works only 4 h per shift) or on a daily basis, where one employee works a certain number of days during the shift cycle while the partner works the remaining days (with each employee working a full 8-h shift on the assigned work day). Some of the major problems with shared work include employee benefits (how to divide insurance and other company benefits) and scheduling problems. If an employee fails to show up for work, the employer might not know which employee was supposed to work that shift (there are also increased opportunities for the two employees to misunderstand who was to work on which days).

Lewis and Swaim (1986) conducted a study at Fast Flux Test Facility near Richland, Washington where the rotating shift schedule was changed from an 8-h-day to a 12-h-day work schedule (i.e., from 5–7 days of 8-h-day to 3–4 days of 12-h-day) using standard laboratory-type measures of performance and alertness, and a questionnaire on sleep patterns. After 7 months' adaptation to the new schedule, an analysis indicated that there were some decrements in alertness, reduction in sleep and disruptions of other personnel activities during 12-h workdays. However, they observed improved gastrointestinal state during night shift and reduced stresses. After 7 months, 84% of the operators favored the change. Results indicated that a 12-h shift schedule was a reasonable alternative to an 8-h schedule at this facility.

One of the issues of concern to companies considering an extended work day (10- or 12-h shifts) with 3 or 4 days off is what employees will do on their days off. In periods of economic stress, the worker might find a second job to work on off days. Any follow-up surveys of worker fatigue and stress should also attempt to look at activities during the day-off periods. However, it may be very difficult to obtain accurate information, especially if the worker fears that the off-duty activity (second job) might have some influence on the principal employment.

16.10 APPLICATIONS AND DISCUSSION

As industry explores options for increasing productivity, the number of workers participating in some sort of shift work is likely to increase also. What may be one person's social problem might be a solution for another. For example, some people might consider an evening shift a hardship because they would like to be a part of their children's after-school activities, such as sports, music, or other recreational activity. However, other individuals might prefer an evening shift so that they would be available to volunteer at their child's school during the day while their child is in school. Also, some parents might have a strong preference for being home in the morning while their children eat breakfast and prepare to go to school, while other parents might be much more concerned about being home when their children return from school in the afternoon.

For whatever reasons, some workers prefer rotating shifts while others prefer fixed shifts. Zedeck *et al.* reported that 32% of the 732 workers they studied at a power-and-gas utility company expressed no desire to change from their rotating shift-work system. This group of workers was satisfied with the rotating shift-work schedule and had fewer health, social, and marital problems than workers who were not satisfied with this shift-work system.

The physiological, psychological, and sociological effects of shift work can impose health, domestic, and social problems on shift workers. Furthermore, these effects are also interactive and aggravate each other. Shift-work education programs should be developed to help shift workers be more able to deal with the complexity of shift-work effects.

Guidelines for coping with shift work are provided in this chapter. These general guidelines may be applied to a majority of shift workers. However, adherence to some of them, such as consumption of caffeine, alcohol, and nicotine, can easily be affected by other factors such as social pressure and simple immediate pleasure. An effective implementation of these guidelines would help the shift worker minimize health, familial, and other problems.

REVIEW QUESTIONS

1. Why is it difficult to compare all shift schedules with each other?
2. Define shift.
3. Define off time.
4. What is a work schedule (with regard to shift work)?
5. What is a work system (with regard to shift work)?
6. What is an elemental cycle or basic sequence of a work-shift pattern?
7. What is the difference between permanent or fixed and rotating shifts?
8. What is the difference between continuous and discontinuous work weeks?
9. What is meant by a shift pattern?
10. What are the lengths of elemental, minor and major cycles of the 2–2–3 shift pattern?
11. In a given shift pattern, the arrangement of morning (M), evening (E) and night (N) shifts, and off days (O) is: [M, M, E, E, N, N, O, O,

O]. Determine the lengths of elemental, minor, and major cycles for this shift pattern.

12. What is the key issue in planning the rotation of shift schedules for 24-h coverage?

13. Describe discontinuous, semicontinuous, and continuous shift systems.

14. What is the difference between forward and backward shift rotations?

15. Write the conventional form of describing the pattern of shift sequence, and describe its notations.

16. Describe circadian rhythms.

17. How can you support the theory that the circadian rhythms follow a 25-h cycle rather than a 24-h one? How can you relate this theory to shift rotation?

18. Describe the typical circadian rhythm in the heart rate during the course of a day.

19. Describe the circadian rhythm in the normal body temperature during the course of a day.

20. Explain why a permanent shift is preferred over a rotating-shift system.

21. List three effects of shift work on circadian rhythms.

22. It is known that shift work disturbs circadian rhythms. What are the consequences of such schedule disturbance?

23. Compare the incidence of illnesses between shift workers and non-shift workers.

24. Compare the sleep periods between shift workers and non-shift workers.

25. List three psychological effects of shift work.

26. List four expected effects of shift work on the worker's domestic and social life.

27. List the three general factors that affect the performance of shift workers.

28. To what can poor night-shift performance be related?

29. Compare the advantages and disadvantages of the morning shift to other shifts.

30. Compare the advantages and disadvantages of the afternoon shift to other shifts.

31. Compare the advantages and disadvantages of the night shift to other shifts.

32. List five guidelines for managing shift work.

33. List seven guidelines for helping shift workers cope with shift-work problems.

34. List some typical criteria that could provide the basis for selecting shift workers.

35. List five characteristics that make an individual a candidate for shift work.
36. List six characteristics that make an individual less suitable for shift work (especially for night shift).
37. Describe the concept of flexible time as an alternative to a fixed-shift schedule.

REFERENCES

Adler, A. and Roll, Y. (1981) Proposal for a standard shift pattern notation. In: *Night and Shift Work: Biological and Social Aspects* (A. Reinberg, N. Vieux and P. Analaver, eds), the Proceedings of the Fifth International Symposium on Night and Shift Work: Scientific Committee on Shift Work of the Permanent Commission and International Association on Occupational Health (PCIAOH), ROUEN. Pergamon Press, Oxford.

Åkertedt, T. (1976) Individual differences in adjustment to shiftwork. In *The Proceedings of the Human Factors Society – 20th Annual Meeting*. Santa Monica, California, pp. 510–514.

Carpentier, J. and Cazamian, P. (1977) *Night Work: Its Effects on the Health and Welfare of the Worker*. International Labour Office (ILO), Geneva.

Edholm, O.G. (1967) *The Biology of Work*. McGraw-Hill, New York.

Emerson, C.P., Jr., and Taylor, J.E. (1946) *Essentials of Medicine*, 15th edn. J.B. Lippincott, Philadelphia.

Folkard, S. and Monk, H.T. (1979) Shift work and performance. *Human Factors* **21**: 483–492.

Fossey, E. (1990) Shiftwork can seriously damage your health. *Professional Nurse* **5**: 576–580.

Ganong, W.F. (1987) *Review of Medical Physiology*, 13th edn. Appleton & Lange, Norwalk, Connecticut.

ILO (1978) Social problems of shift work. Working paper no. 2. In: *Management of Working Time in Industrialised Countries*. International Labour Office, Geneva, pp. 18–35.

Jamal, M. (1981) Shift work related to job attitudes, social participation and withdrawal behavior: A study of nurses and industrial workers. *Personnel Psychology* **34**: 535–548.

Kabaj, M. (1978) Searching for an optimal shift-work pattern. In: *Management of Working Time in Industrialized Countries*. International Labour Office (ILO), Geneva.

Knauth, P., Rutenfranz, J., Karvonen, M.J., Undeutsch, K., Klimmer, F. and Ottmann, W. (1983) Analysis of 120 shift systems of the police in the Federal Republic of Germany. *Applied Ergonomics* **14**: 133–137.

Levi, L. (1984) *Stress in Industry: Causes and Prevention*. Occupational Safety and Health series no. 51. International Labor Organization (ILO), Geneva.

Lewis, M.P. and Swaim, J.D. (1986) Evaluation of a 12-hour/day shift schedule. In *The Proceedings of the Human Factors Society – 30th Annual Meeting*. Santa Monica, California, pp. 885–889.

Monk, T.H. (1986) Advantages and disadvantages of rapidly rotating shift schedules – a circadian viewpoint. *Human Factors* **28**: 553–557.

Monk, T.H. (1988) *How to Make Shift Work Safe and Productive*. American Society of Safety Engineers, Des Plaines, Illinois.

Monk, T.H. (1989) Shift work and safety. *Professional Safety* **34**: 26–30.

Schultz, D.P. and Schultz, S.E. (1986) *Psychology and Industry Today*, 4th edn. Macmillan, New York.

Seward, J.P. (1990) Occupational stress. In: *Occupational Medicine* (LaDou, J., ed.), Appleton & Lange, Norwalk, Connecticut, pp. 467–480.

Tasto, D.L. and Colligan, M.J. (1977) *Shift Work Practices in the United States*. DHEW (NIOSH) publication no. 77–148. National Institute for Occupational Safety and Health, US Department of Health, Education and Welfare.

Teaps, D.I. and Monk, T.H. (1987) Work schedules. In: *Handbook of Human Factors* (Salvendy, G., ed.). John Wiley, New York, pp. 819–843.

Tilley, A.J., Wilkinson, R.T., Warren, P.S.G., Watson, B. and Drud, M. (1982) The sleep and performance of shift workers. *Human Factors* **24**: 629–641.

Urban, N. and Dutta, S. (1995) Shiftwork: strategies for ergonomics intervention. In: *Advances in Industrial Ergonomics and Safety VII* (Bittner, A.C. and Champney, P.C., eds). Taylor & Francis, London, pp. 959–964.

Wedderburn, A. and Scholarios, D. (1993) Guidelines for shiftworkers: trial and errors? *Ergonomics* **36**: 211–217.

Wojtczak-Juroszowa, J. (1977) *Physiological and Psychological Aspects of Night and Shift Work*. US Department of Health, Education and Welfare, NIOSH, Cincinnati, Ohio.

Zedeck, S., Jackson, S.E. and Summers, E. (1983) Shift work schedules and their relationship to health, adaptation, satisfaction, and turnover intention. *Academy of Management Journal* **26**: 297–310.

Office ergonomics 17

17.1 INTRODUCTION

The evolution of the personal computer has resulted in significant changes in the office work environment. Glare, which generally posed no problem with the use of a typewriter, can cause problems at a video display terminal (VDT). Files are stored on computer disks, rather than in file cabinets, thus reducing the need for the operator to move from a desk to filing cabinets. Desks that were not designed for typing activities are now equipped with personal computers having keyboards that are located too high. Because of a lack of suitable space on desks, keyboards and computer displays are often not arranged properly on the desk. All of these situations are common in modern offices, despite the fact that it is known that problems will occur, and there are reasonable approaches available to solve most of them. This chapter discusses the typical problems found in office environments that can be alleviated through ergonomics redesign. Design guidelines for office work will be presented.

17.2 POSTURAL AND HABITUAL PROBLEMS

Most office problems are caused by awkward posture. The postural problems, however, stem from poor workstation design or poor habitual postures. Many computer users sit at their workstations in postures that they think are the most comfortable positions. Thus, office employees may, unknowingly, assume harmful postures and expose themselves to occupational risk factors associated with the development of cumulative trauma disorders (CTDs), such as carpal tunnel syndrome (CTS) or other forms of CTDs at the elbows, shoulders, neck, and back. Poor posture can lead to discomfort and fatigue in the hands, wrists, shoulders, thighs, knees, ankles, and/or feet.

Some computer users find it convenient to rest their wrists on the work-surface while typing on a keyboard. In such a posture the wrist is extended (dorsally bent) so that the fingers can touch the keys. Bending the wrists during keyboarding activities increases tension in the wrist which strains its soft tissues. This causes development of fatigue in the wrists in a relatively short period of time. The reader can experience the strain in the forearm muscles by extension of the wrist. This can be demonstrated by holding the hands in front of the body and extending (or bending dorsally) the wrists, and wiggling the fingers (as done in typing). This activity will induce much more pressure in the forearm muscles than when the wrists are kept straight. The pressure may even reach the shoulders. The amount of wrist fatigue accumulated hour after hour, day after day, becomes tremendous and can lead to CTDs.

When the wrist is placed on a hard surface such as the desk while typing, the weight of the hand and arm places pressure on the underside of the wrist. This pressure interferes with the circulation of blood through the wrists, hands, and fingers, which may cause or aggravate CTS.

As pointed out by Grandjean (1988), the following characteristics of VDT workstations suggest ergonomics analysis:

- working with a VDT for several hours (or perhaps all day) without interruption;
- restricted movements;
- concentrated attention on the screen;
- hands linked to the keyboard and computer mouse;
- VDT operators are more vulnerable to poorly designed workstation problems, such as:
  - strained posture;
  - poor photometric display characteristics;
  - inadequate lighting conditions.

Many office employees develop CTDs, especially CTS. Such disorders have become a problem in service and high-technology industries because of workers performing repetitive motions with fast and forceful movements, coupled with awkward work positions, and insufficient rest breaks over long working periods. The most critical risk factor in office jobs is the requirement of performing the same few motions over and over, without sufficient postural changes to relieve fatigue.

17.3 HEALTH RISKS IN OFFICES

Office employees are usually coping with the risk of CTDs. The most common types of CTDs affecting office employees are the following (see also Chapter 8):

- **Low-back pain**: Low-back pain can be due to muscle spasms or pinched nerves. Chronic low-back pain is usually due to nerve pinch caused as a result of damage to the intervertebral disks in the lumbar region of the spine. Low-back pain among office employees can result from awkward posture during work. An awkward posture may be due to sitting without appropriate back support or improper heights of the seat and work-surface, and poorly positioned equipment, or sitting in the same position for a long period of time.

- **CTS**: This is a disorder affecting the median nerve passing through the carpal tunnel in the wrist. Its symptoms include: numbness, tingling, pain, and aching in the wrist and palmar aspect of the thumb and the first three fingers caused by the compression of the median nerve in the wrist. Extensive, repetitive motions during keyboarding, especially with bent wrists, can cause or contribute to the development of CTS. Flexing, extending, and/or twisting the wrist can strain the muscles, tendons, and median nerve in the wrist, causing CTS. The general work posture influences the hand and wrist postures. When the upper body is flexed (leaning forward), or extended (leaning backward) or slouched, then the hands and wrists become flexed or extended to adapt to the body posture, which may result in increased tension and strain in the wrists and hands.

- **Cervical syndrome**: This is a disorder affecting the disks in the cervical spine. Its symptoms include pain, numbness, and muscular spasms due to pinching of the cervical nerves caused by compressed disks. Awkward body postures induced by improper workstation designs can cause cervical syndrome. When the desk or the keyboard is too low, the user flexes the back, which shifts the weight of the head forward. During such a posture the neck muscles that have to support the head become tense and strained. Repeated and prolonged exposure to such neck tension and strain can lead to cervical syndrome.

- **Tendinitis**: Tendinitis or tendonitis is an injury or inflammation of a tendon, usually around a joint surface (e.g., in the hand, wrist, elbow, and shoulder). Its symptoms include: swelling, tenderness, pain, burning feeling, and weakness in or around the affected joints. Usually, tendinitis caused by office work occurs in the upper extremities.

- **Tenosynovitis**: This injury is the inflammation of tendons and their corresponding sheaths which cover them. Its symptoms are similar to those for tendinitis.

Repetitive movements in restrictive postures can result in nerve pinching, tendon irritation, and blood flow restriction. The occupational ailments are usually the result of a combination of the following factors:

- performing repetitive motions with fast and forceful movements;
- awkward work posture due to poor work habits and workstation design, or improperly adjusted equipment;

- insufficient rest breaks over long working periods, that is, uninterrupted, prolonged use of equipment.

The body strains in response to such conditions include: pain,
tingling, and numbness in the hands; soreness in arms, back, neck, and
legs; headache; and eye strain. These conditions are usually classified as
repetitive strain injuries (RSIs). When the conditions become long-
lasting due to long-term exposures, then they are classified as CTDs.

17.4 OFFICE PROBLEMS

Office jobs present a large number of occupational health hazards that
are often overlooked as ergonomists concentrate on production jobs. In
an investigation of workstations on a university campus (Emanuel and
Tayyari, 1994) and further observations from other offices, the following
problems were consistently found in most office environments.

- There is a lack of knowledge of good posture and proper use of
  workstations.
- Glare is a problem, in most cases, due to the location of overhead
  fluorescent lighting. In some instances, the video displays are even
  facing large windows, resulting in constant glare.
- Space for written material is limited, which forces people to place
  written material on cramped desk-top space or on side chairs, causing
  the operator to bend excessively or strain to read the document.
- Some people raise the backrests too high in order to support the upper
  back, while ignoring support for the lower back (which is usually an
  area for postural problems).
- Computer keyboards and displays are being used on desks that are
  not designed for computers. Typewriters are located on typing desks
  or side pieces that are lower than the desk, but computers are located
  on the desk itself. The height of the keyboard results in the operator's
  hands and arms being placed in a fatiguing configuration.
- Most desks have inadequate space for keyboards and/or display
  monitors. As a result, the displays are set on the desk-top corners or
  on one side of the user. This type of arrangement significantly
  increases strain in the user's neck if sitting perpendicular to the front
  edge of the desk, rotating the head to read documents, using the
  keyboard, and then rotating the head again to view the screen (Figs
  17.1 and 17.2). To alleviate neck strain, the keyboards are often set at a
  30–45° angle to the edge of the desk (Fig. 17.3). This causes the
  operator to sit at an angle at the desk or to sit straight and try to
  operate the keyboard at an angle. Both of these configurations
  eliminate any potential desk-top space for wrist support which can
  result in musculoskeletal strain. Angled seating may also create

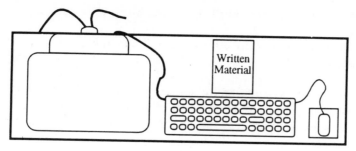

Fig. 17.1 Insufficient desk-top space and size of computer monitor may require the location of the monitor on one side of the user and the keyboard and written material in front. Adapted from Emanuel and Tayyari (1994).

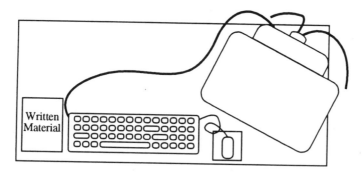

Fig. 17.2 Insufficient desk-top space and size of computer monitor may require the location of the monitor on the desk-top corner. Adapted from Emanuel and Tayyari (1994).

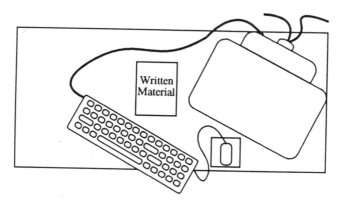

Fig. 17.3 Keyboards are often set at a 30–45° angle to the edge of the desk to prevent neck strain. Adapted from Emanuel and Tayyari (1994).

problems for the operator's legs, which can no longer fit under the desk.

- The seating is either not adjustable or if it is, it is usually not adjusted by the user. Many people tend to use the chair as they found it and make no effort to change the chair height or backrest position. The reasons for this are twofold: some chairs are difficult to adjust and many people seem to be unaware of good seating principles.
- Some seats are too deep, not allowing users to obtain support from the backrests.
- Principles of good seating and footrests are often ignored. Many offices ignore the basic ergonomics principles for designing workstations (e.g., desk, chair, and footrest in combination) to promote a correct working posture.
- The wrists are held in an awkward posture, and elbow and shoulder joints are extended while using the computer mouse.
- The placement of the hands and, particularly, the wrists on the firm desk-surface and its sharp edge is found to be a serious risk factor in developing wrist pain problems.
- The correct computer display height and placement is assumed to be the same for all operators, despite differing visual acuity. The displays are usually located at approximately eye height – a height that would be appropriate for operators with normal visual acuity. For those requiring bifocal corrective lenses, placing the display at eye height may cause eye and/or neck fatigue.
- The workstation designs are not related to the work being done. For example, a desk corner may be an acceptable location for a VDT when a skilled typist spends most of the time viewing the written hard-copy material and only needs to glance at the screen occasionally. On the other hand, tasks such as graphics composition, technical drawing, as well as typing by unskilled operators usually result in almost constant monitoring of the screen.

17.5 SOLUTIONS TO OFFICE HEALTH PROBLEMS

Ergonomics can offer simple and usually inexpensive solutions to many office problems. However, these problems can result in very costly consequences if ignored. In many cases, all it takes is to make some simple adjustments to the office environment to alleviate these work-related problems. Simple adjustments can be made to computer keyboards, monitors, stands, document holders, chairs, illumination, and phones to reduce work-related problems in the office. However, these adjustments must be made properly to achieve desirable results. Sometimes, they require additional equipment, such as back cushions for proper back support or wristrests for the keyboard users. Recently, exercise and relaxation have been explored as important factors for reducing body strain.

Although very specific recommendations can be made to achieve an ergonomically sound workplace, a key issue is to incorporate enough flexibility and adjustability to allow workers to make postural changes easily. Two individuals may have very different preferences in the adjustment of the same workstation and, in fact, the same individual may adjust the workstation differently from day to day, and even from hour to hour during the day.

17.5.1 MANAGEMENT AND EMPLOYEE AWARENESS AND PARTICIPATION

Awareness and participation of both employees and the management are two critical factors required for solving occupational problems. General understanding of ergonomics principles and having good equipment available helps solve many problems. However, this solution requires that, first, management who controls the expenditures must recognize both the severity of the problems and the severity of the consequences of ignoring them. Education becomes a key factor that must be provided to both employees and the management.

The success of any system depends on the comprehensive involvement of its components. Ergonomics programs are no exception. Employees should learn how to arrange, adjust, and use the workstation and equipment so that they can work without straining their body. They also should learn how to release their tensions by simple relaxation and exercise techniques. The following recommendations are based on ergonomics principles for employees to practice.

17.5.2 JOB REDESIGN

The jobs performed by workers in a problem area should be investigated to pinpoint job elements which may be responsible for problems.

- Some repetitive tasks may not be necessary at all, and therefore should be eliminated.
- Some tasks can be made easier to perform.
- If the job requires the same or a few repetitive tasks, the job should be enlarged so that the worker does not perform the same tasks repetitively day after day. For example, if a worker's job is solely data processing using a computer keyboard, other tasks such as answering the telephone, filing documents and copying could be added to enlarge the job (and effectively provide rest breaks from the data entry job).

17.5.3 JOB ROTATION

The workers should be rotated among jobs to avoid continuous exposure to repetitive motions. This should be implemented in daily schedules so that workers perform various tasks during the course of a work shift. For example, if there are four different tasks to be performed by four different employees, each employee should take a turn in performing each task for 2 h.

17.5.4 GOOD WORKING POSTURE

Recommendations for good seating posture were made in Chapter 7 (Fig. 7.2). Employees should be aware of posture at the workstation. They should be directed to practice good posture during work until it becomes natural. The postural recommendations should minimize physical stress on the workers. However, adjustability is the key to giving postural control to the worker to allow for frequent postural shifts during the shift. The goal is to maintain the wrists straight, relax the fingers, and maintain a straight back posture during the task (e.g., keyboarding) so that employees do not feel aches and pain at the end of the working day. The employee at a computer workstation, therefore, should practice the following to achieve this goal:

- Sitting straight, facing the keyboard and computer monitor with the back straight to avoid straining the muscles and the intervertebral disks of the neck and back.
- Holding the head slightly downward to avoid straining the muscles of the neck and shoulders.
- Maintaining the hand and wrist straight and in line with the forearm to avoid irritation of the tendons and their sheaths in the carpal tunnel.
- Keeping the thighs horizontal or the knees slightly raised to avoid pressure underneath the thighs.
- Keeping the lower legs vertical, and the feet flat and pointed forward. If the seat is too high (which is usually the case for short individuals), an inclined footrest should be used to be able to maintain this posture.
- Keeping the elbows as close to the body as possible to avoid straining the muscles and joints of the shoulders. An elbow angle of 90–100° is recommended. Thus, the upper arms should be vertical or slightly flexed and the forearms horizontal or the hands slightly raised.

17.5.5 ADJUSTING WORKSTATIONS

Employees should be able to adjust their workstations easily and quickly to fit them. When the workstation is properly adjusted, the employee is able to maintain a good posture, as described in the previous section. The following simple adjustments to the workstation reduce the odds of encountering many ergonomics-related problems:

- The seat height and backrest should be adjusted so that the user can maintain straight hands and wrists.
- If possible, the desk height should be adjusted to allow the VDT user to maintain a correct posture, in which the torso is upright, the upper arms and the lower legs vertical, the forearms and thighs horizontal and the feet are flat on the floor. The desk height is adjusted according to the adjusted seat height. If the desk height is not adjustable, the seat height should be adjusted to maintain a proper posture and a footrest has to be used for short individuals.
- The keyboard (keyboard tray or desk) height should be adjusted so that the user's wrists are straight during keyboarding.
- The keyboard should be so positioned that the user is facing the keyboard while keeping the wrists straight and the elbows as close to the body as possible.
- The top of the display should be at approximately eye height for use by operators with normal visual acuity. This height for a user requiring bifocal corrective lenses should be at a comfortable level below eye height.

17.5.6 WORKSTATION DESIGN

- The dimensions of the desk-top should provide adequate space for keyboards and displays. If sufficient space cannot be provided, an under-table keyboard holder should be used (Fig. 17.4).
- Use L-shape or cut-in workstations or a special pull-out mouse holder to keep the elbow as close to the body as possible during the use of computer mouse.

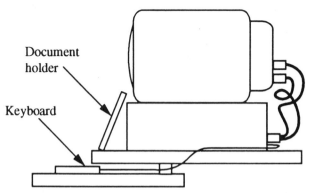

Fig. 17.4 The desk-top should have sufficient space to allow the placement of the computer monitor directly in front of the user.

- Choose chairs with the following characteristics:
  - The seat height should be adjustable. The seat height should be adjusted so that the elbows are not bent more than 90° to reach the keyboard.
  - The chair should provide cushioned support for the lower back.
  - Its backrest should support the lower back, rather than just the upper back. Supporting the lower back keeps the spinal column in its natural double S shape.
  - The seat pan should have a waterfall front design to prevent pressure on the underneath of the thighs.
  - The seat pan and backrest should be made of materials that give way about 2.5 cm (1 in) when pressed. Seat pans that are too soft may interfere with posture.
  - The seat pan, backrest and armrests should be covered with permeable fabric material to absorb the user's body heat and moisture, and provide a sufficient coefficient of friction to prevent slipping and sliding on the chair surface.
  - The seat height should be adjustable to fit a large proportion of employees.
  - The chair should have a backrest which is adjustable up-and-down as well as forward-and-backward. If the latter adjustability is not available, the seat depth should be less than 40 cm (16 in).
- Since the desk heights are usually fixed, adjustable chairs should be used. Operators should raise their chair seats to have their upper arms vertical and forearms horizontal when performing their tasks. Of course, this may require the use of footrests high enough to maintain the thighs horizontal, lower legs vertical and the feet flat on the footrests.
- The workstations should be designed in such a way that the keyboard, computer monitor, and documents are directly in front of the VDT user. It may be possible to place the monitor in front of the user by bringing the desk forward, away from the walls, to provide clearance for the monitor cables. An under-table keyboard holder allows the keyboard to be kept directly in front of the user at a proper height.
- Table heights for computer workstations should be lower than standard desks – 67 vs 73–76 cm (26.5 vs 29–30 in).

17.5.7 GOOD ILLUMINATION

The illumination should be suitable and glare should be eliminated. The glare at office workstations is usually of an indirect, reflective type. Sources are either daylight through the windows or ceiling lights. Glare may be controlled by the following methods:

- Glare can be minimized by better lighting location and diffusion.
- Glare caused by the daylight falling from windows on the computer monitor may be reduced by situating the monitor so that the window is on the side of the user, not on the back or front.
- Daylight can also be prevented from falling on the computer monitor by using curtains and blinds.
- Glare from light fixtures may be controlled using diffusing light screens and/or relocation of the workstation in such a way that the light is not reflected on the monitor. The placement of partitions can block light (Chapter 13).
- If none of the above methods is found effective in preventing light from falling on the computer monitors, glare screen filters that are placed directly on the computer screens or monitor visors should be used.

17.5.8 USE OF WORKSTATION ACCESSORIES

When all necessary adjustments to the workstation are found to be insufficient for the user to maintain a good posture, many workstation accessories are commercially available, or can be made to help the user solve postural problems. Examples of such accessories are wristrests, mouse rests or mouse nests, document holders, and mouse pads. The following are recommended for consideration for VDT workstations:

- Inclined **footrests** should be used by short individuals for whom the seat is usually too high so that they can relieve strain from the legs and back. Footrests should be adjustable to be used by a large number of people rather than a specific person. A type of footrest whose surface rocks in the sagittal plane, not side to side, should be preferred. Such a continuous movement can minimize fatigue development during prolonged sitting.
- **Back cushions** should be attached to the backrests of those chairs that do not properly support the lumbar region of the back. Various types of back cushions are commercially available.
- **Mouse rests**, commercially known as **mouse nests**, can support the hand and help reduce pressure on the palmar side of the wrist and stop the wrist rubbing on the mouse pad. A type of mouse nest is commercially available that encompasses the mouse and glides along with it. It is also contoured to support the hand.
- **Mouse pads** should be friction-free so that the mouse can be used over it with little effort.
- A **document holder** or copy-stand, that is used for data entry and text processing, should be at the same height as the computer screen to help avoid straining the neck muscles.

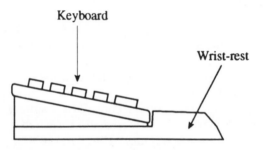

Fig. 17.5 A wristrest for use by keyboard users.

- A **wristrest** (Fig. 17.5), that supports the hands and forearms at the keyboard, not only provides cushion for the wrist against the hard surface of the desk-top or keyboard holder, but also reduces the amount of bend in the wrists. It also reminds the user of the wrist posture and to keep the wrists in a neutral (straight) position. The following are some characteristics of a good wristrest:
  - The wristrest should provide an anatomically neutral position for the wrist. In such a position the wrist is not bent, that is, the hands are in line with the forearms; hence the amount of strain developed in the carpal tunnel structure is minimized.
  - The wristrest should be made of soft materials that minimize the amount of pressure on the palmar side (underside) of the wrist. Wristrests made of hard materials, such as closed-cell foam, which increase pressure on the carpal tunnel and impair circulation, should be avoided.
  - The wristrest should be made of open-cell materials, covered by breathable (permeable) fabric to absorb moisture to minimize the amount of heat and moisture build-up. Wristrests made of closed-cell foams and/or covered by vinyl or other fabric that are impermeable to moisture should be avoided.
  - The wristrest should not be made of abrasive or rough fabric.
  - The wristrest should have a removable and washable cover and be covered with antislip material on its underneath.
  - The wristrest should have a thickness compatible with the height of the keyboard. The thickness of the wristrest should be the same as the height of the keyboard space bar measured from the table-top.
- There are some **spongy plastic grips** commercially available that can be placed on to pens and pencils to provide a soft cushion for helping maintain comfort in prolonged writing or drawing tasks.
- **Telephone headsets should be used to keep the head upright and the upper body straight.**

- **A task light** can be used to provide a higher level of illumination over a specific area.
- An articulated **keyboard tray** enhances the capability to adjust the height of the keyboard.
- A **writing board** (Fig. 17.6) placed over the desk-top is an effective tool for relieving neck and back pain caused by prolonged bending over the desk.

17.5.9 OTHER RECOMMENDED PRACTICES

- Touch the keys lightly with relaxed wrists and fingers. Striking the keys should be avoided.
- The computer screen should be dusted off to improve the visibility of the displayed information.
- The VDT user should take a 15-min break away from the computer every 2 h. This break is not necessarily for resting. The user may perform other job duties, especially those requiring walking away from the workstation, for example to a printer to retrieve printed work, or to a file cabinet to file or retrieve some information.
- The VDT user should perform short exercises and stretches after an hour of work, even while sitting at the workstation. Examples of such exercises are stretching and relaxing fingers and arms, relaxing the hands at the wrists and shaking them, rotating the wrists keeping

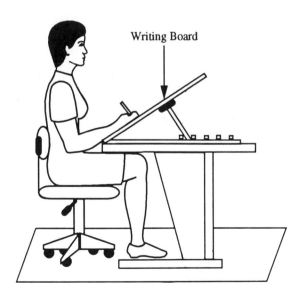

Fig. 17.6 A writing board placed over the desk-top helps the user gain relief from neck pain.

fingers relaxed, rolling shoulders backward in a circular motion, stretching the fingers and the arms upward reaching toward the ceiling, moving the torso to the sides and stretching, and looking away from the screen monitor. Frequently shifting the focus on an object near the eyes to a point across the room and vice versa helps reduce eye strain and headaches. Facial muscles can relax and the neck muscles can stretch by forcing a yawn.

- Employees should be trained that when at home, they should not perform the same repeated wrist and hand motions that they do at work. Instead, they should be involved with some physical activities (especially aerobic exercise such as walking, jogging, and swimming), stretching and relaxation.
- Management should encourage employees to comment on their work-related problems and to request the ergonomic assessment of their workstations.

17.6 EXAMPLES (WORKSTATION DESIGN FOR VDT USERS)

VDTs are now everywhere. They have greatly improved the productivity of office workers. They have created many problems, too! Most problems involve individuals using VDTs for a relatively long period of time, day after day. The problems include wrist, neck and back pain and stiffness, eye strain, headache, discomfort, fatigue, and stress. These problems result from awkward postures due to poorly designed workstations. The application of ergonomic principles in the design of VDT workstations has significantly reduced such problems. A good VDT workstation allows the operator to assume a good posture, reach the keyboard and source document easily, and see the screen clearly.

The recommended dimensions for VDT workstations, as summarized below and illustrated in Figures 17.7 and 17.8, were partially drawn from ANSI/HFS 100-1988 and complemented by the authors.

Work surface and computer

A = Height of the work surface (desk-top) should permit the primary viewing area (θ) to be within 0–60° below the horizontal plane of the eyes. If the work-surface height is not adjustable, its height is determined by the clearance envelope under the desk-top of the 95th-percentile user, given that the line of sight of any user to the top line of the display is not above the horizontal plane through his or her eyes.

B = Height of the display screen: keep the top line of the screen at eye level, for use by operators with normal visual acuity. This height for a user requiring bifocal corrective lenses should be at a comfortable level below eye height.

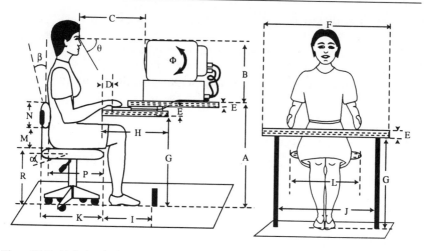

Fig. 17.7 Alphabetical references for recommended video display terminal workstation dimensions.

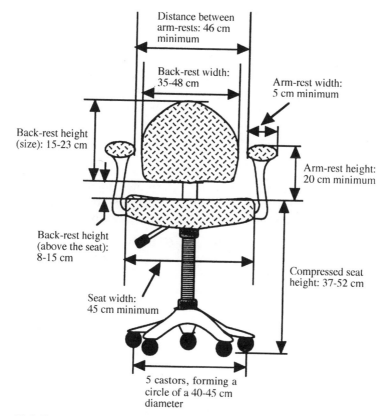

Distance between arm-rests: 46 cm minimum

Back-rest width: 35-48 cm

Arm-rest width: 5 cm minimum

Back-rest height (size): 15-23 cm

Arm-rest height: 20 cm minimum

Back-rest height (above the seat): 8-15 cm

Compressed seat height: 37-52 cm

Seat width: 45 cm minimum

5 castors, forming a circle of a 40-45 cm diameter

Fig. 17.8 Recommended chair dimensions.

C = Viewing distance: between 45 and 50 cm (18–20 in).
D = Handrest: 5 cm (2 in).
E = Thickness of the work surface: 2.5–3 cm (about 1 in).
F = Width of the work-surface: 76 cm (30 in) or more.
Φ = Computer screen should be adjustable (tiltable) between 0 and 20°.

Seat

K = Seat depth: 38–43 cm (15–17 in), front edge relief design (e.g., waterfall contour).
L = Seat width: at least 45 cm (18 in).
M = Backrest height: 8–15 cm (3–6 in) above the seat.
N = Backrest size: 15–23 cm (6–9 in) high and at least 30 cm (12 in) wide.
P = Backrest depth (from front edge of seat): adjustable between 40 and 50 cm (16–20 in) and 42 cm (17 in) if fixed.
R = Seat height (compressed): 37–52 cm (14.5–20.5 in).
α = Seat slope: adjustable 0–10°
β = Backrest tilt: adjustable, permitting a torso-to-thigh angle between 90 and 105°.

The armrest height (above the seat) should be a minimum of 20 cm (8 in), its width should be at least 5 cm (2 in), and distance between the two armrests must be at least 46 cm (18 in). The chair should have at least five castors, forming a circle of 40–45 cm (15.75–17.75 in) in diameter.

Leg-room clearance envelope

The leg-room dimensions should be determined following the procedures recommended in Chapter 7. However, the following figures may be found quite reasonable for such a clearance envelope:

G = Leg-room height: if adjustable: 50–67.5 cm (20–27 in).
      if not adjustable: 67.5 cm (27 in) minimum.
H = Leg-room depth at knee level: 38 cm (15 in) minimum.
I   = Leg-room depth at toe level: 59 cm (23.5 in) minimum.
J   = Leg-room width: 50 cm (20 in) minimum.

17.7 WORKSTATION DESIGN FOR OFFICE WORK

Recommended dimensions and characteristics for office chairs are the same as those given for chairs for use by VDT users. However, the following specific recommendations were made by Grandjean (1988) for an office desk which is used for general office work without a typewriter or VDT:

- The desk-top height should be 74–78 cm for men and 70–74 cm for women (assuming that the chair is fully adjustable and footrests are available).
- The leg-room should permit legs to be crossed without difficulty. Thus, there should be no drawer in the middle, above the knees, or thick edge to the desk-top.
- The desk-top should not be thicker than 3 cm.
- The leg-room under the desk should be at least 68 cm wide and 69 cm high. It should also have a depth of not less than 60 cm at knee level and not less than 80 cm at toe level.

17.8 RELIEF FROM WORK TENSION

As was mentioned previously, prolonged sitting at the workstation, regardless of its design, increases tension in the muscles. They become stiff and sore. The tension can be prevented by performing some simple exercises, right at the workstation without leaving it. Such exercises should be repeated periodically during the day.

The following subsections present some simple exercises that can be performed by office employees and VDT users at their workstations. However, employees should always seek the opinion of a physician about the appropriateness of exercises they want to perform.

17.8.1 ISOMETRIC NECK EXERCISES

The exercises given in Figure 17.9 are useful for maintaining strength and preventing neck pain. They can be performed while seated at the workstation or during driving when stopped behind a traffic light. These exercises should be performed at least once a day. Each contraction should be sustained for about 10 s or a count of 10. The person should breathe out slowly during contractions.

17.8.2 SHOULDER, BACK, AND ARM RELAXATION

The following exercises help prevent tension build-up in the body. They are especially useful for computer users who sit at the workstation for prolonged hours. These exercises should be performed once every hour, with each exercise repeated five times.

- Extend the arms upward and reach toward the ceiling while stretching the fingers. The position should be held for about 5 s, followed by 5 s relaxation.
- Roll the shoulders backward in a wide circular motion.

1

Place both hands against your forehead. Push your head forward against the heels of your palms without moving your head. Hold for a count of 10.

2

Place both hands behind your head. Pull your head backward against your hands while pulling your hands forward so that your head does not move. Hold for a count of 10.

3

Place the palm of your right hand against the right side of your head. Push your head against the heel of your palm without moving your head. Hold for a count of 10.

4

Place the palm of your left hand against the left side of your head. Push your head against the heel of your palm without moving your head. Hold for a count of 10.

5

Place the heel of your right palm against your right temple. Push your chin down against the heel of your palm toward your right shoulder without moving your head. Hold for a count of 10.

6

Place the heel of your left palm against your left temple. Push your chin down against the heel of your palm toward your left shoulder without moving your head. Hold for a count of 10.

Fig. 17.9 Isometric neck exercises, performed in six steps.

- Slouch and swing the body forward and to the sides (in a circular motion) while sitting in the chair, and hold the arms relaxed with flexed elbows.
- Look away from the computer screen. Periodically shifting the focus on a near and far object is recommended, especially for VDT users.

- Hold the hands out in front of the body and spread (abduct) the fingers as much as possible.
- Rotate the wrists while keeping the fingers relaxed.
- Shake the hands up and down, and to the side.

17.9 APPLICATIONS AND DISCUSSION

The frequency with which the term ergonomic appears in office equipment catalogs as a modifier for chairs, desks, anti-glare screens, and footrests seems to suggest that office workers have no reason to continue to suffer from long-term work-related problems. Nevertheless, CTDs in the wrists, back, neck, and shoulders, and vision problems are often dealt with by treating the symptoms rather than addressing the cause. Equipment purchase decisions are often made based primarily on price and ergonomic considerations are frequently ignored.

This chapter addressed various types of problems in office environments and offered many ergonomic-based ideas for resolving such problems. Particular attention needs to be paid to seating, workplace design, and keyboarding activities in offices. The parties involved in equipment acquisition decisions need to be contacted and the information they would need in order to make ergonomically acceptable decisions should be provided. Samples of easy-to-understand guidelines, such as those presented in this chapter, are very helpful when discussing ergonomics with these parties.

REVIEW QUESTIONS

1. With regard to the worker and the workstation, what is the root cause of most office problems?
2. List situations that are common causes of problems in modern offices.
3. What are the two root-causes of the postural problems?
4. In what parts of the body can poor posture cause discomfort and fatigue?
5. When the wrist is placed on a hard surface such as the desk while typing, from what problems would the person probably suffer? Explain.
6. What type of problem is associated with placing the wrist on a hard surface such as the desk under a keyboard?
7. What characteristics of video display terminal (VDT) workstations would call for ergonomic interventions?
8. Occupational cumulative trauma disorders (CTDs) in office environments usually result from exposure to a combination of several factors. Name the factors.

9. What is the most critical risk factor for CTDs among office employees?
10. List the five most common types of CTDs affecting office employees.
11. How can repetitive movements in restrictive postures affect nerves, tendons, and blood flow?
12. What types of body strain (in response to repetitive motions; poor work habits; improperly adjusted equipment; and uninterrupted, prolonged use of equipment) are usually classified as repetitive strain injuries (RSIs)? What is the difference between RSIs and CTDs?
13. The text lists 13 problems that are consistently found in most office environments. List six of these problems.
14. To what level should the employee be aware of and learn about ergonomics principles?
15. List the solutions to office health problems.
16. List three possible cases of job redesign.
17. What kind of jobs should be enlarged? Why? Give a situational example.
18. Why should workers be rotated among jobs?
19. What characteristic of the workstation can be the key to giving postural control to the worker to allow for frequent postural shifts during the shift?
20. List four simple adjustments to the workstation which can reduce the odds of encountering many work-posture-related problems.
21. At what height should the computer display be located?
22. What would you recommend for a computer user to help keep the elbow as close to the body as possible during the use of computer mouse?
23. What are the characteristics of a good office chair?
24. When the desk heights are fixed, what would be your workstation design recommendations for accommodating a large portion of the task-force population?
25. List five possible methods by which glare may be controlled in office environments.
26. Give eight examples of workstation accessories that may help solve ergonomics-related problems in office environments.
27. How often and for how long should a VDT user break away from the computer? What type of break must the user be given?

REFERENCES

ANSI/HFS 100-1988, American National Standard for Human Factors Engineering of Visual Display Terminal Workstations. The Human Factors and Ergonomics Society, Santa Monica, CA.

Emanuel, J.T. and Tayyari, F. (1994) Office ergonomics: workstation problems and solutions. In *The Proceedings of an International Symposium on Ergonomics*. American Society of Safety Engineers, Des Plaines, IL, pp. 157–164.

Grandjean, E. (1988) *Fitting the Task to the Man: A Textbook of Occupational Ergonomics*, 4th edn. Taylor & Francis, London.

Ergonomics assessment of the workplace 18

18.1 INTRODUCTION

An ergonomics assessment is much more than merely an analysis of a job or task. It is a comprehensive investigation of all potential hazards and risks to individuals and exploration of corresponding solutions. The ultimate objectives of an ergonomics assessment program are to increase productivity and to reduce workplace hazards. Therefore, the purpose of an ergonomics assessment program is to evaluate the capabilities and limitations of the workers and to document the demands of the jobs. Once the job demands have been quantified, the ergonomist can use that information for task redesign to accommodate a large percentage of the population, or to select employees for jobs in such a way as to minimize potential job risks to employees.

Reducing workplace hazards, in turn, increases the company's profitability by minimizing the following measurable costs:

- the cost of worker injuries and illnesses;
- the profit lost when production is lost due to worker injuries and illnesses;
- the workers' compensation costs and premiums;
- the health insurance premiums.

In many system design cases, the roles of ergonomists are crucial. For example, workplace and work-space arrangements; determination of types and characteristics of controls and displays, their compatibilities and their layouts; and functional allocations between human operators and machines are assigned to ergonomists to handle. Ergonomists incorporate specific human factor characteristics in the design that, for example, deal with the anthropometrics, psychophysiological and biomechanical capabilities and limitations, and cognitive capacities of the potential users. The expertise of ergonomists is called upon to deal with two types of systems – the design of new systems and the evaluation of existing systems. They play crucial roles in the design, prototype fabrication, testing, and evaluation of a new system. Their roles in the investigation of incompatibilities between the human–equipment interfaces in existing systems and exploration of ergonomics solutions are also critical.

18.2 ERGONOMICS ASSESSMENT APPROACHES

Approaches used for an ergonomics assessment may be either reactive or proactive. **Reactive** ergonomics assessment usually follows an accident or injury which has prompted the investigation. Reactive ergonomics is usually expensive and limited in scope. **Proactive** ergonomics assessment is typically carried out in the design phase or as a planning activity to eliminate workplace problems (accidents, injuries, job stress, etc.) in the future.

18.2.1 REACTIVE-APPROACH ERGONOMICS ASSESSMENT

When accident or injury issues arise, as a response, a post-injury/accident or reactive ergonomics assessment is launched. Such issues are typically due to a mismatch between the job demands and the worker's capability and limitation. Accordingly, a job hazard analysis consisting of the following four steps should be conducted to identify and alleviate the root-cause of the problems.

- A preliminary analysis identifies the problem jobs.
- An indepth ergonomics assessment, using a cause-and-effect analysis, identifies the root-cause of the interface problem. In such an assessment, all activities (tasks) required for performing each problem job are listed.
- The potential hazards associated with each job activity are identified.
- Once the hazardous job activities which cause injuries and accidents have been identified, then ergonomics interventions can be applied to avoid similar hazards in the future.

18.2.2 PROACTIVE-APPROACH ERGONOMICS ASSESSMENT

Surveying the workplace on a regular basis is an effective method of identifying situations for applying ergonomics interventions. The goal of this method should be searching for mismatches between job demands and the capabilities of the workers. Indicators of such potential mismatches include the following:

- jobs involving manual lifting of heavy loads;
- jobs involving lifting, lowering, carrying, twisting, bending, stooping, reaching, stretching, pushing, and pulling;
- jobs involving repetitive motions;
- workers maintaining awkward postures (examples are holding the elbow away from the body, bent wrists, raised elbows or shoulders, bent neck, unnatural spine position, and hanging legs due to lack of proper foot support);
- workers maintaining the same posture for a relatively long period of time;
- workers demonstrating signs of discomfort, such as frequent shaking of their limbs to relieve strain;
- workers making frequent trips to the medical department;
- workers changing their workstations;
- workers modifying their work-tools;
- workers frequently leaving their workstations for a variety of reasons;
- unprotected workers exposed to tool vibration;
- controls that are difficult to reach;
- displays that are difficult to read;
- poor thermal conditions;
- poor lighting conditions (too little or too much light);
- poor ventilation and indoor air conditions.

18.3 ERGONOMICS ASSESSMENT OF THE WORKPLACE DESIGN

As a part of concurrent engineering philosophy, ergonomics principles and concepts must be considered at the initial stage of any designs which are for human use. Thus, systematic ergonomics assessments in designing a workplace should be performed in three stages.

18.3.1 ASSESSMENT AT THE INITIAL DESIGN STAGE

All preliminary activities, or proposal processes, of any system are performed in the initial design stage. This is a crucial stage for incorporating the ergonomics principles and concepts in the design of the workplace. This is in agreement with the "design it right the first time" philosophy, which should save time, money, and employee stress and trauma (Kidd and Van Cott, 1972). At this stage, ergonomists address the following typical questions:

- What is the purpose of the workplace being established?
- What missions (products to be manufactured or services to be rendered) are to be fulfilled in the workplace?
- What are the stages of the workplace missions?

- What functions/activities must be performed in the workplace at each stage? Are these functions/activities divided into specific specialized groups, and each group of functions/activities assigned to a different workstation?
- What numbers and types of employees (i.e., operators, maintenance personnel, etc.) are to use the workplace?
- What work-force population is to occupy the workplace? This information is crucial for determining the corresponding anthropometric dimensions.
- What type and level of physical or mental workload (job stresses and demands) are expected to be placed on the employees?
- What are the environmental conditions (i.e., light, heat, cold, noise, vibration, radiation, etc.) of the workplace?
- What typical hazards are associated with the fulfillment of the job requirements under these environmental conditions?
- What are the alternative technological options available?

In addition to responding to the questions posed above, the ergonomist (or ergonomics team) should begin to quantify and record job demands as early in the process as possible. Quantifiable job demands include weights of objects being handled, physical dimensions of objects handled, push/pull forces required, reach requirements, energy expenditure demands, biomechanical force and torque calculations, environmental considerations (lighting, heat, noise, air-borne contaminants, etc.), and any other job demands that can be identified and quantified.

18.3.2 ASSESSMENT AT THE ADVANCED DESIGN STAGE

Based on the information collected in the preliminary assessment at the initial stages of the design process, the necessary anthropometric data for the type of work-force population, and equipment dimensions are obtained. The ergonomics assessment will then enter the advanced stage of design. At this stage, ergonomists address the following typical questions (Kidd and Van Cott, 1972):

- What functions should be allocated to human operators and support staff? What functions should be performed by machines?
- What condition will impose peak workload on the operators?
- What conditions may cause operator performance to deteriorate?
- What pattern of decision–action will occur at crucial mission stages?
- What information is required by the operators and/or support staff to perform their functions?
- What are the requirements for information flow in the system?
- What is the most useful form (i.e., code, mode, format) of information to the operator?

- How many employees are needed to operate and support the system under the normal and peak load conditions?
- What special skills, capabilities or attributes are required of operators to operate the system effectively?
- What special training is required? Is such training feasible? What resources will be needed to implement the training.
- How should the assigned functions be distributed among operators and support staff?
- How should the workstations be arranged?
- What equipment is required at each workstation? How should the equipment be laid out?
- What specific tools, devices, or controls are appropriate to the pattern of task actions?
- What kinds of aids, instructions, locks, interlocks, cover plates, etc. will be useful to facilitate correct actions and prevent operator errors?
- What means are available to allow quick recovery or to maintain the safety and integrity of the system in the event of operator error or failure?

18.3.3 ASSESSMENT AT THE DESIGN REVIEW STAGE

When the workplace design is semifinalized, it should be reviewed for possible omissions. At this stage the design is assessed on two general bases – engineering anthropometry and biomechanics, as follows:

18.3.3.1 Engineering anthropometric basis

- Does a large-size employee (e.g., the 95th percentile of the population) have sufficient room and clearance for knees, feet, body height and depths, and movements?
- Is a small-size employee (e.g., the 5th percentile of the population) able to reach everything?
- Is the distance between the employee's eyes and work appropriate?
- Is the work-surface suitable for the desired position (sitting or standing)?
- Can the employee vary the working postures?
- Are all frequently used items within the normal reach (of both arms and legs)?
- Does the worker have a suitable chair (i.e., seat, back support, height, etc.)?
- Is (are) an armrest (armrests) necessary and available? If yes, is it (are) they) suitable (i.e., location, position, material, shape)?
- Is a footrest necessary and available? If yes, is it suitable (i.e., height and depth dimensions, slope, shape)?

18.3.3.2 Biomechanical basis

- Has static work been minimized?
- Are jigs, fixtures, vises, conveyors, etc. available and used wherever possible?
- Where prolonged loading of a muscle group is inevitable, is the required muscular force below 15% of the maximum strength?
- Have torques about the body joints been minimized?
- Is the direction or applied force correct to provide the amount of force required?
- Are mechanical assists available and used wherever necessary?
- Has muscular loading been minimized by using counter supports (e.g., balancers for suspending a heavy work-tool)?
- Are objects lifted and carried correctly? Are their weights within the acceptable limits?

18.4 ERGONOMICS WORKSITE ASSESSMENT

Worksite assessments are post-design assessments that are performed in existing workplaces. An ergonomics worksite assessment is conducted to identify existing work-related hazards and conditions, operations that create hazards, and areas where hazards may develop (Occupational Safety and Health Administration (OSHA), 1990). The objectives of such worksite assessments are, therefore, to recognize, identify, and correct workplace risk factors. Workers may already suffer from poor job, workstations, and equipment designs. Ergonomics assessments can uncover problem areas that require methods redesign, workplace rearrangement or redesign, equipment/tool changes for ergonomics reasons, work environmental changes, and/or organizational and social changes to the work environment. A worksite assessment should use systematic procedures to fulfill its objectives and acquire credibility for the ergonomics program. Obviously it is important to have the workers involved in the ergonomics assessment at the earliest stages of the process. Employees who are currently working in the workplace of concern, or similar workplaces, can provide valuable input into the design/redesign process. The following steps are recommended for a systematic worksite ergonomics assessment.

18.4.1 DEVELOPMENT OF AN ERGONOMICS EVALUATION CHECKLIST

Investigation requires a considerable amount of note-taking for documentation and further analysis of the situation and finding the root-cause of problems. An ergonomics evaluation checklist (e.g., similar to the checklist given in Fig. 18.1) can simplify the documentation and help

Ergonomics checklist

A positive response to any of the items in the checklist indicates a need for ergonomics analysis and possible ergonomics intervention.

General ergonomics considerations of the organization

- Is absenteeism too high for this organization?
- Are there jobs with high employee turnover?
- Do employees have trouble meeting production standards?
- Is the organization often trying to find a person to match the demands of the job?
- Has the accident rate been too high?
- Have employees made too many trips to first aid?
- Do workers find excuses to be away from their workstations?
- Has quality deteriorated?
- Has scrap and rework increased?
- Are tools needed to adjust workstations?

Indicators to consider engineering controls

- Do workers have to assume unusual postures to do their job (bowed/bent neck, hunched back, stretching arms, etc.)?
- Are workers reaching high or twisting to get parts?
- Do workers have to assume static postures?
- Do workers have to stand all shift long?
- Are there highly repetitive jobs?
- Are there jobs requiring a great deal of force?
- Are there jobs performed at a rapid pace?
- Do workers have to lift heavy loads?
- Do workers have to push/pull loads often?
- Are work-surface heights too high or too low?
- Do workers have to reach below knee level?
- Do workers have adequate clearances in their workplaces?
- Have workers modified their workstations?

- Have seated operators modified their chairs?
- Are chairs/seats not adjustable?
- Do chairs/seats have no back supports?
- Are footrests needed but not available?
- Are seats not padded?
- Are foot controls required in standing positions?
- Are seated operators using makeshift footrests?
- Do seated operators have non-adjustable chairs?
- Do workstations provide no option to sit or stand while working?
- Do workers have to stand on concrete or other hard surfaces for long periods of time?
- Do the operators have trouble reading displays?
- Do workers have trouble reaching controls?
- Does the use of controls require awkward postures?
- Are workers using controls/tools with sharp edges or pinch points?
- Are workers using controls/tools with very smooth/slippery grips?
- Are workers using controls/tools with handle diameters that are too small or too large?
- Are workers using vibrating tools?
- Do workers use tools that are heavy and difficult to handle?
- Do workers use tools that are inappropriate for their tasks?
- Do workers use tools improperly?
- Is it too light or too dark in the workplace?
- Is there direct or reflected glare?
- Does the noise level in the workplace interfere with communication?
- Is the noise level above 85 dBA?

Indicators to consider administrative controls

☐ Are multiple people required to perform a task?
☐ Do workers complain that they are fatigued at the end of the shift?
☐ Do workers have trouble understanding their job instructions?
☐ Is the organization using too much overtime?
☐ Are workers required to be in a single workstation all shift long?
☐ Do employees have to walk too far to get a drink of water?
☐ Are the workers not adjusting their workstations, but just leaving them as they were when they started the shift?
☐ Have workers modified the standard work practice for their jobs?
☐ Are tools or mechanical assists available, but not used because they are poorly maintained?
☐ Is lifting equipment provided, but not used for some reasons?
☐ Are some workers too isolated from other workers?

☐ Do employees forget to use their personal protective equipment (eye protection, hard hat, ear protection, etc.)?
☐ Do workers use gloves that do not fit properly?
☐ Do employees violate safety rules?

Indicators to consider personal protective equipment

☐ Do workers have to go into noisy areas infrequently?
☐ Are employees exposed to air-borne contaminants?
☐ Are there heat sources in the workplace?
☐ Do workers have to handle hot items?
☐ Do the workers have to work in cold environments?
☐ Do workers handle hazardous chemicals?
☐ Do workers work around high-speed equipment?
☐ Are workers exposed to foreign objects that could injure their eyes?
☐ Are workers exposed to radiation sources?
☐ Are there restricted-entry workstations due to chemical, mechanical, or other hazards?

Fig. 18.1 An example of an ergonomic evaluation checklist.

SYMPTOMS SURVEY QUESTIONNAIRE

Ergonomics programme

Date: _____

Plant _____   Department code/name _____   Job code/name _____

Shift _____   Supervisor _____   Hours worked/week _____ hours   Experience on this job _____ years _____ months

Other jobs you have done during the last year (for more than 2 weeks)

_____   _____   _____
Plant   Dept. code   Job code   Job name

_____   _____   _____
Plant   Dept. code   Job code   Job name

(If more than 2 jobs, include the two on which you worked the most)

Have you had any pain or discomfort during the last year? Yes _____ No _____ (If **No**, stop here)

If **Yes**, carefully shade the area of the following drawing which bothers you the most:

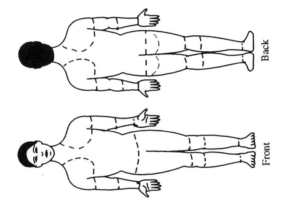

Front   Back

Complete the next page

(Complete a separate page for each area that bothers you)

| Check area: | Neck _____ | Shoulder _____ | Elbow/Forearm _____ | Hand/wrist _____ | Fingers _____ |
| | Upper back _____ | Low back _____ | Thigh/knee _____ | Ankle/foot _____ | |

1. Please mark the symptom(s) that you have experienced during the last year:

_____ Aching
_____ Tingling
_____ Cramping
_____ Loss of control

_____ Burning
_____ Pain
_____ Swelling
_____ Paleness

_____ Numbness (asleep)
_____ Weakness
_____ Stiffness
_____ Other (please name) _____

2. How often have you experienced the problem(s)?

_____ every day
_____ every month

_____ every week
_____ once in a while (longer than a month)

_____ every two weeks

3. How long does each episode last? (Place an 'X' along the line)

1 hour   1 day   1 week   1 month   6 months   1 year

_____

4. How many separate episodes have you had during the last year? _____ times

5. When did you first notice the problem? _____ months ago _____ years ago

6. What do you think causes the problem?

_____ Repetitive motion   _____ Posture   _____ Pressure
_____ Other(s), (describe) _____

7. Have you had this problem during the last week? **Yes** _____ **No** _____

8. How would you rate this problem? (Place an 'X' along the line):
   Now:

   None _____ Unbearable

   When it was the **worst**:

   None _____ Unbearable

9. Have you had medical treatment for this problem? Yes _____ No _____

   9a. If **No**, why not? _____

   9b. If **Yes**, where did you receive treatment?

   1. Company medical _____   Number of times in the past year _____
   2. Personal doctor _____   Number of times in the past year _____
   3. Other _____   Number of times in the past year _____

10. How many work days did you lose in the last year because of this problem? _____ days

11. How many days in the last year were you on restricted or light duty because of this problem? _____ days

12. Please comment on what you think would improve your symptoms: _____

_____

_____

Fig. 18.2 An example of an employee questionnaire. Adapted from OSHA (1990).

remember and record some critical questions, which may otherwise be overlooked. Many ergonomics checklists exist; however, a company or organization may develop its own checklist to reflect particular issues that occur in their workplaces.

18.4.2 DEVELOPMENT OF AN EMPLOYEE'S QUESTIONNAIRE

A questionnaire (e.g., similar to the questionnaire given in Fig. 18.2) for surveying employees should contain the following basic parts:

- assuring employees that the sole purpose of the questionnaire is to help reduce problems (anonymous responses may alleviate employee concerns about potential repercussions from responding to the questionnaires);
- parts of jobs that are most difficult and cause discomfort or pain;
- employees' suggestions for improving the jobs.

18.4.3 STEPS IN CONDUCTING THE INVESTIGATION

The following are recommended steps for an efficient investigation of a company's workplace problems (Manuele, 1991; St John et al., 1993):

- Review the company's accident and injury records (e.g., in the USA review the *OSHA 200 Log* and *101 Accident Report*) and workers' compensation claims to find patterns of work-related injuries. The analyst should look for work-related problems, including musculo-skeletal problems (i.e., back and shoulder pain, carpal tunnel syndrome, tendinitis, tenosynovitis, bursitis, vibration white finger, trigger finger, and tennis elbow–epicondylitis), hearing losses, and heat and cold incidents.
- Review medical claims history and dispensary records, and look for any injuries or illnesses that can be related to workplace and work method design, as listed above.
- Examine the company's accident and injury reports to determine the types of jobs causing injuries and their corresponding departments.
- Obtain data from the human resources and personnel department regarding jobs that are associated with high rates of absenteeism and turnover. Such data may reveal that turnover and absenteeism are due to poor workplace or work method design.
- Interview the human resources and personnel department, and inform them of the intention of the investigation.
- Interview employees and supervisors to identify jobs that require repetitive motions, awkward joint postures and/or excessive force exertion, or which are performed at a rapid pace.

- Observe employees performing the injury-causing jobs, and use the observer's checklist to uncover problem areas. Using a video recorder is highly recommended to document these work areas and to provide a detailed record of the jobs.
- Have the exposed employees fill out survey questionnaires about their work areas.
- Analyze the results of personal observations and video recordings, and the employee questionnaires.
- Discuss the problems with management and employees, and obtain their opinions and suggestions for solving the problems.
- Review all suggestions and their validity based on ergonomics principles.
- Make all necessary modifications in the work methods and/or workstations.

18.5 INJURY INVESTIGATION

The firm's ergonomist should be a member of the safety team that performs a comprehensive investigation of workplace injuries, and prepares a written report of the investigation indicating the job requirements and instances where the job demands exceeded the capacities of the workers. The information extracted from such a report can be used to determine whether job modifications are needed. The ergonomist should take the following steps in investigating a reported injury (Carson, 1993):

- Interview (if possible) injured workers to obtain information about how they perform their job, how the injury occurred and when the injury occurred or first became known to them, and information about their injury history and off-the-job hobbies and activities.
- Perform the job in question (yourself, if possible) to determine motions and forces required by the job.
- Observe other workers performing the job in question. Obtain these workers' opinions and comments about the job.
- Obtain information from the injured worker's supervisor regarding the job history, number of complaints, etc.

18.6 ASSESSMENT OF ERGONOMICS EFFORTS

The final step in the workplace evaluation process is the assessment of the ergonomics efforts. In many cases, the results of ergonomics interventions may not be realized in the short term, but it may require months or even years to realize the full impact of the interventions. However, it is necessary to document the changes and determine the appropriate metrics to monitor the success of the ergonomics program. In general, the following metrics have been successfully used to assess the efforts of an ergonomics program.

- **Improved performance**: Productivity measures should be monitored to note changes after the intervention. The productivity issue must extend beyond the single worker and should also examine productivity at the departmental and location levels. In some cases the productivity of an individual, or even a department, might decline, while the overall location productivity might increase as a result of an ergonomics intervention (a modification that might take additional time at one phase of the operation might save time and effort in several workstations further along the production process).
- **Reduced accident/injury costs**: Both direct and indirect accident/injury costs should be monitored. Direct costs can include medical costs for the injured worker, sick pay, damaged equipment, increased medical premiums and worker compensation costs, and lost production. Indirect costs include lost production (due to less efficient replacements), supervisory time associated with analyzing the incident, lower morale of the employees, and training costs associated with the replacement workers. Accident cost data, especially indirect costs, can be difficult to obtain, but efforts should be made to document the costs involved accurately. Comparisons might be difficult to make with pre-intervention costs due to poor record keeping, or changes in the definitions of direct and indirect costs.
- **Reduced non-productive costs**: Non-productive costs can range from reduced training costs due to lower injury rates and reduced turnover, to reduced trips to first aid, to fellow employees standing around talking about the incident. Again, non-productive costs may be difficult to obtain, but efforts to document such costs would be beneficial.
- **Improved workforce utilization**: Ergonomically designed jobs should accommodate a larger portion of the work-force. Concepts such as work teams and job enlargement would allow employees the flexibility of rotating through several jobs and increasing their range of job skills. Benefits to the company include a more skilled work force that could compensate better for an injured or sick co-worker.
- **Improved employee relations**: Ergonomics assessment of jobs should lead to improved employee relations. Involving workers in the ergonomics assessment of their jobs demonstrates the interest of the company in providing a safe, non-stressful, productive workplace for its employees. Often companies may be reluctant to survey employees for fear of initiating employee complaints (the assumption is that if the issue is not raised, the workers won't know to complain about it).

Such a reluctance may only delay the problem and could result in more serious problems and higher costs in the future.

18.7 APPLICATIONS AND DISCUSSION

A successful assessment program requires a corporate-wide involvement. Creation of an ergonomics task-force (ETF), consisting of representatives from various organizational units, is a major breakthrough for achieving this goal. By tapping the employee knowledge of job demands and desire to have a voice in workplace decision, the ETF can involve employees in the corporate-wide effort to improve the work system (Montante, 1994).

Ergonomics assessments should be performed by a knowledgeable ETF team. Ideally, all team members should be trained in ergonomics principles and analysis methodology. However, if this is not possible, the team should be led by an ergonomist to provide guidance and analysis expertise.

REVIEW QUESTIONS

1. What is an ergonomics assessment, and what are its ultimate objectives and purpose?
2. Reducing workplace hazards increases the company's profitability by minimizing certain measurable costs. List four examples of such costs.
3. What are the roles of ergonomists in dealing with new systems and existing systems?
4. What are the differences between reactive and proactive approaches of ergonomics assessments?
5. List the four steps of reactive ergonomics in a job hazard analysis.
6. List 10 typical indicators of potential mismatches between job demands and the capabilities of workers.
7. List the three design stages at which ergonomics assessments should be performed.
8. List eight typical questions that should be addressed in ergonomics assessments at the initial stage of design of the workplace.
9. List six examples of quantifiable job demands that the ergonomist (or ergonomics team) should begin to quantify and record as early in the design process as possible.
10. List 10 typical questions that should be addressed in ergonomics assessments at the advanced stage of design of the workplace.
11. When the workplace design is semifinalized, it should be reviewed for possible omissions. What are the two general bases on which

ergonomics assessments are performed at this stage of design? List
six typical questions of each.
12. List the steps recommended for a systematic worksite ergonomics
    assessment.
13. What three basic parts should a questionnaire contain for surveying
    employees?
14. What steps should the ergonomist take in investigating a reported
    injury?
15. List five metrics for assessing the efforts of an ergonomics program.

REFERENCES

Carson, R. (1993) How to start a successful ergonomics program. *Occupational
    Hazards* **55**: 122–127.
Kidd, J.S. and Van Cott, H.P. (1972) System and human engineering analyses.
    In: *Human Engineering Guide to Equipment Design* (Van Cott, H.P. and
    Kinkade, R.G., eds). US Government Publishing Office, Washington, DC.
Manuele, F.A. (1991) Workers' compensation cost control through ergonomics.
    Professional Safety **36**: 27–32.
Montante, W.M. (1994) An ergonomic approach to task analysis. *Professional
    Safety* **39**: 18–22.
OSHA (1990) *Ergonomics Program Management Guidelines for Meatpacking Plants.*
    Publication no. OSHA 3123. Occupational Safety and Health Administration
    (OSHA), US Department of Labor, Washington, DC.
St John, D., Tayyari, F. and Emanuel, J.T. (1993) Implementations of an
    ergonomics program: a case report. *International Journal of Industrial
    Ergonomics* **11**: 249–256.

Implementation of ergonomics programs 19

19.1 INTRODUCTION

Ergonomics programs have been implemented for a variety of reasons. At one end of the spectrum, the concern of the company to provide a safe and productive workplace for its workers has been the impetus behind the program, while at the other extreme, fear of non-compliance with regulatory bodies has been the driving force behind ergonomics implementations. Most companies probably fall somewhere between these two extremes. While most organizations are concerned about the welfare of their workers, some form of economic justification is generally the rationale behind an ergonomics program.

Spiraling medical costs often provide the economic rationale for ergonomics efforts. Also, the highly competitive nature of the global marketplace has prompted organizations to review costs at all levels. In the past, many organizations viewed injuries as a "sunk cost" and did little beyond safety posters and occasional safety meetings to address the costs of workplace injuries. The expectations of the "lights-out factory" featuring extensive automation and limited personnel have not been realized. Labor costs continue to be a significant portion of the total costs of many products. Industry is beginning to realize that the unnecessary costs associated with poorly designed workplaces and mismatches between human capabilities and job demands can be controlled through an effective ergonomics program.

19.2 MANAGING AN ERGONOMICS PROGRAM

For an ergonomics program to succeed it must have support throughout the organization. This is why the team approach (ergonomics task-force or ETF) is often successful. An ergonomics team consisting of ergonomists, engineers, medical personnel, safety personnel, first-line supervisors, and hourly employees can explore issues from many perspectives as they use tools such as checklists and surveys to assess the workplace.

Once an ergonomics team or department is formed, it should have a set of well-defined goals. The goals and objectives of the ETF department should be formally developed and presented to both managers and workers. One of the primary goals of such a program should be to promote ergonomics awareness throughout the organization. Typical awareness programs can involve in-house training, short courses, and dissemination of written information. Again, it is important to have the cooperative support of both management and labor for the overall program to be successful.

The ergonomics effort will most likely entail both reactive and proactive components. Ergonomics projects should be conducted both reactively in response to requests from the production area and proactively as self-directed projects based on reviews of safety, productivity, personnel, or other databases that could be used to point out areas that might benefit from ergonomics intervention. It is important that the ergonomics team maintain contact with the production areas and address their needs and issues as they arise. One of the best promotions for any program is the word-of-mouth endorsement of the program by the workers and first-line supervisors. However, the ergonomics team should not rely solely on word-of-mouth support for their programs. Ergonomic interventions should be well-documented (written). A record of interventions can form the basis for a before-and-after analysis of the activity as well as provide a history of ergonomics efforts that can be shared with management, or with ergonomics groups in other departments or locations within the organization, or even with groups outside the organization.

19.3 OBJECTIVES OF ERGONOMICS PROGRAMS

The very basic premise of ergonomics is that job demands should not exceed workers' capabilities and limitations. If this is not the case, then the worker is being exposed to work stresses that can adversely affect safety and health as well as the company's productivity. Therefore, the primary objective of an ergonomics program is to provide a safe and productive workplace to fulfill the goals and objectives of the organization successfully. In meeting the objectives of an ergonomics program, the following steps are recommended:

- providing guidelines and information on the ETF plans:
  - to determine the existence of work-related problems in the workplace that could be alleviated through ergonomics interventions;

- to determine the nature and location of the identified problems;
- to implement ergonomics interventions to reduce or to eliminate the problems.

• forming quantitative and qualitative databases for job requirements, such as postures, forces, and frequencies. Such data are necessary for work system designs or job assignment. The databases can also be useful in establishing baseline data of worker capabilities and limitations for future ergonomics planning efforts;

• identification of job functions for functional allocation between workers and machines;

• identification or design of tasks for assignment to individuals with disabilities;

• reviewing the facility to identify light-duty work activities for alternate-work or return-to-work programs that assign tasks with different or restricted motions to workers who have experienced work-related injuries;

• identification of occupational risk factors associated with each job and the body parts they are affecting.

19.4 COMPONENTS OF AN EFFECTIVE ERGONOMICS PROGRAM

For an ergonomics program to succeed, it should be supported by top management and have a well-defined objective and carefully drawn plan, consistent with the organization's goals and objectives. The following major components or steps are typically necessary for the implementation of successful ergonomics programs.

19.4.1 TOP MANAGEMENT SUPPORT

The first and most crucial step to implement any program, including ergonomics, is obtaining top management support. Management must be fully supportive, otherwise the program will not be successful. The ergonomics programs should be given as much priority as production, quality, and safety. Commitment by management provides the organizational resources and motivating force necessary to deal effectively with ergonomics-related hazards (Occupational Safety and Health Administration (OSHA), 1990). Hence, management's support must be demonstrated at all organizational levels for the program to get credibility and corporate-wide cooperation.

In order to obtain the support of management, it may be necessary to educate them about the benefits of an ergonomics program. A brief seminar on successful ergonomics implementations or of potential ergonomics interventions within the organization is generally very effective. High visibility of a plant manager at the ergonomics meetings and educational programs effectively demonstrates commitment by top management for sustaining the program.

19.4.2 A FORMAL, WRITTEN ERGONOMICS PROGRAM STATEMENT

Once management has approved an ergonomics program, the next step is to develop a written ergonomics program statement which outlines the goals and plans for the program and its support. The program statement should insure that employee well-being and productivity have been addressed. It should also demonstrate management's concern about employee benefits. This written program statement can serve as a medium for familiarizing new employees, supervisors, and managers with the ergonomics philosophy of the organization.

19.4.3 AN ERGONOMICS TASK-FORCE

Since ergonomics is a multidisciplinary movement, an ergonomics program should include an ETF. The membership of the ETF should be composed of representatives from various departments and functions across the organization, such as employee and/or union representatives, first-line supervisors, personnel or human resources representatives, medical personnel, engineering, safety and health (or industrial hygiene), and production management.

The ETF should be viewed as a broad-based resource to aid in the analysis and implementation of ergonomics interventions. Generally, some subset of the ETF, along with necessary support personnel, will be assigned to a particular ergonomics issue. The support personnel might be either internal or external consultants who can lend expertise to solve the problem at hand.

The ETF would be expected to handle both reactive and proactive ergonomics interventions. The task-force should also document all ergonomics activities and develop an ergonomics database for the organization. The ergonomics database should include a documentation of all ergonomics projects, databases of worker capabilities and limitations (e.g., anthropometry, strength, aerobic capacity, etc.) and should be able to serve as a reference for justification of ergonomics efforts in the organization.

19.4.4 A SYSTEMATIC ERGONOMICS ASSESSMENT PROGRAM

The purpose of the systematic ergonomics assessment program is to evaluate the capabilities and limitations of workers and to document the demands of the jobs. Once the job demands have been quantified, the ergonomist can use that information for task redesign to accommodate a large percentage of the population, or to select employees for jobs in such a way as to minimize potential job risks to employees. It is important to establish baseline data on employees and job demands, and to update the data on a regular basis.

19.4.5 A HAZARD PREVENTION AND CONTROL PROGRAM

Once the ergonomics assessment program identifies the work-related hazards, the next step is to apply appropriate measures to prevent and control them. These hazards are alleviated by design/redesign of the workstations, tools/equipment, and jobs. In order for an ergonomics program to be effective, the program should use appropriate engineering and administrative controls, work practices and personal protective equipment to correct and control work-related hazards. As usual, the preferred hierarchy is first to find an engineering solution to the problem. If an engineering solution is not possible or feasible, administrative controls become the next order for implementation. As a last resort, personal protective equipment is recommended as a short-term solution until engineering or administrative controls can be implemented.

- **Engineering controls**: As mentioned earlier in this book, the fundamental principle of ergonomics is to fit the work to the person, as opposed to fitting the person to the work. This can be accomplished by designing/redesigning or modifying workstations, tools, and work methods to eliminate the occupational risk factors.
  - **Workstation design**: Workstations should be so designed to accommodate the population that will be assigned to the workstation. The design should not be for the average-size worker; reaches and strength should be designed for smaller, weaker workers, while clearances should be designed to accommodate larger workers.
  - **Work method design**: Static, awkward and extreme postures, repetitive movements, and excessive forces should be minimized. The strength and endurance requirements of the jobs should be within the abilities of the workers.
  - **Tool and handle design**: Ergonomically designed tools and handles reduce the risk of cumulative trauma disorders (CTDs). Also, so far as possible the concept of one-size-fits-all should be avoided, and various sizes of tools should be provided for a proper fit. Special-purpose tools should be used where needed for certain jobs.
- **Administrative controls**: Administrative controls are applied to reduce the duration, frequency, and/or severity of exposures to work stressors. Examples of administrative controls are:

- Providing rest breaks to recover from work-induced fatigue. This is a very technical strategy since the length of time needed for recovery depends on the task workload and endurance. However, remember that rest does not have to mean sitting down in a break room. Rest can be accomplished by performing a task that utilizes different muscle groups from those being fatigued. Filling out paper work, restocking supplies, and housekeeping are examples of tasks that most likely would be providing rest from repetitive job activities.
- Increasing the number of employees assigned to a task to distribute the overall load over a larger number of individuals. This reduces the severity of exposure to the work stressors, especially in lifting heavy objects.
- Using job rotation. This should be a preventive measure, not a remedial response to symptoms. The purpose of job rotation is to alleviate physical fatigue and stress of a particular group of muscles and tendons by rotating employees among various tasks that use different muscle–tendon groups. Job rotation should be used with caution to insure that the same muscle–tendon groups are not used. A job analysis is necessary to determine what muscle–tendon groups are used in each task.
- Job enlargement. Various tasks may be added to the employee's job. For example, several office employees can share tasks, so that each answers the telephone, files documents, uses the copy machine, etc., instead of assigning each separate task to a different employee.
- Establishing an effective mechanism to insure that facilities, equipment, and tools are well-maintained so that they are in proper working order and within the original manufacturer's specifications.
- Conducting a methods analysis to insure that each job is being performed in the best manner, using proper tools and techniques.
- Developing an effective training program that provides training for new employees and follow-up training for experienced employees. The training program should address issues of acclimatization and "work hardening" to insure that workers are prepared for their jobs prior to permanent job assignment.
- **Personal protective equipment**: Personal protective equipment should be utilized as a last resort for safeguarding employees against work-related hazards. Personal protective equipment should be selected according to the physical characteristics of workers and the jobs. Factors that should be considered in selecting personal protective equipment include proper fit, and protection against physical stressors (i.e., cold, heat, vibration, etc.). It should be noted that

braces, splints, back belts, etc. are not personal protective equipment
– they are medical devices.

19.4.6 AN ERGONOMICS TRAINING AND EDUCATION PROGRAM

Employee training is an effective way of increasing awareness of
ergonomics issues (e.g., CTDs and back injuries) and resolving
problems before injuries occur. For example, many ergonomists and
experts in biomechanics prefer pushing over pulling and pivoting over
twisting. The employees should be trained to practice these recommended procedures in their work and personal daily activities.

Ergonomics education allows managers, supervisors, and employees
to understand work-related hazards with a job, their avoidance, and
their harmful consequences. For example, employees should know that
an adjustable work height minimizes bending, that reducing weight
minimizes the required exertion, and that a special-purpose tool makes
the task easier. By reducing medical claims, training lowers insurance
premiums and increases the organization's profitability, and it helps
prevent liability suits.

19.4.7 MEDICAL MANAGEMENT PROGRAM

This program requires periodic physical examinations and preventive
programs, such as fitness testing, hypertension screening, smoking
cessation, etc. A medical management protocol should be developed for
handling work-related injuries. It should describe the exact procedures
to be followed by the company when such cases arise. Copies of this
protocol should be made available to employees. This is done to avoid
any misunderstandings, and to improve communications between
management and employees. The medical management protocol should
cover the following basic steps (St John *et al.*, 1993):

- Injury is reported to the foreman, human resources department and
  the ETF.
- The injury (illness) is recorded.
- The ETF investigates the injury with the injured employee and
  supervisor. The ETF examines potential solutions, including process/
  workplace redesign and job rotation.
- The injured employee is referred to an occupational injury specialist to
  determine whether the injury is work-related (e.g., CTDs), and for
  observation, X-rays, tests and conservative treatments.
- The injured employee is referred to an orthopedist (or hand specialist)
  by the occupational injury specialist if the injury is determined to be a
  CTD-type (e.g., carpal tunnel syndrome), and for a second opinion, if
  necessary.

- The orthopedist (or hand specialist) will examine the worker to determine the severity or stage of the injury.
- If necessary, the injured employee will receive conservative treatment, such as therapy, splints, injections and prescriptions of anti-inflammatory drugs, exercise and rest through job rotation, light duty, or time off under the care of the orthopedist (or hand specialist).
- An orthopedist performs surgery if the injury is severe, followed up with the physician.
- The injured employee may be given a recovery break, followed up with the physician for determination of recovery and returning to work or light duty.
- The human resources/personnel department notifies the physician of the light-duty tasks available.
- Light-duty or modified work is made available to the employee when the physician approves performing such restricted work activities, followed up with the physician.
- The employee returns to regular duty upon physician's approval, followed up with the physician.

19.4.8 EXERCISE PROGRAMS

As a part of fitting the person to the task, a before-work warm-up exercise program, such as those performed by athletes and Japanese employees, could be considered. Although the concept of warm-up exercises has not become widely accepted in the western workplace, this practice minimizes the risk of back injuries and other CTDs by stretching and loosening up the muscles and, perhaps, warming up the intervertebral disks. Lutz and Hansford (1987) reported that incorporation of daily exercise in their ergonomics program had both physiologic and psychologic benefits for their employees. Among the exercise benefits were:

- improved posture and breathing;
- improved joint flexibility and muscle extensibility;
- balanced muscle tone;
- improved blood circulation;
- reduced risk of inflammation;
- reduced stress.

Of course, there are drawbacks in implementing an exercise program due to worker attitudes and union resistance. Therefore, the exercise programs should be offered on a voluntary basis. While the basic techniques of the exercise should be consistent for all units throughout the company, its contents should suit the nature of the work being performed in each unit, and changed periodically.

19.5 JUSTIFICATION FOR AN ERGONOMICS PROGRAM

The most difficult part of most ergonomics programs is getting them established. Once started, an ergonomics program will build the momentum while its benefits are revealed one after another. An ergonomics program achieves high credibility if it presents other benefits in addition to safety and health benefits, such as improved productivity and quality, and reduced costs. Smith and Smith (1977, 1982) presented three methods for justifying ergonomics activities:

- justification based on increased productivity (benefit–cost);
- justification based on reduced non-productive time and overhead expenses;
- justification based on social (legal) responsibility.

19.5.1 JUSTIFICATION BASED ON INCREASED PRODUCTIVITY

Justification on the basis of increased productivity is probably the first approach that comes to the mind. This approach uses a benefit–cost analysis. Most organizational decisions are made based on benefit–cost analysis. The benefits and costs of implementing ergonomics programs are listed below:

Benefits of implementing ergonomics

Ergonomists cite a variety of indicators of jobs that could benefit from ergonomics intervention. Therefore, the benefits of implementing an ergonomics program may be some or all of the following (LaBar, 1991; Mahone, 1993):

- reduced direct and indirect medical costs associated with accidents and injuries;
- reduced worker complaints, especially related to musculoskeletal problems (e.g., back pains and other CTDs), and high frequency of visits to the medical department;
- reduced medical insurance premiums;
- reduced workers' compensation premiums;
- reduced first-aid cases and costs;
- reduced workers' compensation claims;
- reduced costly citations (e.g., by OSHA in the USA);
- reduced rate or probability of liability litigation;
- good system reliability;
- decreased employee absenteeism;
- lower employee turnover rates;
- decreased lost time and need for overtime;

- fewer lost work-days;
- fewer days of restricted duty;
- decreased product reworks and material costs;
- improved productivity, especially when accompanied with higher quality;
- improved product quality and customer satisfaction;
- better capacity to deal with emergencies;
- improved morale among the workforce;
- improved worker–management relations.

Costs of implementation ergonomics

The following is a list of common cost items incurred to implement and sustain an ergonomics program:

- cost of workplace redesign when deemed necessary;
- cost of tool replacement;
- fees paid to ergonomics consultants;
- salaries and benefits paid to internal ergonomists.

It should be remembered that ergonomics solutions do not necessarily produce immediate results; rather they tend to have long-term results. Nevertheless, the solutions are usually simple and inexpensive. Manuele (1991) reported that ergonomically sound work environments can reduce workers' compensation claims by as much as 50%.

19.5.2 JUSTIFICATION BASED ON REDUCED NON-PRODUCTIVE TIME AND OVERHEAD COSTS

Reduction in non-productive time and overhead costs is accomplished through reduction in absenteeism; labor turnover; time lost due to occurrence of situations that interfere with production operations; supervisor's time in preparing injury reports, causal investigation, and taking part in legal litigation; injured worker's non-working paid time when getting medical attention and recovery; and productivity lost during the new or returning workers' conditioning periods. Increased workers' compensation costs associated with job-related injuries also add to non-productive and overhead costs.

19.5.3 JUSTIFICATION BASED ON SOCIAL RESPONSIBILITY

This method uses the improvement in the quality of work-life of the employees as the basis for justification of ergonomics activities. Such improvement is a part of the employer's social responsibility. The employer is also legally obliged to provide a place of employment free of recognized hazards for the employees. The potential liabilities associated with violation of occupational safety and health standards/regulations and equal employment laws – as well as potential product liability litigations – encourage the employer to be concerned with ergonomics efforts.

19.6 IMPLEMENTATION OBSTACLES

The implementation of a major project, such as a proactive ergonomics program, faces a number of problems. These problems should be anticipated to prevent the downfall of the project. The following are four common problems associated with the implementation of ergonomics projects (St John *et al.*, 1993).

- **Resistance to change**: People are usually resistant to change and more comfortable with the existing routine, with which they are familiar. A natural human reaction to change is to express doubt about any new process or procedure. Ergonomics is a relatively new idea to many people, so resistance against its implementation should be expected. People may become convinced to cooperate with the implementation of an ergonomics project by demonstrating the potential benefits from the change (let the worker try a new tool or modified workplace).
- **Ergonomics as "soft-engineering"**: Ergonomics is occasionally referred to as a "soft-engineering" discipline due to its involvement primarily with people. Few hard numbers exist for its justification, thus ergonomics is difficult to sell to management on the basis of future cost savings or other quantitative bases. Although tools such as biomechanics, physiology, and anthropometry are used, many ergonomics principles are implemented because they "feel better" to the worker. The basic idea to remember or justification factor to use, is that we deal with the safety and welfare of people – an asset that is hard to replace.
- **Cost–benefit justification**: The economic benefits of ergonomics may not be seen immediately, but its cost may easily be determined. This fact often inhibits the commitment of upper management and makes ergonomics hard to sell. The immediate benefits of a newly implemented ergonomics program are the safety, comfort, and well-being of the employees. The long-term benefits which will eventually be realized include those listed in the previous section.
- **New cases**: When a new ergonomics program is initiated, more reported cases of work-related disorders should be expected. With education of the workforce, previously unrecognized occupational health problems will be realized and more injuries will be reported. This movement will eventually diminish and a true decrease in the

number of cases will be attained along with the long-term benefits of ergonomics implementation.

19.7 APPLICATIONS AND DISCUSSION

Ergonomics expertise is often called upon after a work system produces undesirable outcomes, such as work-related injuries (e.g., CTDs), frequent operator errors, and poor productivity and product quality. In such reactive cases, the ergonomist must analyze the work system and tasks allocated to employees. The analysis identifies ergonomics risk factors and those tasks whose requirements are beyond the employee's capabilities and limitations. Then, the system may be redesigned and those tasks can be reallocated, redesigned, or automated. However, if redesign efforts, administrative controls, and personal protective equipment cannot effectively resolve the issues, selection of workers with the required capabilities is the last resort. This strategy is based on the "fitting the person to the task" philosophy, which should always be used as the last remedy.

By a proactive ergonomics approach, ergonomics principles and concepts are incorporated into the original design, and the dynamics of work systems are routinely reviewed and monitored to identify problems before they occur.

Obtaining approval of funds to implement ergonomics programs is usually very difficult. Even more difficult is to gain approval for capital expenditures for improving existing systems. The justification of such expenditures is usually based on benefit–cost analysis. Examples of benefits and costs of ergonomics were presented in this chapter. They can be used as clues for performing a benefit–cost analysis.

Commitment and involvement of both the employer and employees are complementary and crucial elements of a successful ergonomics program. A commitment by the employer encourages employee involvement in the program. This commitment is also essential in obtaining funding for new or improved tools and equipment, to get worker cooperation to try new methods, and to acquire support from employees at all levels. Employees are invaluable sources of information since they are the ones facing the related problems. The progress of the program must be regularly reviewed and evaluated in order to continue its momentum. As OSHA (1990) suggested, management, supervisors, and employees should participate in reviewing the program and re-evaluating its goals and objectives and discuss possible changes that may be necessary.

REVIEW QUESTIONS

1. What are the employer's driving forces for ergonomics implementations?
2. Why is an ergonomics task-force often successful?
3. How can an ergonomics program be promoted within an organization?
4. How can the ergonomics effort entail both reactive and proactive components?
5. Why should ergonomic interventions be well-documented (written)?
6. Based on what basic premise of ergonomics can it be determined if the worker is being exposed to work stresses?
7. What steps are recommended for meeting the objectives of an ergonomics program?
8. List the eight major components or steps necessary for implementing a successful ergonomics program.
9. What should be done in order to obtain the support of top management for implementing an ergonomics program?
10. What is the purpose of the systematic ergonomics assessment program?
11. Briefly explain the engineering controls for fitting the work to the person, as opposed to fitting the person to the work.
12. Briefly explain the purpose of applying administrative controls for controlling work-related hazards. List four examples of such controls.
13. What basic steps should a medical management protocol cover?
14. What are the three methods for justifying ergonomics activities?
15. List 10 examples of benefits of implementing ergonomics programs.
16. List four examples of costs of implementing and sustaining an ergonomics program.
17. List the four common problems associated with the implementation of ergonomics projects.

REFERENCES

LaBar, G. (1991) Building ergo land. *Occupational Hazards* **53**: 29–33.
Lutz, G. and Hansford, T. (1987) Cumulative trauma disorder controls: the ergonomics program at Ethicon, Inc. *Journal of Hand Surgery* **12A**: 863–866.
Mahone, D.B. (1993) Review of system design employs ergonomics prior to work injuries. *Occupational Health and Safety* **62**: 89–92, 105.

Manuele, F.A. (1991) Workers' compensation cost control through ergonomics. *Professional Safety* **36**: 27–32.

OSHA (1990) *Ergonomics Program Management Guidelines for Meatpacking Plants*. Publication no. OSHA 3123. Occupational Safety and Health Administration, US Department of Labor, Washington, DC.

St John, D., Tayyari, F. and Emanuel, J.T. (1993) Implementations of an ergonomics program: a case report. *International Journal of Industrial Ergonomics* **11**: 249–256.

Smith, J.L. and Smith, L.A (1977) Justification of a human factors program in industry. In: *The Proceedings of the Human Factors Society – 21st Annual Meeting*. San Francisco, pp. 524–527, Human Factors & Ergonomics Society, Santa Monica, California.

Smith, L.A. and Smith J.L. (1982) How can an IE justify a human factors activities program to management? *Industrial Engineering* **14**: 39–43.

Appendix A: Ergonomics-related periodicals

JOURNALS AND MAGAZINES

AGING TITLES

Age and Aging
Journal of Gerontology

BIOMECHANICS TITLES

ASME Transaction: Journal of Biomechanics
Biomechanics
Clinical Biomechanics
Journal of Biomechanics

ENGINEERING AND PHYSICS TITLES

American Journal of Physics
Automatica
Color Research and Application
Computers and Industrial Engineering
Datamation
IEEE Transactions on Systems, Man, and Cybernetics
International Lighting Review
Journal of the Acoustical Society of America
Journal of the Illuminating Engineering Society
Journal of Navigation
Lighting Research and Technology
Noise Control Engineering Journal

ERGONOMICS TITLES

Applied Ergonomics
Ergonomics
Human Factors
Human Movement Science
International Journal of Industrial Ergonomics
International Journal of Man–Machine Studies
Japanese Journal of Ergonomics
Journal of Human Ergology
Occupational Ergonomics
Office Ergonomics
Office Ergonomics Review
Scandinavian Journal of Work, Environment and Health
Work and Occupations
Work and People
Work and Stress

PSYCHOLOGY TITLES

Applied Psychology: An International Review
Behavior and Information Technology
Cognitive Science
European Journal of Applied Physiology and Occupational Physiology
International Review of Applied Psychology
Journal of Applied Psychology
Journal of General Psychology
Journal of Managerial Psychology
Psychologist
Quarterly Journal of Experimental Physiology
Quarterly Journal of Experimental Physiology and Cognate Medical Science
Perception
Perceptual and Motor Skills
Perceptual and Psychophysics

SAFETY AND HEALTH TITLES

Accident Analysis and Prevention
American Industrial Hygiene Association Journal
American Journal of Public Health and the Nation's Health
Annals of Occupational Hygiene
Australian Safety News
Canadian Journal of Public Health

Health and Safety (Journal)
Health and Safety at Work
Health Laboratory Science
Health Physics
Industrial Health
Journal of Occupational Accidents
Journal of Safety Research
Occupational Hazards
Occupational Health and Safety
Occupational Health in Ontario
Professional Safety

SCIENCE AND MEDICAL TITLES

Acta Physiologica Scandinavica
Acta Physiologica Scandinavica, Supplementum
Advances in Biological and Medical Physics
American Journal of Physiology
American Journal of Science
American Journal of the Medical Science
Annual Review of Physiology
Australian Journal of Experimental Biology and Medical Science
Australian Journal of Science
Aviation Space and Environmental Medicine
Canadian Journal of Research (Section E–Medical Sciences)
Canadian Medical Association Journal
International Journal of Rehabilitation
JAMA – the Journal of the American Medical Association
Journal of Applied Physiology
Journal of Occupational Medicine
Journal of Physiology
Journal of Science and Labor
Journal of Sport Medicine and Physical Fitness
Lancet
Life Sciences
Metabolism, Clinical and Experimental
New England Journal of Medicine
Physiological Reviews
Respiration Physiology
Science
Science Abstract
Science and Medicine in Sports
Sports Medicine

OTHER TITLES

Human Relations
IIE Transactions
Industrial Engineering
Management Services
New Technology, Work, and Employment
Personnel Management
Research Quarterly

CONFERENCE PROCEEDINGS

Advances in Industrial Ergonomics and Safety (the Proceedings of the Annual International Industrial Ergonomics and Safety Conference, since 1989). Taylor & Francis, London.

Proceedings of the Ergonomics Society's Conferences. Taylor & Francis, London.

Proceedings of Human Factors Society Annual Meetings. Human Factors and Ergonomics Society, Santa Monica, CA.

Proceedings of Human Factors and Ergonomics Society Annual Meetings. Human Factors and Ergonomics Society, Santa Monica, CA.

Trends in Ergonomics/Human Factors (the Proceedings of the Annual International Industrial Ergonomics and Safety Conference, prior to 1989). Elsevier, North-Holland, Amsterdam.

Appendix B: Units of measurement and conversion factors

ABBREVIATIONS

The following abbreviations are used in conversion tables:

atm	atmospheres (pressure)
BTU (or Btu)	British Thermal Unit (work and energy)
cm	centimeters (length)
dyn	dynes (force)
ft	foot or feet (length)
ft-lb	foot · pounds (work and energy)
g	gravitational acceleration = 9.80665 m · s² $=$ 32.174049 ft · s^{-2} (acceleration)
gal	gallons (volume)
h	hours (time)
hp	horsepowers (power)
in	inches (length)
J	Joules (work and energy)
kcal	kilocalories = Calories (or large calories) = 10^3 calories (work and energy)
kg	kilograms (1 kg = 1000 g; mass)
kJ	kilojoules (1 kJ = 1000 J; work and energy)
km	kilometers (1 km = 1000 m; length)
kp	kiloponds
kpm	kilopond · meter = kilogram · meter
l	liters (volume)
lb	pounds (weight)
mi	miles (length)
min	minutes (time)

mmHg	millimeters of mercury (pressure)
mph	miles per hour
N	newtons (force)
oz	ounces (weight)
Pa	pascals (pressure)
psi	pounds per square inch (lb · in^{-2}; pressure)
s	seconds (time)
W	watts (power)
yd	yards (length)

SI UNITS OF MEASUREMENTS

Basic SI units (le Système International d'Unités):
   Length: meters (m)
   Mass: kilograms (kg)
   Time: seconds (s)
   Quantity: moles (mol)
Other SI units (Table B.1)

Table B.1 Other SI units of measurement

	Unit	Symbol	Calculation expression
Force	Newton	N	$kg \cdot m \cdot s^{-2}$
Work or energy	Joule	J	$kg \cdot m^{-2} \cdot s^{-2} = N \cdot m$
Power	Watt	W	$kg \cdot m^{-2} \cdot s^{-3} = J \cdot s^{-1}$
Pressure	Pascal	Pa	$N \cdot m^{-2}$

CONVERSION FACTORS

LENGTH

1 in = 2.54 cm = 0.0254 m
1 ft = 30.48 cm = 0.3048 m
1 yd = 3 ft = 36 in = 91.44 cm = 0.9144 m
1 mi = 5280 ft = 1.609344 km
1 cm = 0.3937 in = 0.01 m
1 m = 100 cm = 39.37 in = 3.28084 ft = 1.09360 yd
1 km = 1000 m = 0.621371 mi

AREA

$1 \text{ in}^2 = 1/144 \text{ ft}^2 = 6.4516 \text{ cm}^2$
$1 \text{ ft}^2 = 144 \text{ in}^2 = 929.0304 \text{ cm}^2 = 0.092903 \text{ m}^2$
$1 \text{ acre} = 43\,560 \text{ ft}^2$
$1 \text{ mi}^2 = 27\,878\,400 \text{ ft}^2 = 640 \text{ acre} = 2.589988 \text{ km}^2$
$1 \text{ cm}^2 = 0.155 \text{ in}^2 = 0.0001 \text{ m}^2$
$1 \text{ m}^2 = 10\,000 \text{ cm}^2 = 1{,}550 \text{ in}^2 = 10.7639 \text{ ft}^2$
$1 \text{ km}^2 = 10^6 \text{ m}^2 = 0.3861022 \text{ mi}^2$

VOLUME

$1 \text{ l (or 1 cc)} = 10^3 \text{ ml} = 10^{-3} \text{ m}^3$
$\qquad\qquad = 0.0353147 \text{ ft}^3 = 61.0237 \text{ in}^3 = 0.2641721 \text{ gal (US fluid)}$
$1 \text{ m}^3 = 10^3 \text{ l (or cc)} = 35.3147 \text{ ft}^3 = 6.10237 \times 10^4 \text{ in}^3 = 264.1721 \text{ gal}$
$\qquad$ (US fluid)
$1 \text{ gal (US fluid)} = 3.7854118 \text{ l}$
$1 \text{ gal (British Imperial and Canadian fluid)} = 1.201 \text{ gal (US fluid)}$
$\qquad\qquad\qquad\qquad\qquad\qquad\qquad\qquad\quad = 4.546 \text{ l}$

SPEED

$1 \text{ m} \cdot \text{s}^{-1} = 3.6 \text{ km} \cdot \text{h}^{-1} = 3.28084 \text{ ft} \cdot \text{s}^{-1} = 2.23694 \text{ mph}$
$30 \text{ mph} = 44 \text{ ft} \cdot \text{s}^{-1} \text{ (exact)}$

MASS

$1 \text{ slug} = 14.59 \text{ kg}$
$1 \text{ kg} = 0.06852 \text{ slug}$

FORCE AND WEIGHT

$1 \text{ N} = 10^5 \text{ dyn} = 0.2248 \text{ lb}$
$1 \text{ lb} = 16 \text{ oz}$
1 kg mass weighs 9.80665 N (at standard **g**)
$1 \text{ kg} \propto 2.2046 \text{ lb}$ ($\propto$ means equivalent, i.e., 1 kg mass weighs 2.2046 lb at standard **g**)
$1 \text{ lb} \propto 0.45359 \text{ kg mass}$

Table B.2 Conversion factors for the units used in photometry

Measurement	Unit	System of measurement	Conversion factor
Candlepower	Candela (cd)	USCS and SI	$1\ cd = 4\pi$ or $12.566371\ lm$
	Lumen (lm)	USCS and SI	$1\ lm = 1/4\pi$ or $0.0795775\ cd$
Illumination level (illuminance)	Foot-candle (fc)	USCS	$1\ fc = 1\ lm \cdot ft^{-2} = 10.76391\ lx$
	Lux (lx)	SI	$1\ lx = 1\ lm \cdot m^{-2} = 0.092903\ fc$
Luminance	Foot-Lambert (fL)	USCS	$1\ fL = 3.4262591\ cd \cdot m^{-2}$
	Candela per square meter (cd · m^{-2})	SI	$1\ cd \cdot m^{-2} = 0.2918635\ fL$

USCS = US Customary System; SI = International System

Table B.3 Interrelationship among units of pressure

Unit	Pa	mbar	psi	atm	torr
Pa = N · m^{-2}	1	10^{-2}	1.45×10^{-4}	9.8692×10^{-6}	7.5006×10^{-3}
mbar	100	1	1.45×10^{-2}	9.8692×10^{-4}	0.75006
psi	6895	68.95	1	0.06805	3.519250
atm	101 325	1013	14.7	1	760
torr	133.322	1.33322	0.28415	1/760	1

1 Pa = 1 N · m^{-2} = 0.001 kPa (kilopascal) = 10 dyn · cm^{-2} = 10^{-5} bar = 0.0075 mmHg.
1 torr = 1 mmHg at 0°C = 3.93701 inHg at 32°F.

Table B.4 Interrelationship among units of work and energy

Unit	J	kW·h	kcal	BTU	hp·h	ft-lb	kpm
Joule (J)	1	2.7778×10^{-7}	2.3885×10^{-4}	9.4782×10^{-4}	3.7251×10^{-7}	0.73756	0.101972
kW·h	3.6×10^{6}	1	859.85	3412.15	1.341	2.6552×10^{6}	3.6710×10^{5}
kcal	4186.78	1.163×10^{-3}	1	3.9684	1.5596×10^{-3}	3088.0	426.924
BTU	1055.05	2.9307×10^{-7}	0.252	1	3.9292×10^{-4}	778.17	107.585
hp·h	2.6845×10^{6}	0.7457	641.19	2545.0	1	1.98×10^{6}	2.7374×10^{5}
ft-lb	1.35582	3.76617×10^{-7}	3.2383×10^{-4}	1.2851×10^{-3}	5.0505×10^{-7}	1	0.13826
kpm	9.80665	2.72407×10^{-6}	2.3423×10^{-3}	9.29494×10^{-3}	3.6531×10^{-6}	7.23300	1

Table B.5 Interrelationship among units of power or workload

Unit	W	kcal·h⁻¹	BTU·h⁻¹	hp	ft-lb·min⁻¹	kpm·min⁻¹
Watt (J·s⁻¹)	1	0.85985	3.4121	$1.341 \; 10^{-3}$	44.2536	6.1183
kcal·h⁻¹	1.163	1	3.9684	$1.5596 \; 10^{-3}$	51.4674	7.1154
BTU·h⁻¹	0.2930	0.252	1	$3.93 \; 10^{-4}$	12.9696	1.7931
hp	745.7	641.19	2545	1	33 000	4562
ft-lb·min⁻¹	0.022597	0.01943	0.0771	30.303×10^{-6}	1	0.13826
kpm·min⁻¹	0.16345	0.140538	0.5574	0.2192×10^{-3}	7.233	1

DEFINITIONS AND FORMULAS

ACCELERATION

Rate of change in velocity (speed), calculated as:

$$a = \frac{V_t - V_o}{t}$$

where V_0 and V_t = initial and final velocities, respectively; and t = time elapsed reaching final velocity from initial velocity.

BEER'S LAW

The intensity of light (or color) transmitted through a liquid is inversely proportional to the depth of the liquid.

BTPS

Body temperature and ambient barometric pressure saturated with water vapor.

$$\text{BTPS}_{\text{correction}} = \left[\frac{273°K + 37°C}{273°K + T°C} \right] \times \left[\frac{P_B \ (mmHg) - P_{H_2O} \ (mmHg)}{P_B \ (mmHg) - 47.1 \ (mmHg)} \right]$$

where $273°K$ = freezing temperature of water on Kelvin scale equivalent to 0°C;

$37°C$ = normal body temperature (constant);

$T°C$ = temperature of collected gas sample;

P_B = ambient barometric pressure (in mmHg);

P_{H_2O} = water vapor tension (in mmHg) at $T°C$, the pressure exerted by water vapor in collected gas sample, which is obtained from the empirical tables (Table B.6);

$47.1 \ mmHg$ = water vapor tension at 37°C, i.e., $P_{37°C}$ (constant).

CARDIAC OUTPUT

The volume of blood ejected per unit of time (usually in $ml \cdot min^{-1}$):

$$\text{Cardiac output} = SV \times HR$$

where SV = stroke volume (ml), and HR = heart rate (bpm).

CENTER OF GRAVITY

The point at which the weight of a mass is concentrated.

DALTON'S LAW

The total pressure of a mixture of gases is equal to the sum of the partial pressures of the individual gases in the mixture; i.e., for a mixture of n gases

$$P_{Total} = P_1 + P_2 + P_3 + \ldots + P_n$$

DENSITY

The ratio of mass or weight to volume of an object.

$$D_m = m/v \text{ or } D_w = w/v$$

where D_m = mass density; m = mass; v = volume; D_w = weight density; and w = weight.

EFFICIENCY

The ratio of energy (work) output to energy (work) input:

$$\text{Gross efficiency} = \frac{\text{energy output}}{\text{energy input}} \times 100$$

$$\text{Net efficiency} = \frac{\text{energy output}}{\text{net energy input}} \times 100$$

Net energy input = gross energy input − resting energy for a comparable time period

ENERGY

The capacity for doing work or overcoming tension:

$$\text{Potential energy (PE)} = mgh = wh$$

$$\text{Kinetic energy (KE)} = \frac{1}{2}mv^2$$

where m = mass; g = gravitational acceleration; h = vertical distance; w = weight; and v = velocity.

ENERGY COST OF WORK

- **Gross energy cost** is the amount of oxygen (or kilocalories of energy) required for performing and recovering from the activity.

- **Net energy cost** is the amount of oxygen (or kilocalories of energy) required for performing and recovering from the activity minus resting oxygen uptake (or energy consumption) for a comparable time period.

ENERGY EXPENDITURE (WEIR'S FORMULA):

$$E(\text{kcal} \cdot \text{min}^{-1}) = 4.92\,(F_{IO_2} - F_{EO_2}) \cdot \dot{V}_E$$

where $\dot{V}_E$ = volume of expired air in $1 \cdot \text{min}^{-1}$ at STPD;

F_{IO_2} = fraction of oxygen in the inspired air (usually assumed to be 0.2093);

F_{EO_2} = fraction of oxygen in the expired air (a fraction of the volume of expired air);

4.92   = kilocalories of energy equivalent to 1 l of oxygen consumed.

NEWTON'S LAWS OF MOTION

- **First law (inertia)**: An object continues in a state of rest or uniform motion (with constant velocity), unless acted upon by unbalanced forces that change its state.
- **Second law (acceleration)**: An object is accelerated in the direction of and in direct proportion to the force exerted on it and in inverse proportion to its mass.
- **Third law (interaction)**: For every action there is an equal and opposite reaction.

POWER

The rate of performing work:

$$P = W/t$$

where W = work, and t = elapsed time.

RELATIVE HUMIDITY

The ratio of the water vapor saturation of a given volume of air to its maximum possible saturation (expressed as a percentage).

STPD

Standard temperature pressure dry. The gas volume is converted to an equivalent volume of dry gas at sea level $P_B = 760$ mmHg and $T = 0°C$.

$$STPD_{correction} = \frac{273}{273 + T\,(°C)} \times \frac{P_B\,(mmHg) - P_{H_2O}\,(mmHg)}{760}$$

where P_B = ambient barometric pressure (mmHg):

P_{H_2O} = vapor tension of water (mmHg) at the temperature of the gas-meter, which is obtained from the following empirical tables (Table B.6);

T = temperature of the gasmeter or collected gas sample, in °C.

VAPOR TENSION OF WATER (OR VAPOR PRESSURE)

That part of total atmospheric pressure attributable to its water vapor content.

Table B.6 The vapor tension of water (mmHg) at temperatures between 0 and 49°C

°C	P_{H_2O}	°C	P_{H_2O}	°C	P_{H_2O}	°C	P_{H_2O}	°C	P_{H_2O}
0	4.6	10	9.2	20	17.5	30	31.8	40	55.3
1	4.9	11	9.8	21	18.7	31	33.7	41	58.3
2	5.3	12	10.5	22	19.8	32	35.7	42	61.5
3	5.7	13	11.2	23	21.1	33	37.7	43	64.8
4	6.1	14	12.0	24	22.4	34	39.9	44	68.3
5	6.5	15	12.8	25	23.8	35	42.2	45	71.9
6	7.0	16	13.6	26	25.2	36	44.6	46	75.7
7	7.5	17	14.5	27	26.7	37	47.1	47	79.6
8	8.0	18	15.5	28	28.3	38	49.7	48	83.7
9	8.6	19	16.5	29	30.0	39	52.4	49	88.0

VELOCITY

The rate of displacement:

$$v = d/t$$

where v = average velocity; d = displacement (distance); and t = elapsed time.

WORK

The product of force acting upon an object and the distance the object is moved:

$$W = F \cdot d$$

where F = force; and d = distance the object is moved.

Appendix C: Standard normal distribution

Table C.1 Standard normal distribution

$$F(z) = \frac{1}{\sqrt{2\pi}} \int_{-\infty}^{z} e^{-\frac{1}{2}t^2} dt$$

z	0.00	0.01	0.02	0.03	0.04	0.05	0.06	0.07	0.08	0.09
0.0	0.5000	0.5040	0.5080	0.5120	0.5160	0.5199	0.5239	0.5279	0.5319	0.5359
0.1	0.5398	0.5438	0.5478	0.5517	0.5557	0.5596	0.5636	0.5675	0.5714	0.5753
0.2	0.5793	0.5832	0.5871	0.5910	0.5948	0.5987	0.6026	0.6064	0.6103	0.6141
0.3	0.6179	0.6217	0.6255	0.6293	0.6331	0.6368	0.6406	0.6443	0.6480	0.6517
0.4	0.6554	0.6591	0.6628	0.6664	0.6700	0.6736	0.6772	0.6808	0.6844	0.6879
0.5	0.6915	0.6950	0.6985	0.7019	0.7054	0.7088	0.7123	0.7157	0.7190	0.7224
0.6	0.7257	0.7291	0.7324	0.7357	0.7389	0.7422	0.7454	0.7486	0.7517	0.7549
0.7	0.7580	0.7611	0.7642	0.7673	0.7704	0.7734	0.7764	0.7794	0.7823	0.7852
0.8	0.7881	0.7910	0.7939	0.7967	0.7995	0.8023	0.8051	0.8078	0.8106	0.8133
0.9	0.8159	0.8186	0.8212	0.8238	0.8264	0.8289	0.8315	0.8340	0.8365	0.8389
1.0	0.8413	0.8438	0.8461	0.8485	0.8508	0.8531	0.8554	0.8577	0.8599	0.8621
1.1	0.8643	0.8665	0.8686	0.8708	0.8729	0.8749	0.8770	0.8790	0.8810	0.8830
1.2	0.8849	0.8869	0.8888	0.8907	0.8925	0.8944	0.8962	0.8980	0.8997	0.9015
1.3	0.9032	0.9049	0.9066	0.9082	0.9099	0.9115	0.9131	0.9147	0.9162	0.9177
1.4	0.9192	0.9207	0.9222	0.9236	0.9251	0.9265	0.9279	0.9292	0.9306	0.9319
1.5	0.9332	0.9345	0.9357	0.9370	0.9382	0.9394	0.9406	0.9418	0.9429	0.9441
1.6	0.9452	0.9463	0.9474	0.9484	0.9495	0.9505	0.9515	0.9525	0.9535	0.9545
1.7	0.9554	0.9564	0.9573	0.9582	0.9591	0.9599	0.9608	0.9616	0.9625	0.9633
1.8	0.9641	0.9649	0.9656	0.9664	0.9671	0.9678	0.9686	0.9693	0.9699	0.9706
1.9	0.9713	0.9719	0.9726	0.9732	0.9738	0.9744	0.9750	0.9756	0.9761	0.9767
2.0	0.9772	0.9778	0.9783	0.9788	0.9793	0.9798	0.9803	0.9808	0.9812	0.9817
2.1	0.9821	0.9826	0.9830	0.9834	0.9838	0.9842	0.9846	0.9850	0.9854	0.9857
2.2	0.9861	0.9864	0.9868	0.9871	0.9875	0.9878	0.9881	0.9884	0.9887	0.9890
2.3	0.9893	0.9896	0.9898	0.9901	0.9904	0.9906	0.9909	0.9911	0.9913	0.9916
2.4	0.9918	0.9920	0.9922	0.9925	0.9927	0.9929	0.9931	0.9932	0.9934	0.9936
2.5	0.9938	0.9940	0.9941	0.9943	0.9945	0.9946	0.9948	0.9949	0.9951	0.9952
2.6	0.9953	0.9955	0.9956	0.9957	0.9959	0.9960	0.9961	0.9962	0.9963	0.9964
2.7	0.9965	0.9966	0.9967	0.9968	0.9969	0.9970	0.9971	0.9972	0.9973	0.9974
2.8	0.9974	0.9975	0.9976	0.9977	0.9977	0.9978	0.9979	0.9979	0.9980	0.9981
2.9	0.9981	0.9982	0.9982	0.9983	0.9984	0.9984	0.9985	0.9985	0.9986	0.9986
3.0	0.9987	0.9987	0.9987	0.9988	0.9988	0.9989	0.9989	0.9989	0.9990	0.9990
3.1	0.9990	0.9991	0.9991	0.9991	0.9992	0.9992	0.9992	0.9992	0.9993	0.9993
3.2	0.9993	0.9993	0.9994	0.9994	0.9994	0.9994	0.9994	0.9995	0.9995	0.9995
3.3	0.9995	0.9995	0.9995	0.9996	0.9996	0.9996	0.9996	0.9996	0.9996	0.9997
3.4	0.9997	0.9997	0.9997	0.9997	0.9997	0.9997	0.9997	0.9997	0.9997	0.9998

Note

p values are given in the body of the table, and Z_p is the sum of the margin values. Example: by linear interpolation $Z_{0.95} = (1.64 + 1.65)/2 = 1.645$ since 0.9500 is not found.

Appendix D: Guidelines for systematic workload estimation

INTRODUCTION

A systematic workload estimation (SWE) method was developed by Burford *et al*. (1984) for evaluating the metabolic cost of work performed in underground mines (Ramsey and Burford, 1985; Ramsey *et al*., 1986). The SWE is a coding method that can be used by individual observers who have little or no previous experience in measuring/estimating metabolic rate (Tayyari *et al*., 1989). The method was validated in the field (Burford *et al*., 1985) by simultaneous measurements of oxygen consumption, and documented in the final report of a National Institute for Occupational Safety and Health (NIOSH) project (Ramsey and Burford, 1985).

SWE DATA RECORDING FORMS

An SWE data-recording form (Fig. D.1) includes three sections:

- A heading which provides spaces for the data-sheet number, observer's number (or name), worker's (ID) number, and worker's physical characteristics. This heading is actually to be prepared by users to incorporate their own required information.
- The SWE coding schema which provides codes for 45 combinations of body activities, body members, and work intensities.
- The SWE time-and-motion-study work-sheet which provides space for recording a short description of the work activity (motion) performed by the subject during each minute (time) of the study, and the SWE metabolic rate-estimation code for the task performed during

that specific minute. This work-sheet also provides a space for the minute caloric value of the SWE code which is used to calculate the time-weighted average of energy expenditure in performing the work in question.

Sheet No.: _____ Date: _____ Observer: _____ Worker's ID: _____ Wt: _____ Ht: _____ Age: _____

		(Kcal/min)	1	2	3	4	5	6	7	8	9	10	11	12
STATIONARY S	Sitting Standing Stooping	Hand Work												
		1-Arm Work												
		2-Arm Work												
		Whole-Body Work												
WALKING W	Slow & Medium	Hand Work												
		1-Arm Work												
		2-Arm Work												
		Whole-Body Work												
EXTRA EFFORT E	Fast Grade Soft	Hand Work												
		1-Arm Work												
		2-Arm Work												
		Whole-Body Work												

STATIONARY — Hand Work 0 1 2 3; 1-Arm Work 4 5 6; 2-Arm Work 7 8 9; Whole-Body Work 10 11 12 13 14

WALKING — Hand Work 0 1 2 3; 1-Arm Work 4 5 6; 2-Arm Work 7 8 9; Whole-Body Work 10 11 12 13 14

EXTRA EFFORT — Hand Work 0 1 2 3; 1-Arm Work 4 5 6; 2-Arm Work 7 8 9; Whole-Body Work 10 11 12 13 14

(Watts) 70 140 210 280 350 420 490 560 630 700 770 840

Starting Time Hour _____ : 00/30 Min

Start Min	Task/Work Description	SWE Code	Min Kcal	End Min
00/30				01/31
01/31				02/32
02/32				03/33
03/33				04/34
04/34				05/35
05/35				06/36
06/36				07/37
07/37				08/38
08/38				09/39
09/39				10/40
10/40				11/41
11/41				12/42
12/42				13/43
13/43				14/44
14/44				15/45
15/45				16/46
16/46				17/47
17/47				18/48
18/48				19/49
19/49				20/50
20/50				21/51
21/51				22/52
22/52				23/53
23/53				24/54
24/54				25/55
25/55				26/56
26/56				27/57
27/57				28/58
28/58				29/59
29/59				30/00

Fig. D.1 Systematic workload estimation (SWE) data form.

GUIDELINES FOR THE USE OF THE SWE TECHNIQUE

- Enter the heading data before the state of observation (e.g., date, subject's ID, page number, etc.).
- Enter the starting hour and circle 00 for the first half-hour or 30 for the second half-hour (e.g., if the starting time is 9:40, enter 9 as the starting time hour and circle 30).
- Observe the worker while he or she performs the tasks.
- Assign an appropriate letter-code (first part of coding) according to the activity class as follows:
  - **S** = stationary (sitting, standing, stooping, etc.);
  - **W** = walking (slow or medium);
  - **E** = extra effort or exertion (e.g., walking fast, walking on a grade, walking on soft materials, climbing a ladder, walking with a heavy load or working with high exertion).
- Assign an appropriate numerical code (second part of coding) according to the activity subclass and work intensity as follows:
  - 0 = No hand/arm work (e.g., resting is rated as S–0, and slow walking as W–0);
  - 1–3 = hand work (1 = light; 2 = moderate; 3 = heavy);
  - 4–6 = one-arm work (4 = light; 5 = moderate; 6 = heavy);
  - 7–9 = two-arm work (7 = light; 8 = moderate; 9 = heavy);
  - 10–14 = whole-body work (10 = light; 11 = slightly light; 12 = moderate; 13 = heavy; 14 = very heavy).
- Record the desired notes in the task/work description column and enter the assigned codes in steps 4 and 5 in the SWE code column.
- Repeat steps 3–6 for each minute of work, unless the worker is performing the same task for several minutes. In the latter case make a note of the situation and enter the new coding at the time the code must change. This is required to calculate the time-weighted average of the metabolic rate of the work.
- At the end of observation, convert the SWE codes into their corresponding caloric values (in kcal · 70 kg^{-1}, where 70 kg is the body weight of an arbitrary chosen reference man) from the top of the SWE scheme or from Table D.1 for a more accurate result, and enter the values in the "Min kcal" column.
- Calculate the total amount of energy consumed during the observed time by summing all caloric values.
- Calculate the time-weighted average of the metabolic rate of work (kcal · min^{-1} · 70 kg^{-1}) by dividing the sum of all caloric values by the number of minutes in the observation time.
- Calculate the time-weighted average of the energy expenditure (metabolic rate) for the observed worker by dividing the rate calculated in step 10 by 70 and multiplying by the worker's body weight (kg).

Table D.1 Systematic workload estimation (SWE) schema values (kcal · min^{-1} · 70 kg^{-1})

Code	0	1	2	3	4	5	6	7	8	9	10	11	12	13	14
S	1.20	2.00	2.20	2.50	2.60	3.00	3.40	3.10	3.60	4.10	3.60	4.85	6.10	7.35	8.60
W	3.25	3.65	3.85	4.15	4.25	4.65	5.05	4.75	5.25	5.75	5.25	6.50	7.75	9.00	10.25
E	4.50	4.90	5.10	5.40	5.50	5.90	6.30	6.00	6.50	7.00	6.50	7.75	9.00	10.25	11.50

Notes
S = Stationary (sitting, standing, stooping, etc.).
W = Walking (slow or medium).
E = Extra-exertion (or walking fast, walking on a grade, walking on soft materials, climbing a ladder, walking with a heavy load or working with high exertion).

GUIDELINES FOR TRAINING THE SWE USER

Observers who make metabolic rate estimations using the SWE technique should be trained in assigning the SWE codes. They must study the SWE scheme to learn the classes and subclasses of activities, and should personally perform some physical activities representing each activity class and subclass to have a better understanding of the workload.

To calibrate the observer's subjective coding, workers should perform several different tasks while their pulmonary ventilation rate and metabolic rate for each task are measured. The observer maintains a time log of each task, along with the SWE code on the data-recording worksheets. The observer will compare these ratings with the measured metabolic rates to calibrate the ratings.

The training requires a respiratory gas-meter to measure the pulmonary ventilation rate ($\dot{V}_e$), a barometer to convert the observed pulmonary ventilation rate into standard temperature pressure dry (STPD), SWE data-recording forms, air sampler pump and air sample bags to collect expired air for each task, and an oxygen analyzer to measure the oxygen concentration of the collected sample air. The air sample bags and oxygen analyzer can be eliminated if not available, and the oxygen consumption may then be estimated from pulmonary ventilation rate, using the following equations (Bernard *et al.*, 1979):

For men:

$$\dot{V}_{O_2} = (0.69 - 0.002 \cdot A) \, \dot{V}_E^{0.44} - 1.47 \qquad (D.1)$$

For women:

$$\dot{V}_{O_2} = (1.64 - 0.002 \cdot A) \, \dot{V}_E^{0.25} - 2.34 \qquad (D.2)$$

where: $\dot{V}_{O_2}$ = the rate of oxygen consumption in $l \cdot min^{-1}$ (STPD);

$\dot{V}_E$ = the rate of pulmonary ventilation (expired air flow) in $l \cdot min^{-1}$ (STPD);

A = the subject's age in years.

REFERENCES

Bernard, T.E, Kamon, E. and Franklin, B.A. (1979) Estimation of oxygen consumption from pulmonary ventilation during exercise. *Human Factors* **21**: 417–421.

Burford, C.L., Ramsey, J.D., Tayyari, F., Lee, C.H. and Stepp, R.G. (1984) A method for systematic workload estimation (SWE). In: *Proceedings of the Human Factors Society – 28th Annual Meeting*. Human Factors Society, Santa Monica, CA, pp. 997–999.

Burford, C.L., Ramsey, J.D., Akbar-Khanzadeh, F., Lee, C.H., Tayyari, F. and Dukes-Dobos, F. (1985) Field validation of a procedure for systematic workload estimation. In: *The Abstract Proceedings of American Industrial Hygiene Conference*, May 24, Las Vegas, Nevada, American Industrial Hygiene Association, Akron, Ohio.

Ramsey, J.D. and Burford, C.L. (1985) *Development of a Simple Procedure for Predicting the Effects of Heat on Underground Miners*. Final report. NIOSH contract no. 210–81–6104. US Department of Health and Human Services, NIOSH, Cincinnati, Ohio.

Ramsey, J.D., Burford, C.L., Dukes-Dobos, F.N., Tayyari, F. and Lee, C.H. (1986). Thermal environment of an underground mine and its effect upon miners. *Annals of American Conference of Governmental Industrial Hygienists* **14**: 209–223.

Tayyari, F., Burford, C.L. and Ramsey, J.D. (1989) Guidelines for the use of systematic workload estimation. *International Journal of Industrial Ergonomics* **4**: 61–65.

Appendix E: Relative humidity

Table E.1 A psychrometric chart for estimation of relative humidity from dry-bulb and wet-bulb temperatures (in °F)

Wet-bulb depression	Dry-bulb temperature (°F)														
	60	65	70	75	80	85	90	95	100	105	110	115	120	125	130
0.5	97	97	98	98	98	98	98	98	98	98	98	98	98	98	98
1.0	94	95	95	96	96	96	96	96	96	97	97	97	97	97	97
1.5	91	92	93	93	94	94	94	95	95	95	95	95	95	95	95
2.0	89	90	90	91	91	92	92	93	93	93	93	94	94	94	94
2.5	86	87	88	89	89	90	91	91	91	92	92	92	92	92	92
3.0	83	85	86	86	87	88	89	89	89	90	90	91	91	91	91
3.5	81	82	83	84	85	87	87	87	88	89	89	90	90	90	90
4.0	78	80	81	82	83	84	85	86	86	87	87	88	88	88	89
4.5	75	77	79	80	81	82	83	84	85	85	86	87	87	87	88
5.0	73	75	77	78	79	81	81	82	83	84	84	85	85	86	86
5.5	70	72	74	76	77	79	80	81	82	82	83	84	84	85	85
6.0	68	70	72	74	75	76	78	79	80	80	81	82		82	83
7.0	63	65	68	70	72	73	74	76	77	78	78	79	80	80	81
8.0	58	61	64	66	68	70	71	72	73	75	75	76	77	78	78
9.0	53	56	59	62	64	66	68	69	70	72	73	74	74	75	76
10.0	48	52	55	58	61	63	65	66	68	69	70	71	72	73	73
11.0	42	48	51	54	57	59	61	63	65	66	67	68	69	70	71
12.0	39	44	48	51	54	56	58	60	62	64	65	66	67	68	69
13.0	34	39	44	47	50	52	55	57	59	61	62	64	65	66	67
14.0	30	35	40	44	47	50	52	54	56	58	60	61	62	63	64
15.0	26	31	36	40	44	47	49	51	54	56	57	59	60	61	62
16.0	21	27	32	37	41	44	47	50	51	53	55	57	58	59	60
17.0	17	24	29	34	38	41	44	47	49	51	52	54	55	57	58
18.0	13	20	26	30	35	38	41	44	46	49	50	52	53	55	56
19.0	9	16	22	27	32	36	39	42	44	46	48	50	51	53	54
20.0	5	12	19	24	29	33	36	39	41	44	46	48	49	51	52
21.0	1	9	16	21	26	30	34	37	39	42	44	46	47	49	50

Adapted from a psychrometric table developed by Weksler Instruments Corporation, Freeport, L.I., NY.

Table E.2 A psychrometric chart for estimation of relative humidity from dry-bulb and wet-bulb temperatures (in °C)

Wet-bulb depression	Dry-bulb temperature (°C)														
	17	20	23	25	28	30	33	35	38	40	45	48	50	53	55
0.5	95	96	96	96	96	96	96	96	96	97	97	97	97	97	97
1.0	90	91	92	92	92	92	92	93	93	94	95	95	95	95	95
1.5	85	86	87	88	89	89	89	90	90	91	91	92	92	92	93
1.7	83	85	86	87	87	88	89	89	89	89	90	91	91	91	92
2.0	81	82	83	85	86	86	87	87	87	88	89	90	90	90	90
2.2	79	81	82	83	84	85	85	86	86	87	88	88	88	89	89
2.5	76	78	80	81	82	83	83	84	84	85	87	87	87	88	88
2.8	74	76	77	79	80	80	82	82	83	83	84	85	85	86	86
3.0	71	73	75	77	78	79	80	81	82	82	83	84	84	85	85
3.5	67	70	72	73	73	74	78	76	79	79	79	82	82	82	83
4.0	62	66	68	70	71	72	75	75	77	77	78	79	80	80	81
4.5	60	63	65	67	69	70	72	73	73	75	76	77	78	78	79
5.0	55	58	60	63	65	67	68	69	70	72	74	74	75	76	76
5.5	50	54	57	60	62	64	65	67	68	70	71	72	73	74	74
6.0	46	50	53	56	59	61	62	64	66	67	68	69	70	71	71
6.7	41	45	49	52	55	57	59	60	62	64	65	67	67	68	69
7.0	39	44	47	50	52	54	57	58	60	62	64	65	66	67	68
7.2	37	42	46	49	51	53	55	57	59	61	63	64	66	67	67
7.5	35	40	44	47	50	52	54	56	57	59	62	63	64	65	65
7.8	32	37	42	45	48	50	53	54	56	58	60	62	62	63	64
8.0	32	36	40	44	47	49	52	53	55	57	59	61	62	63	63
8.3	29	34	38	42	46	48	50	51	54	56	58	59	61	62	62
8.5	27	32	37	41	45	47	49	51	54	55	57	58	60	61	62
8.8	24	30	35	39	43	46	48	50	52	54	56	57	59	60	61
9.0	23	29	34	38	42	45	47	49	51	53	55	57	58	59	60
9.5	21	27	32	34	39	42	45	46	49	50	53	55	56	58	58
10.0	17	23	28	33	37	40	42	45	46	49	51	53	54	56	56
10.5	13	19	25	30	34	37	40	52	44	46	49	51	52	54	55
11.0	9	16	22	27	32	35	39	40	41	44	47	49	50	52	53
11.5	6	13	19	24	29	32	36	38	40	42	46	48	49	50	51
12.0	4	11	15	21	26	29	34	37	38	40	43	46	47	48	49

Adapted from a psychrometric table developed by Weksler Instruments Corporation, Freeport, L.I., N.Y.

INSTRUCTION FOR THE USE OF THE RELATIVE HUMIDITY PSYCHROMETER

The use of Tables E.1 and E.2 requires two similar thermometers, a dry-bulb and wet-bulb thermometer. The dry-bulb thermometer measures the ambient air temperature (T_{db}). The sensing portion (mercury reservoir) of the wet-bulb thermometer is covered with a clean cotton wick, whose free end is immersed in water and thereby kept wet. The wet-bulb is cooled by evaporation, at a rate depending on the relative humidity of the air being evaluated. To maintain the accuracy of the measurement of the relative humidity, the wick must be kept clean and wet, and it is desirable to fan the wet bulb until there is no further drop in the wet-bulb temperature (T_{wb}) before reading the thermometers. The following measurements should be made:

- Read both the ambient air (dry-bulb) temperature (T_{db}) and the wet-bulb temperature (T_{wb}) simultaneously.
- Calculate the wet-bulb depression (i.e., $T_{db} - T_{wb}$).
- Use Table E.1 if temperature is measured in °F, or Table E.2 if temperature is measured in °C. Read the relative humidity at the intersection of dry-bulb temperature (column) and the wet-bulb depression (row).

For example, if T_{db} is 40°C and T_{wb} is 34.5°C, the wet-bulb depression is 5.5°C and, from Table E.2, the relative humidity (rh) is found to be 70%. If T_{db} is 90°F and T_{wb} is 85°F, the wet-bulb depression is 5.0°F and, from Table E.1, rh is read as 81%.

Index